Dahlem Workshop Reports
Life Sciences Research Report 48
Organic Acids in Aquatic Ecosystems

Goal of this Dahlem Workshop:
to identify major processes
that cause spatial and temporal
variability in properties and
concentrations of organic
acids in aquatic ecosystems

Life Sciences Research Reports

Series Editor: Silke Bernhard

Held and published on behalf of the
Stifterverband für die Deutsche Wissenschaft

Sponsored by:
Senat der Stadt Berlin
Stifterverband für die Deutsche Wissenschaft
Deutsche Forschungsgemeinschaft

Organic Acids in Aquatic Ecosystems

E.M. Perdue and E.T. Gjessing, Editors

Report of the Dahlem Workshop on
Organic Acids in Aquatic Ecosystems
Berlin 1989, May 7–12

Rapporteurs:
W.H. Glaze, D.M. McKnight, P.J. Mulholland, M.S. Shuman

Program Advisory Committee:
E.M. Perdue and E.T. Gjessing, Chairpersons
F.H. Frimmel, J.R. Kramer, J.R. Sedell,
C.E.W. Steinberg, E.M. Thurman

A Wiley–Interscience Publication

John Wiley & Sons 1990
Chichester · New York · Brisbane · Toronto · Singapore

Copy Editors: J. Lupp, K. Klotzle, M. Leskin

Photographs: E. Thonke

With 4 photographs, 58 figures, and 18 tables

Library of Congress Cataloging-in-Publication Data

Dahlem Workshop on Organic Acids in Aquatic Ecosystems (1989 : Berlin, Germany)
Organic acids in aquatic ecosystems : report of the Dahlem Workshop on Organic Acids in Aquatic Ecosystems, Berlin 1989, May 7–12 / E.M. Perdue and E.T. Gjessing, editors : rapporteurs, W.H. Glaze . . . [et al.].
p. cm.—(Life sciences research report ; 48) (Dahlem workshop reports)
"A Wiley-Interscience publication."
Includes bibliographical references.
ISBN 0 471 92631 0
1. Organic water pollutants—Environmental aspects—Congresses. 2. Organic acids—Environmental aspects—Congresses. I. Perdue, Edward M. II. Gjessing, Egil T., 1932– . III. Glaze, William. IV. Title. V. Series. VI. Series: Dahlem workshop reports.
TD427.07D34 1989
574.5′263—dc20 89–29728
CIP

British Library Cataloguing-in-Publication Data

Dahlem Workshop on Organic Acids in Aquatic Ecosystems (1989 : Berlin, Germany)
Organic acids in aquatic ecosystems.
1. Aquatic ecosystems. Pollution by chemicals
I. Title II. Perdue, E.M. III. Gjessing, Egil T. 1932–
IV. Series
574.5′263

ISBN 0 471 92631 0

Typeset by Photo·graphics, Honiton, Devon
Printed and bound in Great Britain
by Biddles Ltd, Guildford, Surrey

Table of Contents

The Dahlem Konferenzen

Founders

Recognizing the need for more effective communication between scientists, the Stifterverband für die Deutsche Wissenschaft*, in cooperation with the Deutsche Forschungsgemeinschaft**, founded Dahlem Konferenzen in 1974. The project is financed by the founders and the Senate of the City of Berlin.

Name

Dahlem Konferenzen was named after the district of Berlin called *Dahlem*, which has a long-standing tradition and reputation in the sciences and arts.

Aim

The task of Dahlem Konferenzen is to promote international, interdisciplinary exchange of scientific information and ideas, to stimulate international cooperation in research, and to develop and test new models conducive to more effective communication between scientists.

The Concept

The increasing orientation towards interdisciplinary approaches in scientific research demands that specialists in one field understand the needs and problems of related fields. Therefore, Dahlem Konferenzen has organized workshops, mainly in the Life Sciences and the fields of Physical, Chemical, and Earth Sciences, of an interdisciplinary nature.

The Dahlem Workshops provide a unique opportunity for posing the right questions to colleagues from different disciplines who are encouraged to state what they do not know rather than what they do know. The aim is not to solve problems or to reach a consensus of opinion, the aim is to define and discuss priorities and to indicate directions for further research.

---

* The Donors Association for the Promotion of Sciences and Humanities, a foundation created in 1921 in Berlin and supported by German trade and industry to fund basic research in the sciences.

** German Science Foundation.

Topics

The topics are of contemporary international interest, timely, interdisciplinary in nature, and problem oriented. Dahlem Konferenzen approaches internationally recognized scientists to suggest topics fulfilling these criteria. Once a year, the topic suggestions are submitted to a scientific board for approval.

Program Advisory Committee

A special Program Advisory Committee is formed for each workshop. It is composed of 6–7 scientists representing the various scientific disciplines involved. They meet approximately one year before the workshop to decide on the scientific program and define the workshop goal, select topics for the discussion groups, formulate titles for background papers, select participants, and assign them their specific tasks. Participants are invited according to international scientific reputation alone. Exception is made for younger German scientists. Invitations are not transferable.

Dahlem Workshop Model

Since no type of scientific meeting proved effective enough, Dahlem Konferenzen had to create its own concept. This concept has been tested and varied over the years. It is internationally recognized as the *Dahlem Workshop Model*. Four workshops per year are organized according to this model. It provides the framework for the utmost possible interdisciplinary communication and cooperation between scientists in a period of $4\frac{1}{2}$ days.

At Dahlem Workshops 48 participants work in four interdisciplinary discussion groups. Lectures are not given. Instead, selected participants write background papers providing a review of the field rather than a report on individual work. These papers, reviewed by selected participants, serve as the basis for discussion and are circulated to all participants before the meeting with the request to formulate written questions and comments to them. During the workshop, each of the four groups prepares reports reflecting their insights gained through the discussion. They also provide suggestions for future research needs.

Publication

The group reports written during the workshop together with the revised background papers are published in book form as the Dahlem Workshop Reports. They are edited by the editor(s) and the Dahlem Konferenzen staff. The reports are multidisciplinary surveys by the most internationally distinguished scientists and are based on discussions of advanced new concepts, techniques, and models. Each report also reviews areas of priority interest and indicates directions for future research on a given topic.

The Dahlem Workshop Reports are published in two series:

1. Life Sciences Research Reports (LS), and
2. Physical, Chemical, and Earth Sciences Research Reports (PC).

Director

Silke Bernhard, Dr med., Dr phil. h.c.

Address

Dahlem Konferenzen
Tiergartenstr. 24–27
D-1000 Berlin (West) 30

Tel.: (030) 262 50 41

THE DAHLEM WORKSHOP MODEL

MONDAY	TUESDAY	WEDNESDAY	THURSDAY	FRIDAY
A. Opening (P) B. Introduction (P) C. Selection of Problems for the Group Agendas (S) ① ② ③ ④	① ② ③ ④	① ③	F. Report Session ① ② ③ ④	G. Distribution of the Reports H. Reading Time I. Discussion of the Group Reports (P)
D. Presentation of Group Agendas (P) E. Group Discussions (S) ① ②		② ④		J. Groups Meet to Revise their Reports (S) ① ② ③ ④

KEY: (P) = Plenary Session;
(S) = Simultaneous Sessions;
○ = one discussion group

Explanation of the Dahlem Workshop Model

A. Opening
Background information is given about Dahlem Konferenzen and the Dahlem Workshop Model.

B. Introduction
The goal and the scientific aspects of the workshop are explained.

C. Selection of Problems for the Group Agenda
Each participant is requested to define priority problems of his choice to be discussed within the framework of the workshop goal and his discussion group topic. Each group discusses these suggestions and compiles an agenda of these problems for their discussions.

D. Presentation of the Group Agenda
The agenda for each group is presented by the moderator. A plenary discussion follows to finalize these agendas.

E. Group Discussions
Two groups start their discussions simultaneously. Participants not assigned to either of these two groups attend discussions on topics of their choice.
The groups then change roles as indicated on the chart.

F. Report Session
The rapporteurs discuss the contents of their reports with their group members and write their reports, which are then typed and duplicated.

G. Distribution of Group Reports
The four group reports are distributed to all participants.

H. Reading Time
Participants read these group reports and formulate written questions/ comments.

I. Discussion of Group Reports
Each rapporteur summarizes the highlights, controversies, and open problems of his group. A plenary discussion follows.

J. Groups Meet to Revise their Reports
The groups meet to decide which of the comments and issues raised during the plenary discussion should be included in the final report.

Organic Acids in Aquatic Ecosystems
eds. E.M. Perdue and E.T. Gjessing, pp. 1–3
John Wiley & Sons Ltd

Introduction

E.M. Perdue[1] and E.T. Gjessing[2]

[1]*School of Geophysical Sciences*
Georgia Institute of Technology
Atlanta, GA 30332, U.S.A.

[2]*Norwegian Institute for Water Research*
NIVA
P.O. Box 68
0808 Oslo 8, Norway

On a global scale, dissolved organic carbon (DOC) and dissolved inorganic carbon (DIC) are present at comparable concentrations in fresh surface waters (average riverine concentrations of about 7 and 12 mg C/l, respectively). Unlike DIC, whose composition and properties are well understood, DOC occurs primarily as an unresolvably complex mixture of organic acids whose strongly interdependent biological, chemical, and geological roles in the aquatic environment are poorly understood. Efforts to understand or possibly anticipate the effects of increased utilization of the Earth's water resources have only recently begun to consider the role of dissolved organic acids.

Environmental scientists have found that organic acids must often be considered in studies of the impact of human activities on the aquatic environment. Their efforts are greatly facilitated by familiarity with the research of scientists who are conducting basic studies of the biological, chemical, and geological roles of dissolved organic acids in surface waters. Similarly, much of the new data that environmental scientists have collected on the temporal and spatial variations of dissolved organic acids in aquatic ecosystems are of great value to scientists in the other disciplines. This Dahlem Workshop has provided an ideal forum for an exchange of such knowledge and for fostering the kind of interdisciplinary perspective that is needed to integrate the results of separate studies into a comprehensive conceptual view of the general role of dissolved organic acids in aquatic ecosystems.

As we prepared for this Workshop, we identified several major research areas in which a greater understanding of the role of dissolved organic acids is needed. First, dissolved organic acids are both a natural background source of acidity and a pH buffer in low alkalinity waters, and they must be considered in assessments of the long-term acidification of surface waters by acidic deposition, including the potential impact of proposed pollution control strategies. Second, the ability of dissolved organic acids to form soluble complexes with many metal ions (including nutrients and toxicants) must be considered in studies and mathematical models of the mobilization, transport, and chemical and biological reactivities of metal ions in surface waters. Next, because the entire food chain in many rivers begins with microbial utilization of dissolved organic acids, the effects of dissolved organic acids on the productivity of aquatic ecosystems and, conversely, the effects of microbial processes on the chemical properties of dissolved organic acids must be better understood. Finally, the riverine flux of organic carbon from the continents to the oceans must be better understood to predict the long-term effects of deforestation and fossil fuel combustion on the global carbon cycle, e.g., on CO_2 levels in the Earth's atmosphere and associated climatic effects.

It was clearly impossible to incorporate such a broad spectrum of interesting topics into a five-day workshop. We chose instead to identify a more narrowly defined, yet generally applicable, topic area to serve as the focal point of our discussions. The main goal of this Dahlem Workshop was "to identify major processes that cause spatial and temporal variability in properties and concentrations of organic acids in aquatic ecosystems." To achieve this goal, nearly fifty scientists from ten countries in Europe and North America were invited to participate in the workshop. Internationally recognized scientists in aquatic chemistry, soil chemistry, organic geochemistry, environmental chemistry, water treatment technology, microbiology, forest ecology, and stream ecology collectively addressed four specific questions:

1. What is the composition of organic acids and how are they characterized?
2. How are acid-base properties of "DOC" measured and modeled and how do they affect aquatic ecosystems?
3. How do organic acids interact with solutes, surfaces, and organisms?
4. What are the temporal and spatial variations of organic acids at the ecosystem level?

Each of these questions is addressed in this volume by several background papers and by a Group Report that summarizes the technical discussion sessions. The background papers were used to provide all the participants with basic technical information that would be needed to fully appreciate the complexity of the problems being addressed in the workshop. The author of each background paper has thus tried to write an acceptably rigorous technical review of a given topic without "losing" other scientists who are not familiar with that discipline. Accordingly, the reader should benefit from reading all the background papers, even those in unfamiliar disciplines. The Group Reports are the best source of new ideas for future research and for recognition of the major areas of agreement and disagreement that were identified during the workshop.

We were very pleased with the extent to which the goals of this workshop were achieved. We were particularly satisfied with the lively exchanges between the ecologists and the other scientists which clearly identified the fundamentally different scientific philosophies that are manifested in their respective approaches to environmental research. The need for close cooperation between analytical chemists, microbiologists, and ecologists was repeatedly emphasized during the technical discussion sessions.

Participants at a Dahlem Workshop are able to spend several days engaging in intense formal and informal discussions on topics of common interest. This shared experience with scientists from a variety of academic backgrounds is not only intellectually quite enriching but it can result in the establishment of professional relationships that lead to productive collaborative research (e.g., between European and American scientists). We hope to have conveyed to the reader some sense of the spirit and intensity of our experiences at the workshop, and we hope that you will be sufficiently intrigued to join the group of scientists who attended this workshop in our efforts to understand the role of organic acids in aquatic ecosystems.

Finally, on a less optimistic note, even as we are writing this Introduction, the future of the Dahlem Konferenzen is being debated. Some of the differences of opinion regarding the sponsorship, financial support, and management of the Dahlem Konferenzen have recently been examined in the prestigious scientific journals *Science* (Lewin 1989) and *Nature* (Dickman and Altman 1989). On behalf of those of us who have participated in a Dahlem Workshop and those who have not yet had this unique opportunity, we sincerely hope that Dr. Silke Bernhard and her staff will be given the strong mandate and financial support that is needed to continue to provide such an excellent forum for scientific discourse.

REFERENCES

Dickman, S., and J. Altman. 1989. The end of a scientific era? *Nature* 340:86.
Lewin, R. 1989. Dahlem Conferences face ax. *Science* 245:122.

Organic Acids in Aquatic Ecosystems
eds. E.M. Perdue and E.T. Gjessing, pp. 5–23
John Wiley & Sons Ltd

Characterization of Organic Acids in Freshwater: A Current Status and Limitations

F.H. Frimmel

Lehrstuhl für Wasserchemie
Engler-Bunte-Institut der Universität Karlsruhe
7500 Karlsruhe 1, F.R. Germany

Abstract. Most organic acids in surface waters (on a mass basis) are classified as humic substances (HS). Due to their complicated structures and high molecular weights, these compounds cannot be identified and quantified in classic chemical terms. Well-defined operations and careful interpretation of the results are necessary to obtain meaningful data from elemental analysis, such as the determination of molecular size distribution, optical properties, resonance behavior, and functional groups. Special information can be derived from degradative studies. A serious impediment is the lack of authentic, well-defined samples for calibration of the methods. This can be partly compensated for by characterizing identical samples through a broad variety of analytical procedures and through biological testing, which provides the basis for new concepts of method development and correlations of analytical data to biological observations.

INTRODUCTION

Aquatic reactions of organic acids can be best understood on the basis of the Bronsted concept of conjugated acid-base pairs. Most organic acids are weak acids. Perdue (1985) showed that the pK_a values for all organic acids containing only carbon, hydrogen, and oxygen have a Gaussian distribution with a maximum at about 4.0. In aquatic systems it is important that the anionic base typically has a much better solubility than the corresponding nondissociated acid.

Small alkyl carboxylic acids with only a few C atoms per molecule, simple aromatic acids, and amino acids are significant products of chemical and biological degradation of organic matter. They can be identified and quantified, e.g., by gas chromatography/mass spectrometry (Cordt and Kußmaul 1989) or HPLC (Fuhrmann and Ferguson 1986). Because most of the simple acids are readily biodegradable, they occur in aquatic systems in fairly low steady-state concentrations. However, as information on the pattern of the acids increases, conclusions on their precursors and the aquatic reaction conditions can be expected. Small acids such as citric acid, lactic acid, salicylic acid, and phthalic acid have been used in numerous experiments as well-defined surrogate compounds for complicated high molecular weight matter (Öhman and Sjöberg 1988). It is obvious that a generalization of the results must be made very carefully. In addition to the biogenic small acids, an increasing number of synthetic organic acids can be detected in aquatic systems. For example, sulfonic acids, nitrilotriacetic acid (NTA), and ethylenediaminetetraacetic acid (EDTA) were determined by Schaffner and Giger (1984). Liquid chromatography and gas chromatography are well suited to separate these relatively small acids and their derivatives. Identification by UV-visible spectroscopy and mass spectrometry and the use of authentic compounds for calibration lead to reliable results on the distribution and environmental significance of many of these acids with well-defined structures.

This article focuses on the organic acids whose structures are poorly known and which occur as mixtures of refractory, relatively stable, yellow compounds with an average molecular weight of about 1,000 daltons. They can be considered as examples of humic substances (HS), and from this they derive their environmental significance (Frimmel and Christman 1988). The variety of their precursors in aquatic systems and the numerous possibilities of their formation lead to mixtures without any regularity in chemical composition.

There have been numerous attempts to supply composite structures for "typical" HS (Stevenson 1982). All approaches could not depict reality and are strongly dependent on the limitations of the methods applied.

The elemental composition and the hypothetical molecular weight of ca. 2500 daltons infers an average formula of $C_{100}H_{109}O_{69}N_2$ for the acid and base soluble fraction (fulvic acids [FA]) of an isolated aquatic HS sample. It is interesting to note that there are several thousand possible isomers for this general formula. In addition, the averaged molecular weight is a result of a broad distribution of molecular weights of the humic species. This further increases the number of possible compounds. This again raises doubts about the prospects of using an analytical approach to identify individual fulvic molecules.

The goal of this chapter is (*a*) to outline the analytical methods commonly used for the characterization of dissolved organic acids in freshwater; (*b*) to discuss the limitations of the methods with respect to their application to substances with poorly defined structures; and (*c*) to point out some feasible ways which lead to meaningful results for the characterization of organic acids.

ISOLATION PROCEDURES

When dealing with complicated environmental systems, it is advisable to simplify the samples by separation steps. This can lead to subsamples (fractions) which are better defined. In the case of high molecular weight organic acids, isolation procedures have two main purposes:

1. separation of the organic acids from inorganic and organic water constituents which are not humic-like in nature;
2. concentration of the organic acids to gain suitable samples for defined reactions and structural investigations.

A wide variety of isolation procedures has been elaborated using membranes, sorbents, and liquid extractions (review by Aiken 1985). Due to the activity of the International Humic Substances Society, the use of nonionic macroporous resins (e.g., Amberlite XAD type) has become the most commonly employed method worldwide. The availability of standard and reference materials obtained with a defined method from specific surface water samples has led to a sound basis for interlaboratory comparisons of methods and results. The different steps of the procedure, however, include some possibilities for uncontrolled fractionation or reactions (Fig. 1).

The protocol outlined in Fig. 1 starts with the separation of dissolved and particulate matter by means of 0.45 μm membrane filtration. This step is somewhat ambiguous because samples may contain colloidal particles that are less than 0.45 μm and because filter pore size decreases as a filter becomes clogged. Adsorption on XAD resin and subsequent desorption are conducted at pH 2 and 13, respectively. This treatment can result in the uncontrolled production of artifacts, especially when oxygen and light are present. The exclusion of light and oxygen and the minimization of reaction times at extreme pH values lead to a further complication of the whole procedure, but seem to be necessary in order to avoid major artifacts in the samples. There are very few data available concerning the structural changes due to the isolation procedure. The aim of the methods is to produce low-ash humic material regardless of how significant the inorganic components are for the structure of the humic substance itself. The striking similarity of many samples from different origins leads to the question of whether this similarity reflects natural conditions or is an artifact of the method of isolation.

Because *in situ* measurements of humic substances and their reactions are presently restricted to the few cases of highly colored water with dissolved

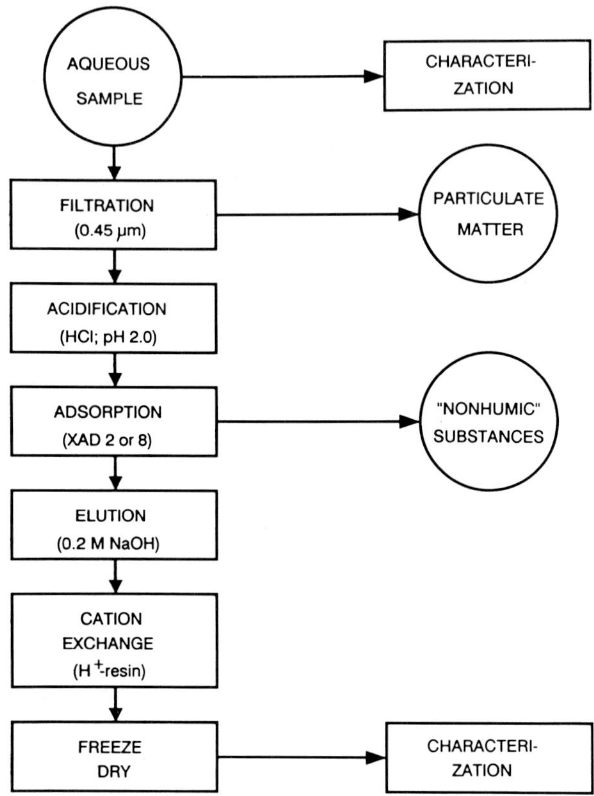

Fig. 1—Scheme of the isolation procedure for aquatic humic substances (based on Mantoura and Riley 1975).

organic carbon (DOC) concentrations of more than 10 mg/l, there is an urgent need for isolation procedures, even though they can only lead to operationally defined substances. The limitations of this approach should clearly be seen by identifying and minimizing the structural changes of the material and by comparing the products gained from different isolation procedures. In addition to the XAD methods, reverse osmosis appears to be promising and has the advantage of being a fast method utilizing mild process conditions. The membrane technique also leads to a higher yield of organic acids from water samples compared to the XAD methods.

DETERMINATION OF ELEMENTAL COMPOSITION

Methods for elemental analysis are among the oldest yet still most powerful tools for obtaining information on the constitution of organic compounds.

The elemental composition of humic substances is most important because we do not know their molecular structure. However, meaningful results can only be gained for samples whose isolation and treatment are well defined and in cases where sufficient material is available. Generally a complete elemental analysis requires at least 100 mg of dry substance. The moisture content, usually around 7%, needs to be measured carefully. Water determination according to the method used by Karl Fischer may be tedious and is often limited by the small amount of sample available but generally leads to good results. Another way is to vacuum-dry the samples at close to ambient temperature. The analysis of an equilibrated sample should be followed by the correction derived from a separate determination of the water content.

In many analyses, oxygen was not determined experimentally but was calculated as the difference between the total weight of the sample and all elements determined. This of course precludes a mass balance control. In addition, such oxygen contents include all the errors made in the determination of the other elements and the ash content. The ash content is often used as a criterion for judging the quality of the isolation procedure. There is an aim to lower the ash content of samples as far as possible; a "good" aquatic humic sample should have less than 2% ash. The determination of carbon and hydrogen is usually accomplished with a precision range of $\pm 0.3\%$. For all other elements, the relative standard deviation is much higher. This is due either to a number of possible interferences (e.g., for oxygen) or to small contents in the humic samples (e.g., sulfur, nitrogen, and phosphorus). An interlaboratory study in which four laboratories determined three samples is reported by Huffman and Stuber (1985). In the case of the sample with the lowest ash content (0.85%), the relative standard deviations for oxygen, nitrogen, sulfur, and ash were approximately 10, 20, 54, and 62%, respectively. The results suggest that great care should be taken in using data from elemental analyses. This also applies to the calculation of atomic ratios such as H/C, O/C, and N/C, and to the conclusions drawn from these data. Due to the satisfactory precision in determining H and C, this ratio can be used reliably. It is often close to 1.0. The O/C ratio is beyond doubt the most sensitive indicator of acid type (Thurman and Malcolm 1983). Therefore, more effort should be put into improving the precision of O-determination.

It is also possible to determine DOC in solution. Especially the method using photodegradation and quantification of the produced CO_2 by means of nondispersive IR spectrometry is sufficiently sensitive, reaching concentrations of 0.1 to 0.3 mg/l. Thus, DOC has become a fundamental parameter in aqueous humic chemistry, and it is suited for balances and as a basis for comparing results. In general, FA, which clearly represent the dominant fraction of aqueous HS, have a carbon content of about 50% on the basis of ash-free dry matter. The transformation of the DOC into the total mass of FA or HS is therefore fairly easy.

In addition to the main elements, there is an increasing interest in trace elements, especially phosphorus, halogens, and metals. Table 1 gives typical concentrations for trace elements in isolated surface water FA. Details on the kind of bonding and their influence on the reactivity of FA are largely unknown. The combination of trace analytical digestion methods and atomic spectroscopy guarantees reliable results for most of the trace elements.

MOLECULAR SIZE AND WEIGHT DETERMINATION

Molecular size and weight are fundamental characteristics of HS. The ratio of the weight-average molecular weight (M_w) and the number-average molecular weight (M_n) can be used as an indication of polydispersity. However, there are many uncertainties and operational definitions attached to the values. Depending on the isolation procedure and the analytical protocol applied, different molecular sizes can be determined. The most commonly used methods and some of their limitations are given in Table 2.

In addition to the general problem of characterizing a polydisperse "Gemisch" such as HS by average molecular size or weight numbers, there is the fact that different methods applied to HS always lead to fairly different results. This can be explained by the polyfunctionality of the substances concerned and by the influence of the matrix on their tertiary structure. Therefore meaningful results for molecular size or weight can only be gained

TABLE 1. Metal concentrations in fulvic acids isolated using the XAD method (values are ppm of dry weight).

Metal	FA (BM4)[a]	FA (BM6)[b]	FA(BM10)[c]
Al	210	160	113
Cd	0.4	0.4	0.7
Co	1.3	1.8	0.6
Cu	72	21	17
Cr	25	25	36
Fe	160	107	123
Mn	130	68	7.6
Mo	1.3	1.8	1.4
Ni	3.2	4.6	1.6
Zn	13	32	19

[a] Brownwater lake, March 1983
[b] Brownwater lake, February 1985
[c] Brownwater lake, September 1986

TABLE 2. Commonly used methods for molecular size and weight measurements and main pitfalls.

Method	Information	Pitfall
EM[a]	size, shape	solid sample
Ultrafiltration	size	coagulation, membrane clogging
GPC[b]	size	adsorption, electrostatic interaction, lack of standards
Radiation scattering	size	aggregation, fluorescence
Ultrafiltration	weight	polydispersity, lack of standards
Viscometry	weight	lack of standards
Colligative properties	weight	dissociation, polydispersity

[a] Electron microscopy
[b] Gel permeation chromatography

after careful corrections of the raw data, and from the application of several independent methods for the same sample. The lack of suitable standard materials for the calibration of the methods is a severe drawback.

Recently, flow-field-flow fractionation was used to characterize the molecular weight distribution of several FA and HA (Beckett et al. 1987). The M_n and M_w were determined to be 1150 and 1910, respectively, for stream FA. The distribution maxima were in good agreement with measurements based on vapor pressure osmometry and low-angle X-ray scattering.

Another attractive approach is the determination of molecular weight distribution by liquid chromatography. Despite several pitfalls, size exclusion chromatography can also provide useful information. Calibration is one of the major problems of these methods. Therefore results should be used only for comparative purposes. In combination with a DOC detector, useful fractionations can be achieved (Fuchs 1985a, 1985b, 1986), and the fractions can be integrated to a total mass of the sample as shown in Fig. 2.

The use of preparative columns leads to molecular size fractions large enough to be investigated further. This also seems to be the only way to obtain appropriate standards for the calibration of other methods.

MEASUREMENTS OF OPTICAL PROPERTIES

There is much information available on the spectroscopic properties of high molecular weight organic acids in the infrared (IR), visible (vis), and ultraviolet (UV) regions. Spectroscopic methods are well suited for the detection of specific functional groups in both pure substances and mixtures

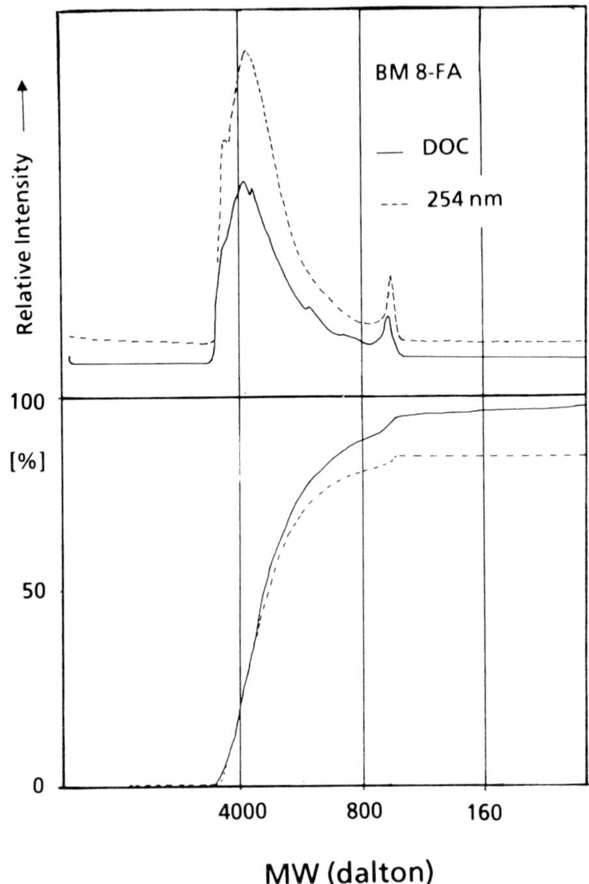

Fig. 2—Gel chromatogram and total carbon mass integration of BM8-FA (brown water, Brunnensee), Gel: TSK-40; eluent: pH 7.0 phosphate buffer; flow rate = 35 ml/h; detection: DOC and UV absorption at 254 nm; molecular weight (MW) calibration: polyethylenglycol, maltose, glucose, glycin.

of polyelectrolytic molecules. However, detailed insight into molecular structures cannot be achieved, because these spectroscopic methods measure the superimposed responses of all different species.

Infrared spectrometric characterization of organic acids has been enhanced by using Fourier transformation (FT). As a result, the poorly resolved bands and the hydrogen bonds are more clearly resolved. From the shape and relative intensity of the CH stretching bands around 2900 cm^{-1} and the double band at 1720 and 1630 cm^{-1}, the content of hydrocarbon and the carboxyl groups can be deduced. However, there is no way to quantify aromaticity or carbohydrate content by means of FTIR.

Spectral absorbance in the UV-vis ranges does not contribute much to structural information. Nonetheless, the absorbances at 254 nm and 436 nm are often used for characterization of HA and FA and quantification of their yellow color. Fluorescence is a typical property of high molecular weight organic acids. It is not known which structural regions in the molecules are responsible for this effect. Again, the summation of the signals from different molecules precludes any detailed interpretation. The use of synchronous fluorescence spectroscopy (Frimmel and Bauer 1987) supplies more sophisticated spectra which can be used as fingerprints for different samples without any obvious correlation to typical structures.

Information on the kinetics of excited states of organic acids can be deduced from laser flash experiments (Frimmel et al. 1987). Figure 3 shows the typical decay of excited FA measured by absorption for an FA isolated from a brownwater lake. From the absorption data, at least two transients with different lifetimes on the order of 10^{-5} and 10^{-4} s can be postulated. The results from quenching experiments on well-characterized samples can help to increase the understanding of the photochemical reactions on a molecular basis.

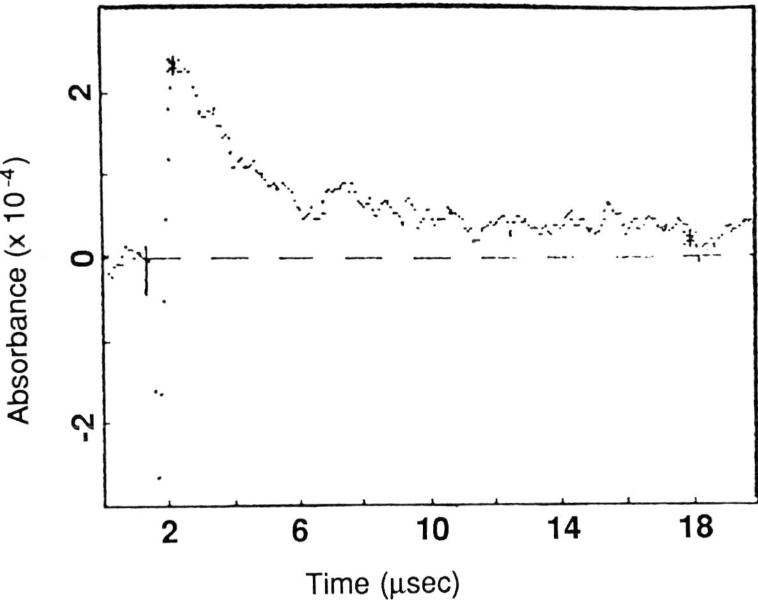

Fig. 3—Decay of the transient absorption of BM10-FA in a laser flash photolysis experiment. Sample origin: brown water, Brunnensee; DOC = 45 mg/l; pH = 7.0; normal atmosphere; λ_{exc} = 532 nm; pulse = 15 ns; λ_{abs} = 900 nm.

APPLICATION OF RESONANCE SPECTROSCOPY

The recent developments in nuclear magnetic resonance (NMR) spectroscopy have had a tremendous impact on structural investigations of organic acids. Methods such as magic angle spinning (MAS) and the use of cross polarization (CP) have enhanced sensitivity so much that relatively high resolution solid-state ^{13}C NMR spectra can be obtained for many samples, even fairly small ones (Wilson 1987). Two-dimensional spectra have been introduced and are an attractive way for representing the results. In addition, ^{1}H NMR spectra of liquid samples are very useful for comparison with and more realistic characterization of aquatic samples. One of the most important results of NMR investigations is the finding that aliphatic structures in HS are obviously present to a much higher extent than the results from degradative studies showed. In addition, specifically bound atoms involved in aromatic structures, olefinic bonds, aliphatic-type regions, carboxylic groups, hydroxyl carbohydrate-like structures, or ether and ester bonds can be assigned. However, it has to be kept in mind that the results are interpreted by correlating the resonance signals with those of defined molecules whose structure is known. There is no direct proof of the validity of this approach.

Electron spin resonance (ESR) spectroscopy works for unpaired electrons in the molecule. However, relatively little data can be used for structural assignment (Blough 1988). The presence of semiquinone moieties has been suggested but not yet proved. Resonance spectroscopy of the Mössbauer type is mainly applicable for the study of iron. Beyond the evidence of at least three distinguishable binding sites in the organic macroligand, there is little detailed structural information (Griffith et al. 1980).

CHARACTERIZATION BY PYROLYSIS AND MASS SPECTROMETRY

The application of mass spectrometry (MS) for structural investigations of high molecular weight organic acids is severely limited by the low volatility of these compounds. For the same reason, gas chromatography (GC) can only be used successfully after derivatization or degradation of the substances. The overall yield of the methods is low and the chemical reactions of the samples are poorly understood.

Pyrolysis (PY) is a special form of degradation. Even though the thermal processes involved in pyrolysate formation are not well known, the methods provide information on the main part (60% or more) of the samples applied. There are two major possibilities for analytical pyrolysis: (*a*) Curie point PY in combination with either low voltage electron impact MS or GC/MS (Saiz-Jimenez and de Leeuw 1986) and (*b*) time-resolved PY combined with field ionization MS (Schulten et al. 1987). There have been several papers on both approaches, and a review of the application of thermal degradation for structural studies of humic substances was published by Bracewell et al. (1989).

The advantage of the Curie point PY-GC/MS approach is the identification of the chromatographed products by their mass spectra whereas the PY has to be accepted as a fast and unresoluble step. Time- or temperature-resolved PY gives information on the thermal stability of HS. However, the identification of PY products is limited to their mass ions. A typical field ionization mass spectrum of PY products of a stream FA is given in Fig. 4.

Generally, there are several hundred mass ions formed as products of a single PY experiment. A large amount of information is available, especially when the specific conditions of the instrument are included and time and temperature resolution are achieved.

High resolution of the nominal masses, which can be achieved by the photoplate technique, leads to a split into several molecular ions which can be characterized by exact mass determination. With the help of literature data, structural suggestions can be made. For a reliable identification, MS/MS is needed.

The major pitfall of this method is the lack of knowledge about the details and reaction pathways of the PY step of macromolecular polydisperse substances such as aquatic acids. The limited identification of the PY products is another drawback which can only be partly compensated for by the application of statistical pattern recognition methods.

Recently the alternative use of advanced soft ionization methods, such as field ionization, field desorption, or fast atoms bombardment, in combination with sophisticated mass spectrometers, has led to mass fragments of up to several thousand daltons. In this case the spectra library has not been able to supply a sufficient selection of data for complicated substances like aquatic acids. Pure and defined authentic samples are also not available in this range.

DETERMINATION OF FUNCTIONALITY

As a consequence of the different approaches with highly sophisticated instrumental methods, the question remains whether it is possible to describe high molecular weight acids according to their exact molecular structure or whether it is more rewarding to characterize their reactivity. Many humic chemists would agree that both should be attempted. Without doubt the determination of functional groups, which is essential for understanding the role of organic acids in the environment, has been a more successful area of research than the pursuit of molecular structures.

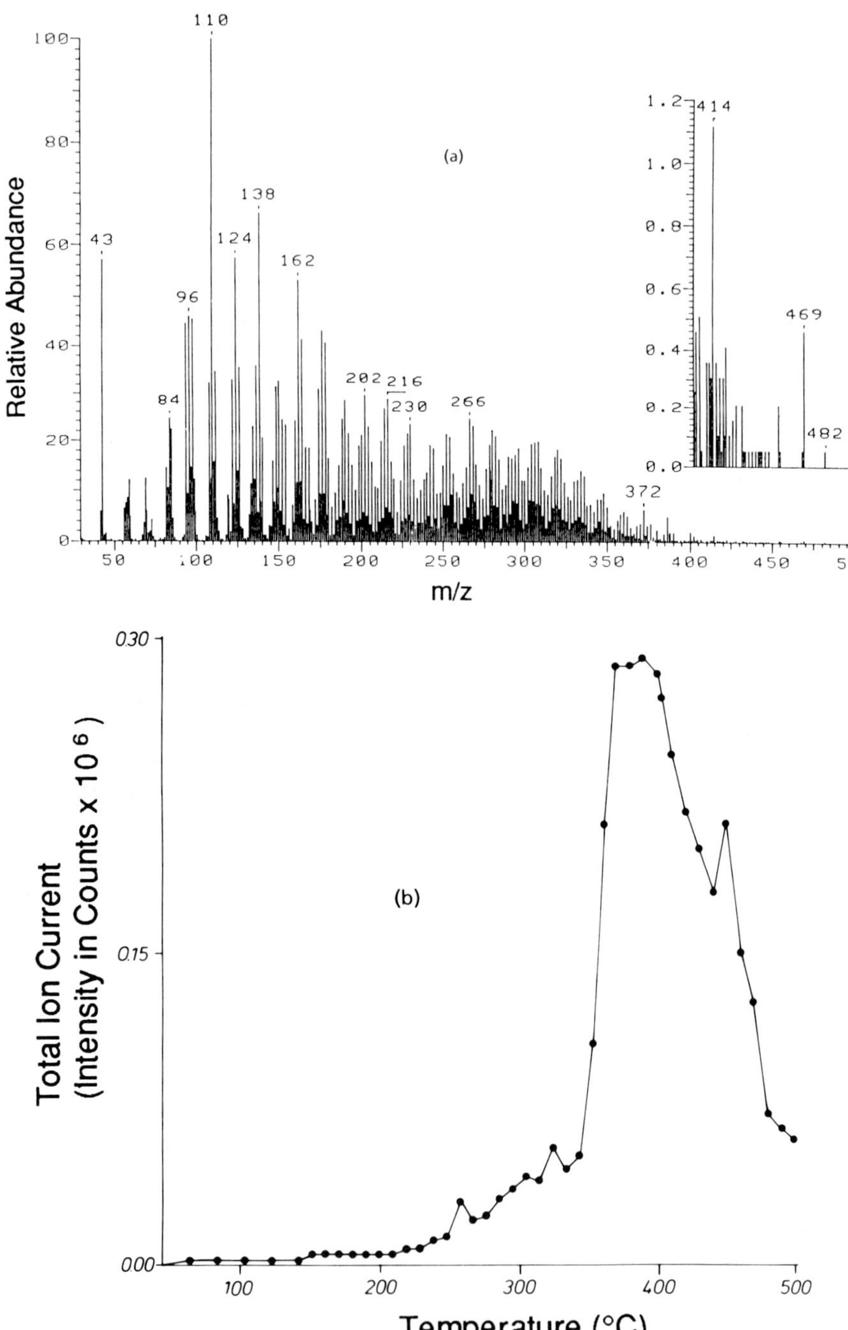

Fig. 4—Integrated field ionization mass spectrum (a) and total ion chromatogram (b) of PY products of BM8-FA (brown water, Brunnensee). Experimental details from Schulten et al. (1987).

There are several reviews and experimental papers on the quantification of functional groups. Schnitzer and Khan (1972) have systematically elaborated chemical procedures for the determination of carboxyl, hydroxyl carbonyl, and other functional groups. From a titration with strong acids or base, the acid functional groups can be quantified. In the pH range between 4 and 12, values on the order of 10^1 μeq/mg DOC are typical for aquatic FA.

The binding of heavy metals by organic acids (complexation) can be studied by several methods which were reviewed recently (Weber 1988). Separation techniques often use chromatography or membranes to quantify complexed and hydrated metal ions. The determination of uncomplexed metal ions can also be done by ion-selective electrodes, polarography, and voltammetry. All methods lead to results which are operationally defined. Because most complexation reactions are equilibria which have to be reversible by definition, care must be taken that the procedure itself does not shift the equilibria significantly. Results of Shuman (1988) show that some of the Cu complexes of FA dissociate with a rate constant of about 2 s^{-1}, which proves that some methods work on a much slower time scale than the complexation reactions. The control of pH is another important factor. Protons compete with metals for available ligands, and most metal ions are involved in the formation of hydroxides and basic salts with small solubility products, especially in case of pH values typical for natural aquatic systems. It is not surprising that the various competing reactions can hardly be corrected. The application of independent methods often leads to different results.

Another approach for studying metal complexation uses the fluorescence of HA and FA, which can be quenched by paramagnetic metal ions (Ryan et al. 1983). On the hypothesis of a 1:1 complex formed between the metal ion and a functional group of the organic molecule, conditional stability constants can be calculated. It is also possible to derive from titration experiments stability functions which reflect the decreasing stability of the metal-ligand bond as the occupation of ligand functional groups by metals proceeds. Laser flash experiments show that the excited states of HA and FA can be effectively quenched by paramagnetic metal ions (Frimmel et al. 1987). The Stern–Volmer plot proves that these results are due to a static quench mechanism from which a defined coordinative bond formation can be deduced. The limitation of fluorescence quenching experiments is that they work only with paramagnetic metal ions (e.g., VO^{2+}, Cr^{3+}, Mn^{2+}, Fe^{3+}, Co^{2+}, Ni^{2+}, Cu^{2+}). The influence of diamagnetic metal ions which compete for the ligand functional groups can be seen indirectly. However, the different equilibria which exist simultaneously are generally too complicated to be calculated precisely. Another experimental pitfall arises from the formation of insoluble products when polyvalent metal ions are introduced. This includes uncontrolled adsorption and coprecipitation reactions.

There is very little information on the formation of specific covalent bonds between metals and elements other than oxygen. This is surprising, for it is known from coordination chemistry that ligand-bound N and S carry preferable electron donor groups for metal complexation, e.g., amino compounds of Cu^{2+} and thio compounds of Hg^{2+}, Cd^{2+}, and Pb^{2+}. This difficulty of their specific identification in HA/FA complexes arises from the relatively low natural abundance of the ligand sites concerned. Because it is clear that these coordination bonds are of high significance for many biochemical reaction sequences, more work should be done to understand these reactions in the aquasphere.

It is not surprising that the problems involved in structural investigations of organic acids lead to the more practical approach of determining their functionality. This is crucial for the understanding of the environmental fate of humic substances as well as their behavior during water treatment. In addition to chlorination and oxidation reactions, there is much interest in the role organic acids play in photochemistry and in the photic zone (Zepp et al. 1987). The sensitized formation of hydrated electrons, singlet oxygen, and other reactive species has been reported. Another significant function of organic acids in the aquatic environment is the transportation of metals and organic micropollutants such as pesticides, polychlorinated biphenyls, and polycyclic aromatic hydrocarbons. An enhanced water solubility of the pollutants in the presence of FA and HA was shown by Chiou et al. (1987) and Gauthier et al. (1987). The different steps of water treatment can also be influenced by dissolved organic acids. There is clear evidence that the flocculation processes can be disturbed and that the effectiveness of activated carbon treatment in the removal of organic micropollutants is decreased by preadsorbed acids (Baldauf 1986).

The biochemical reactions which lead to biodegradation and biotransformation are still unknown to a great extent. The characterization of organic acids in this respect is most important. Our limited understanding of reactions on the molecular level has to be compensated for through the application of practically significant reaction conditions (e.g., control of temperature, pH, salinity, and oxygen content). Comparison of results on different samples is facilitated if measured properties are normalized to the DOC concentration of a water sample (e.g., metal complexation capacity should be μeq/mg C).

DERIVATIZATION

As has already been pointed out for the pyrolysis step, derivatization is one of the necessary tools for obtaining information on the original unknown substance. The common aim of these reactions is to convert a nonanalyzable substance into products which can be identified easily. To make this process meaningful, we have to understand the reaction pathway and know the yields in order to construct (from the finally identified building blocks) a model which is close to the original sample. This process is of course filled with many pitfalls.

Common chemical degradative methods can be adopted from humic chemistry of soils and have been reviewed for aquatic organic acids by Thurman (1985). There are several oxidative and reductive reactions and hydrolysis at different pH values. The pitfalls in the degradation approach are discussed by Norwood (1988). In general, all methods have to be carefully selected and adjusted according to the aim of the investigation. At that point, a conflict can already arise between the necessity of clear and well-understood reactions on the one hand and the information which can be achieved by their application on the other. Experimental limitations such as detection limits and the lack of adequate standard and reference materials have been discussed earlier and have to be faced as well in the degradative approach. For example, exhaustive oxidation leads to CO_2 as a final product in the case of organic acids. This reaction can be performed quantitatively and is therefore well suited for analytical purposes (e.g., determination of DOC). However, all structural information is lost.

A more selective and gentle oxidation is chlorination, which also includes substitution reactions. As long as the reaction is performed with a relatively high concentration of starting material, chloroform is formed as one of the main products. From the difference between chloroform and the amount of total organically bound chlorine (TOCl), an active search for other identifiable reaction products was stimulated (Christman et al. 1983).

As one advances from the simple end products to the more complicated molecules, there is a decrease in the concentrations of the individual compounds and an increase in their structural complexity. There is also generally less literature data for confirmation of structural assignments. An increase in hydrophilicity is another factor which leads to analytical problems. A milestone in chlorination chemistry was the identification of the two mutagens, MX and EMX (Kronberg et al. 1988). It was a clear example of how investigations can be successfully guided by always relating the individual compounds and their effect to the total amount of substance.

Despite the numerous products which have been identified after degradation of organic acids, there remains a many-sided uncertainty in interpretation. Experimental pitfalls must be recognized in the generally poor understanding of degradation reactions on substances with complicated polyfunctional structures. The descriptions of experimental conditions, including yield measurements, are often imprecise and incomplete. This adds to the limitation of the interpretability and comparability of the results.

Artifacts and their formation are another severe pitfall which must be discussed, together with possible influences due to the origin, isolation, and storage of the samples. Finally, conclusions drawn from well-defined derivatization reactions with model compounds should be applied to the complex system of aquatic organic acids only with the utmost care.

CONCLUSIONS

The complexity of the structure and functions of aquatic organic acids leads to one of the most stimulating challenges for environmental scientists. It is hopeless to deal with these high molecular weight polydisperse substances as one would with classically pure compounds which have a defined molecular structure. As a consequence, all analytical approaches to the investigation of the structure of organic acids and their reactions must encounter more or less severe pitfalls. To make the situation acceptable and to improve our knowledge of these environmentally and technologically significant compounds, it is advisable to:

1. start from homogeneous samples which are as well defined as possible in terms of origin, treatment, isolation, and storage;
2. apply as many methods as possible for analyzing the same sample and cross-check the reliability of the results;
3. perform system control experiments to minimize errors and quantify artifacts;
4. interpret *all* results together with the limitation of the methods used;
5. exchange standard and reference material with other laboratories and discuss the results.

An overview of the commonly available methods and their use in the characterization of organic acids is given in Fig. 5.

There is no question that we need additional methods. Analytical and preparative fractionation is still not satisfactory. (HP)LC seems to be the most promising method. High resolution ion exchange chromatography and supercritical fluid chromatography (SFC) are methods which have proved their potential but have not been applied extensively for the fractionation of water-derived organic acids. New generations of various detection systems (e.g., fluorescence, diode array) are available or are being developed (e.g., MS, MS/MS, elemental specific plasma analyzer). With these tools more information can also be gained on the nature of hydrophilic organic acids. As the methods become more and more sensitive, the size of the samples will decrease. This not only has the advantage that the number of experiments will increase, but it will also bring the reactions closer to reality.

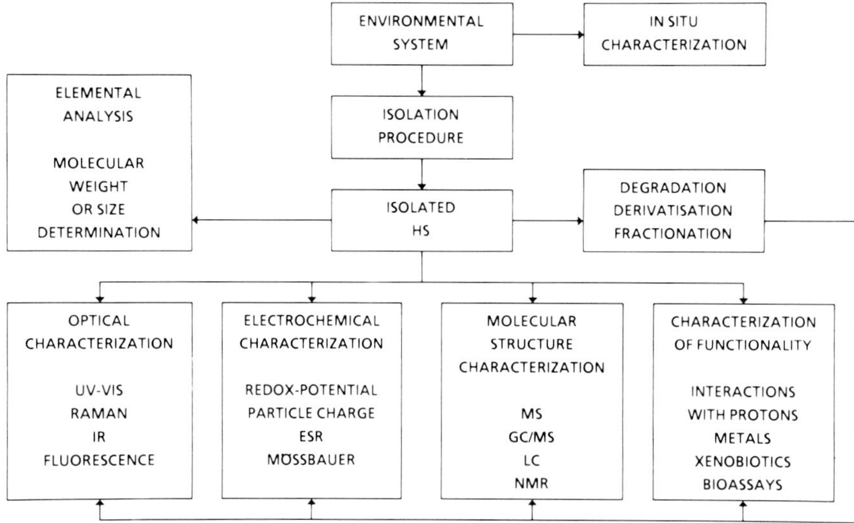

Fig. 5—Analytical multiple method approach for characterization of aquatic organic acids.

As is similar to other natural reference and standard materials, we shall have to live with the lack of identity of standard samples. This problem can only be resolved as in other comparable cases—by careful investigations with independent methods and a precise recording of the operations. Only then will the information result which is needed for an appropriate model of organic acids and an understanding of their role in aquatic systems.

REFERENCES

Aiken, G.R. 1985. Isolation and concentration technique for aquatic humic substances. In: Humic Substances in Soil, Sediment and Water, ed. G.R. Aiken, D.M. McKnight, R.L. Wershaw, and P. MacCarthy, pp. 363–385. New York: Wiley.

Baldauf, G. 1986. Einfluß natürlicher organischer Wasserinhaltsstoffe auf die Adsorption von Spurenstoffen in Aktivkohlefiltern. *Vom Wasser* 67:11–21.

Beckett, R., J. Zhang, and J.C. Giddings. 1987. Determination of molecular weight distributions of fulvic and humic acids using flow field-flow fractionation. *Env. Sci. Tech.* 21:289–295.

Blough, N.V. 1988. Electron paramagnetic resonance measurements of photochemical radical production in humic substances. I. Effects of O_2 and charge on radical scavenging by nitroxides. *Env. Sci. Tech.* 22:77–82.

Bracewell, J.M., K. Haider, S.R. Laster, and H.-R. Schulten. 1989. Thermal degradation relevant to structural studies of humic substances. In: Humic Substances II: In Search of Structure, ed. M.H.B. Hayes, R.L. Malcolm, and R.S. Swift. Chichester: Wiley, in press.

Chiou, C.T., D.E. Kile, T.I. Briton, R.L. Malcolm, and J.A. Leenheer. 1987. A comparison of water solubility enhancements of organic solutes by aquatic humic materials and commercial humic acids. *Env. Sci. Tech.* 21:1231–1234.

Christman, R.T., D.L. Norwood, D.S. Millington, J.D. Johnson, and A. Stevens. 1983. Identity and yields of major halogenated products of aquatic fulvic acid chlorination. *Env. Sci. Tech.* 17:625–628.

Cordt, T., and H. Kußmaul. 1989. Short chain organic acids in soil solution, seepage and groundwater. In: Progress in Hydrogeochemistry—Organics, Carbonates, Silicates, Microbiology, Models, ed. G. Mattheß, F. Frimmel, P. Hirsch, H.D. Schultz, and E. Usdowski. Berlin: Springer, in press.

Frimmel, F.H., and H. Bauer. 1987. Influence of photochemical reactions on the optical properties of aquatic humic substances gained from fall leaves. *Sci. Total Env.* 62:139–148.

Frimmel, F.H., H. Bauer, J. Putzien, P. Murasecco, and A.M. Braun. 1987. Laser flash photolysis of dissolved aquatic humic material and the sensitized production of singlet oxygen. *Env. Sci. Tech.* 21:541–545.

Frimmel, F.H., and R.F. Christman, eds. 1988. Humic Substances and Their Role in the Environment. Dahlem Konferenzen. Chichester: Wiley.

Fuchs, F. 1985a. Gelchromatographische Trennung von organischen Wasserinhaltsstoffen. I. Durchführung von Messungen. *Vom Wasser* 64:129–144.

Fuchs, F. 1985b. Gelchromatographische Trennung von organischen Wasserinhaltsstoffen. II. Ergebnisse der Trennungen bei Oberflächenwässern und Abwässern. *Vom Wasser* 65:93–105.

Fuchs, F. 1986. Gelchromatographische Trennung von organischen Wasserinhaltsstoffen. III. Untersuchungen zu den Wechselwirkungen zwischen Gelmatrix, Probensubstanz und Elutionsmittel. *Vom Wasser* 66:127–136.

Fuhrmann, J.A., and R.L. Ferguson. 1986. Nanomolar concentrations and rapid turnover of dissolved free amino acids in seawater: agreement between chemical and microbiological measurements. *Mar. Ecol. Prog. Ser.* 33(10):237–242.

Gauthier, T.D., W.R. Seitz, and C.L. Grant. 1987. Effects of structural and compositional variations of dissolved humic materials on pyrene K_{oc} values. *Env. Sci. Tech.* 21:243–252.

Griffith, S.M., J. Silver, and M. Schnitzer. 1980. Hydrazine derivatures at Fe^{3+} sites in humic materials. *Geoderma* 23:299–302.

Huffman, E.W.D., and H.A. Stuber. 1985. Analytical methodology for elemental analysis of humic substances. In: Humic Substances in Soil, Sediment and Water, ed. G.R. Aiken, D.M. McKnight, R.L. Wershaw, and P. MacCarthy, pp. 433–455. New York: Wiley.

Kronberg, L., B. Holmbom, M. Reunanen, and L. Tikkanen. 1988. Identification and quantification of the Ames mutagenic compounds 3-chloro-4-(dichloromethyl)-5-hydroxy-2(5H)-furanone and of its geometric isomer (E)-2-chloro-3-(dichloromethyl)-4-oxobutenoic acid in chlorine-treated humic water and drinking water extracts. *Env. Sci. Tech.* 22:1097–1103.

Mantoura, R.F.C., and J.P. Riley. 1975. The analytical concentration of humic substances from natural waters. *Anal. Chim. Acta* 76:97–106.

Norwood, D.L. 1988. Critical comparison of structural implications from degradative and nondegradative approaches. In: Humic Substances and Their Role in the Environment, ed. F.H. Frimmel and R.F. Christman, pp. 133–148. Dahlem Konferenzen. Chichester: Wiley.

Öhman, L.-O., and S. Sjöberg. 1988. Thermodynamic calculations with specific reference to the aqueous aluminium system. In: Metal Speciation: Theory,

Analysis and Application, ed. J.R. Kramer and H.E. Allen, pp. 1–40. Chelsea, MI: Lewis.

Perdue, E.M. 1985. Acidic functional groups of humic substances. In: Humic Substances in Soil, Sediment and Water, ed. G.R. Aiken, D.M. McKnight, R.L. Wershaw, and P. MacCarthy, pp. 493–526. New York: Wiley.

Ryan, D.K., C.P. Thompson, and J.H. Weber. 1983. Comparison of Mn^{2+}, Co^{2+}, and Cu^{2+} binding of fulvic acid as measured by fluorescence quenching. *Can. J. Chem.* 61:1505–1509.

Saiz-Jimenez, C., and J.W. de Leeuw. 1986. Chemical characterization of soil organic matter fractions by analytical pyrolysis-gas chromatography-mass spectrometry. *J. Anal. Appl. Pyr.* 9:99–119.

Schaffner, C., and W. Giger. 1984. Determination of nitrilotriacetic acid in water by high resolution gas chromatography. *J. Chromat.* 286:413–421.

Schnitzer, M., and S.U. Khan. 1972. Humic Substances in the Environment. New York: Marcel Dekker.

Schulten, H.-R., G. Abbt-Braun, and F.H. Frimmel. 1987. Time-resolved pyrolysis field ionization mass spectrometry of humic material isolated from freshwater. *Env. Sci. Tech.* 21:349–357.

Shuman, M.S. 1988. Comparison of anodic stripping voltammetry speciation data with empirical model predictions of pCu. In: Metal Speciation: Theory, Analysis and Application, ed. J.R. Kramer and H.E. Allen, pp. 125–133. Chelsea, MI: Lewis.

Stevenson, F.J. 1982. Humus Chemistry: Genesis, Composition, Reactions. New York: Wiley.

Thurman, E.M. 1985. Organic Geochemistry of Natural Waters. Dordrecht: Nijhoff/Junk.

Thurman, E.M., and R.L. Malcolm. 1983. Structural study of humic substances: new approaches and methods. In: Aquatic and Terrestrial Humic Materials, ed. R.F. Christman and E.T. Gjessing, pp. 1–23. Ann Arbor, MI: Ann Arbor Science.

Weber, J.H. 1988. Binding and transport of metals by humic materials. In: Humic Substances and Their Role in the Environment, ed. F.H. Frimmel and R.F. Christman, pp. 165–178. Dahlem Konferenzen. Chichester: Wiley.

Wilson, M.A. 1987. NMR Techniques and Application in Geochemistry and Soil Chemistry. Oxford: Pergamon.

Zepp, R.G., A.M. Braun, J. Hoigné, and J.A. Leenheer. 1987. Photoproduction in aquatic environments. *Env. Sci. Tech.* 21:485–490.

Organic Acids in Aquatic Ecosystems
eds. E.M. Perdue and E.T. Gjessing, pp. 25–42
John Wiley & Sons Ltd

The Application of Novel Approaches for Characterizing Organic Acids from Aqueous Matrices Focusing Biological Systems on Environmental Problems

D.C. White, D.E. Nivens, and M.W. Mittelman

Institute for Applied Microbiology
University of Tennessee
Knoxville, TN 37932-2567, U.S.A.

Abstract. This paper proposes that the spatial and temporal variability in the distribution of humic-derived organic acids in waters, sediments, and soils is primarily a function of the resident microbial biofilms which are able to process or generate these complex organics. These microbiota represent the largest biomass and have the metabolic versatility to control the humic-derived organic acid distribution. Mechanisms for the formation of humic organic acids include at least two major processes. First, microbes are involved in processing the products of primary photosynthesis such as soluble photosynthetate, cell bodies, litter, and other elaborations in the rhizosphere. Second, the microbes themselves liberate organic acids during microbial metabolism under conditions of growth-limiting concentrations of high potential electron acceptors. These microbially derived organic acids can be further processsed to form components of the humic organic acids. The microbial consortia responsible for the variability in the distribution of these humic organic acids are examined. In addition, the novel methods for characterization of both the biomass and community structure as well as the nutritional status and metabolic activities of the microbial biofilm consortia are described. New methods are required because the traditional methods for microbial community analysis are inadequate. Traditional methods require quantitative removal of the microorganisms from their habitat for cultural enumeration. Many organisms that can be removed and detected microscopically will not grow on the laboratory media. A biochemical methodology which utilizes the analysis of cellular components found universally in all microbes can be utilized as a measure of cellular biomass. If the components of the cells utilized in these measurements are sufficiently unusual in their distribution amongst the various types of microbes, then they can be utilized as "signature" biomarkers for specific types of microbes. If a sufficient catalog of "signatures" can be developed, then changes in community structure can be defined. This study will show that the membrane phospholipids are sufficiently universal to be used as biomass indicators. The distribution of specific groups of phospholipid ester-linked fatty acids (PLFA) amongst specific groups of microorganisms is sufficiently asymmetric to provide "signatures" which can be used to define the community structure. Other lipids, such as the endogenous storage polymer, poly beta-hydroxyalkanoate (PHA), together with the accumulation of specific PLFA, can be used to indicate the community nutritional status. The powerful methods of molecular biology have added a new dimension to microbial ecology. Nucleic acids can be quantitatively recovered from environmental samples and labeled oligonucleotide probes prepared against specific enzymes, antigens, etc. Particularly valuable probes can be prepared against the 16S RNA of the ribosomes. The differentially conserved portion of the sequence can be utilized for the preparation of oligonucleotide probes directed at the kingdom: either phylogenetically related groups or individual species. These labeled oligonucleotide probes readily permeate intact cells and are able to define individual cells in consortia or biofilms. However, the nucleic acids require that the organisms be isolated and the sequences be determined. Once determined, the probes can act in undisturbed consortia. Their primary difficulty is that the activity of the specific gene or cell line is not as yet readily determined. Since the methods reported herein involve the chromatographic fractionation and mass spectral detection of the "signature" PLFA patterns, it is possible to define metabolic rates with mass-labeled (^{13}C) components utilizing enrichments into specific microbes. New methodologies involving nondestructive electrochemical and/or spectroscopic techniques can be used to detect the formation of organic acids by living biofilms. These analyses have shown that microbial biofilms respond to changes in nutrient and electron donor/acceptor concentration and composition, predation, bioturbation (of sediments), and contamination of the sediments and soils. The richness and diversity of the microbiota coupled with its potential for rapid response to changing conditions make possible its use as a means to assess the impacts of toxicants on specific environmental sites quantitatively. Microbiota can also be utilized to begin to define the mechanisms involved in the temporal and spatial variability of humic organic acid distribution in the environment.

INTRODUCTION

The primary focus of this workshop was to identify the major processes that cause spatial and temporal variability in the properties of humic organic acids in the environment. This chapter deals primarily with one of the biological bases for the generation, modification, and fate of organic acids.

Hypothesis

The hypothesis of this presentation is that the critical factor in the spatial and temporal variability of humic organic acid distribution lies in the distribution and activities of the microbiota. The microbiota in the aquatic or terrestrial environments are primarily found attached to the surfaces in the soils and sediments. The microbes provide the largest biomass in nearly every environmental niche. The microbial component of a community is considerably greater in metabolic diversity than the metazoa. For example, the whole complex of anaerobic life is almost exclusively a microbial characteristic. Microbes have the greatest diversity in terms of physiological types. Nearly every possible chemical reaction with sufficient free energy has been exploited by a microbe. In addition, microbes have the capacity for an extremely rapid response to selective conditions in the environment. These microbes have the most rapid biomass doubling times in the soils and sediments with favorable conditions yet have extraordinary metabolic mechanisms for survival in compromising conditions.

The microbes influence the spatial and temporal distribution and structure of organic acids in several ways:

1. *Microbial catabolic production*. Under conditions of limited high potential electron acceptors, such as oxygen or nitrate, the metabolism of the obligate or facultative aerobes shifts so that organic acids rather than carbon dioxide and water become the primary metabolic end products. These anaerobic conditions are often the result of microbial metabolism itself. The anaerobic microbial communities routinely produce organic acids from biopolymers. A portion of these organic acids are further catabolized by the sulfate-reducing bacteria (in the presence of oxidized sulfur) or organic acids and/or hydrogen plus carbon dioxide with the production of methane. In more limited niches, nitrogen oxides and possibly even some oxidized metallic elements (such as Fe^{III}) can be reduced as electron acceptors.
2. *Microbial growth*. Viable microbes are surrounded by a lipid membrane consisting of hydrocarbon esters or ethers with a great structural diversity. The polar components of these membranes are enzymatically hydrolyzed following cell death. The hydrolysis results in the creation of neutral lipids that could form the raw materials for a portion of the organic carbon found in sediments.
3. *Microbial processing of photosynthetic products*. In the course of mineral recycling, the polymeric and simple products of photosynthesis are processed by the microbiota. These may represent detritus on the forest floor or stream bed, algal photosynthetate in the water column and sediments, or organic materials elaborated in the rhizosphere of higher plants. In many cases, organic material is depolymerized but not subsequently mineralized to carbon dioxide. This material may then be repolymerized or otherwise modified into products such as kerogen, which becomes increasingly resistant to biodegradation. The elegant work of W. Michaelis and his group has shown that selective depolymerization in the presence of mass-labeled isotopes allows the

recovery of labeled recognizable bacterial components such as bacterial hopanoids or archaebacterial ethers which are covalently bound to the polymers (Mycke et al. 1987).

Requisite Analysis

In addition to the analysis of the specific organic acids themselves, the determination of the heterogeneous distribution of the microbes, as well as their metabolic activities, is required. Analysis of the localization and activities of microbes in soil and sediments that could be responsible for organic acid distribution and structure requires special techniques. The classical microbiological techniques that proved so extraordinarily effective in the control of epidemic infectious diseases unfortunately are not as effective in helping to understand the distribution and activity of microbes in the environment. Classical microbiological methods that involve the detachment and subsequent culturing of organisms on petri plates can lead to gross underestimations of the numbers of organisms (White 1983, 1988). Assays to define microbial consortia have been developed so that the bias of cultural selection associated with classical methods has been eliminated. Since the total community is examined in these procedures without the necessity of removing the microbes from surfaces, the microstructure of multispecies consortia is preserved. The method involves the measurement of biochemical properties of the cells and their extracellular products. Those components generally distributed in cells are utilized as measures of biomass. Components restricted to subsets of the microbial communities can be utilized to define the community structure. The concept of "signatures" for subsets of the community based on the limited distribution of specific components has been shown for many monocultures (Ratledge and Wilkinson 1988). Phospholipids seem an ideal component for estimating active or potentially active microbial biomass (Fig. 1).

ANALYSES OF THE MICROBES IN SEDIMENTS AND SOILS

Biomass and Community Structure

Phospholipids, intracellular adenosine nucleotides, and cell wall aminosugars are biochemical components of cells that have been utilized to estimate microbial biomass (Balkwill et al. 1988; White 1983, 1988). Of these, phospholipids have proven the most useful when examining predation effects on microbial biofilms (White and Findlay 1988). Phospholipids are found in the membranes of all cells. Under the conditions expected in natural communities, bacteria contain a relatively constant proportion of

CELL COMPONENTS	PERSISTENCE AFTER CELL DEATH
Cell Walls	
Muramic Acid	+
KDO	±
Lipid A	+
Nucleic Acids	
DNA (Genes)	+
16S RNA	?
mRNA	?
Exopolymers	
Uronic Acids	+
Nucleotides	
ATP	+
Enzymes	+
Lipids	
Neutral Lipids	+
Phospholipids	-

Fig. 1—Cellular components for estimation of potentially active microbial biomass.

their biomass as phospholipids (White, Bobbie et al. 1979). Phospholipids are not found in storage lipids and have a relatively rapid turnover in some sediments, so the assay of these lipids gives a measure of the "viable" cellular biomass when compared to enzyme activities, total intracellular adenosine nucleotides, and cell wall muramic acid (White, Davis et al. 1979). As shown by Smith et al. (1986), there appears to be a unique microbial community in uncontaminated subsurface sediments from below the root zone. The microbiota are sparse and have an identical coccobacillary morphology. In these subsurface sediments, the biomass and cell numbers estimated from direct cell counts after acridine orange staining agree with the numbers and biomass estimated from the extractable phospholipid phosphate and total fatty acids, the total adenosine triphosphate, the fatty acids from the lipopolysaccharide lipid A, and the cell wall muramic acid content (Balkwill et al. 1988).

The ester-linked fatty acids in the phospholipids (PLFA) are presently both the most sensitive and the most useful chemical measures of microbial biomass and community structure thus far developed (Guckert et al. 1985). The specification of fatty acids that are ester-linked in the phospholipid fraction of the total lipid extract greatly increases the selectivity of this assay, as many of the anthropogenic contaminants as well as the endogenous storage lipids are found in the neutral or glycolipid fractions of the lipids. For example, by isolating the phospholipid fraction for fatty acid analysis, it proved possible to show bacteria in the sludge of crude oil tanks. The extraction of the lipids provides both a purification and a concentration of the components. The transesterification of the PLFA to form the volatile esters necessary if the resolution of capillary gas-liquid chromatography (GLC) is to be utilized can be performed under conditions in which free fatty acids will not be derivatized. Utilizing the exquisite sensitivity of mass spectral detection with soft chemical ionization and selective ion monitoring, the sensitivity can be in the femtomolar range. At these sensitivities, contaminants result in the solvents; this determines the signal-to-chemical noise ratio. The specificity of the PLFA assay has been greatly increased by the determination of the configuration and position of double bonds in monoenoic fatty acids (Nichols et al. 1986) and also by the formation of electron capturing derivatives, which after separation by capillary (GLC) can be detected by chemical ionization mass spectrometry as negative ions at femtomolar sensitivities (Odham et al. 1985). This allows the detection of specific bacteria with signature PLFA in the range of 100 to 1000 organisms. This sensitivity makes it possible to extract specific areas in biofilms possibly differentiated by microelectrodes to localize microbes. If the biofilms or consortia have been exposed to nonradioactive, mass-labeled (^{13}C) precursors, then the activity of specific microorganisms can be quantitated. Since many environments such as marine sediments often yield 150 ester-linked fatty acids derived from the phospholipids, a single assay provides a large amount of information. Combining a second derivatization of the fatty acid methyl esters, to provide information on the configuration and localization of the double bonds in monounsaturated components, provides even deeper insight. Figure 2 provides a diagram of the methods used in the signature biomarker analysis. These methods could be applied to the analysis of the humic organic acids as well.

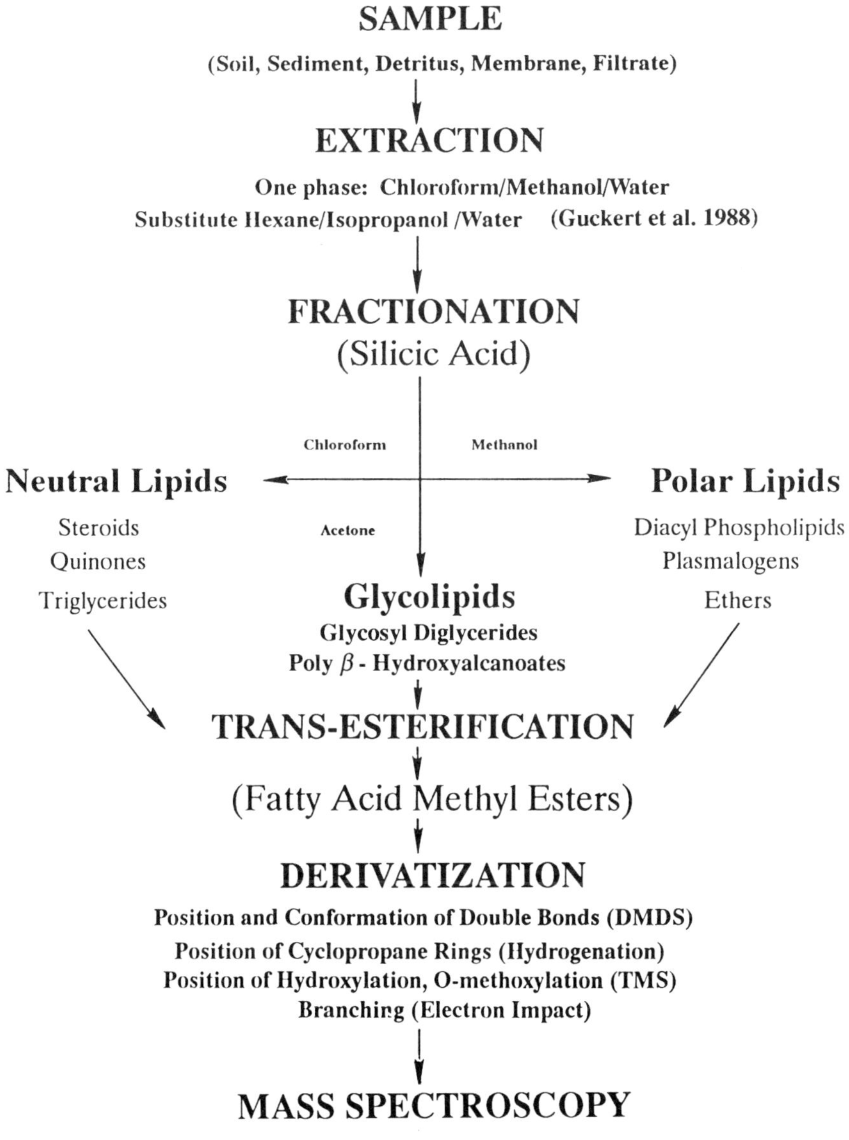

Fig. 2—Processing for lipid analysis.

Despite the fact that the analysis of PLFA cannot provide an exact description of each species or physiologic type of microbes in a given environment, the analysis provides a quantitative description of the microbiota in the particular environment sampled. With the techniques of statistical pattern recognition analysis it is possible to provide a quantitative estimate of the differences between samples with PLFA analysis.

Potential problems with defining community structure by analysis of PLFA come with the shifts in fatty acid composition of some monocultures with changes in media composition or temperature. There is as yet little published evidence for such shifts in PLFA in nature, where the growth conditions that allow survival in the highly competitive microbial consortia would be expected to restrict severely the survival of specific microbial strains to much narrower conditions of growth. The shifts in microbial PLFA patterns with changes in physiological conditions can be utilized to gain insight into the nutritional status of the organisms in a particular biofilm so long as the behavior of the specific groups of microbes under consideration is carefully validated by studies of cultures under defined conditions. A potentially very useful finding is the detection of increased proportions of *trans* monoenoic fatty acids in the minicells which result from the starvation of some marine bacteria (Guckert et al. 1986). This biomarker appears to indicate starvation with attachment to biofilms in the initial microfouling community.

The techniques of molecular biology have provided new tools to examine the distribution and community structure of microbial consortia in soils and sediments. Techniques have been developed which enable the quantitative recovery of DNA and ribosomal RNA from sediments. It is particularly difficult to recover nucleic acids from soils or sediments rich in clay, since the nucleic acids recovered are typically fragmented. This emerging technology is enormously powerful as oligonucleotides made with sequences of more than 10–20 bases provide a specificity that is virtually absolute. Once the sequence is known and the appropriate fragment defined, the synthetic DNA probe is made automatically in substantial quantities. This is then tagged most often with ^{32}P by nick-translation. The environmental samples are treated with detergents to lyse the bacteria and nucleic acids recovered and purified. Once the nucleic acid is recovered from the environment, it is placed in appropriate ionic strength buffers and heated to denature (form single strands). The single strands are then allowed to anneal (hybridize) with the probe and the unhybridized single-stranded nucleic acid, and excess probe removed by various techniques. The degree of homology can be controlled by the hybridization conditions. Under proper conditions it is possible to detect single gene copies.

The proper sequence must be selected for the probe, and the enzyme (or gene) must have been sequenced to provide the sequence. With access to gene libraries it is possible to select for a wide variety of genes. Genes that are involved in functional groups of bacteria can be selected if the distribution of the enzyme is known. Of all the naphthalene degrading aerobes isolated from several contaminated soils, less than 25% hybridized with the DNA oligonucleotide probe in the best-studied pathways. Clearly, there are other enzymes involved in naphthalene degradation. As more of the sequences of genetic determinants of specific processes become known, the probes can be modified or mixed probes utilized. The probe defines the presence of the specific gene, not the enzyme of the metabolic activity. G.S. Sayler has established that certain genes can be detected outside of bacteria adsorbed to the sediment (Ogram and Sayler 1988). Even with these provisos, the technique is powerful and can give insight into the enzyme distribution in a community. Problems may exist in referring the presence of the gene to the presence of a specific bacteria. Some genes are widely distributed; for example the APS-reductase enzyme, which activates sulfate and is essential to all the sulfate-reducing bacteria, was also found to occur in sulfur-oxidizing bacteria as well. These DNA probes are particularly useful in the detection of genetically engineered microbes in the environment where the specificity of the probe can be controlled. The use of DNA gene probes has been reviewed (Jain et al. 1988).

An alternative or complementary method that allows detection of the ribosomal RNA (rRNA) with probes has some additional properties. The rRNA are remarkably conserved molecules that are involved in protein synthetic systems common to all life. Initial work concentrated on the 5S rRNA which consists of about 120 nucleotides. This short sequence was sufficiently invariate in that it showed a paucity of independently varying nucleotide positions. The 16S rRNA ($\sim$1600 nucleotides) was ideal for phylogenetic comparisons and, with the advent of DNA cloning and sequencing methods, provides exciting possibilities. There are invariate sequences universal to all life and sequences common to the major kingdoms—eukaryotes (plants, animals, and microeukaryotes) and prokaryotes (eubacteria and archaebacteria). There are sequences which can be used to define closely phylogenetically related groups (which may not be functionally obvious, such as the plant mitochondrial rRNA, the methane-oxidizing bacteria, and the plant tumor-inducing agrobacterium) and individual species or strains. The fact that there are 10^4 copies of the rRNA in each cell, that the analytical systems can detect 50 or so molecules, and that the probe nucleotides can penetrate the intact cells of bacteria in environmental samples provides a system in which appropriately labeled probes can be used to identify specific bacteria or groups of bacteria in biofilms. When fluorescent probes are used on biofilms the localization of specific cells can be readily determined. This exciting new technology has been reviewed (Olsen et al. 1986). An elegant use of this technology in an examination of the effects of a commonly used antibiotic on the community structure in the bovine rumen has been reported (Stahl et al. 1988).

The disadvantage of the rRNA probe technology is that for maximum effectiveness the sequence of the 16S RNA must be known. This means the organism must be culturable and the sequencing performed. Determination of the sequence is a procedure involving several difficulties, not the least of which comes in defining the best positioning (placement of "skipped" bases) to use in phylogenetic matching. There is also a constant problem of almost ubiquitous RNAase contamination which can easily ruin the experiments. Nonetheless, the application of nucleic acid probe technology to environmental systems is a powerful means for defining the distribution of microbes in samples and their community structure.

Nutritional Status and Metabolic Activity

A major problem with the nucleic acid probes is in the detection of metabolic activities and nutritional status. DNA probes provide an elegant definition of the presence of the gene but do not indicate the activity of the enzyme in question. G.S. Sayler is in the process of using molecular manipulations to insert the LUX operon (the genes from a phosphorescent bacteria for its light generating system) into the operon for specific biodegradative enzymes (pers. comm.). When this is done and the bacteria are induced to form the enzymes, they phosphoresce, thus indicating their activity. Individual metabolically active cells can be detected. This requires considerable metabolic insight into the specific microbes and the ability to generate stable specific genetic constructs. Techniques are developing to allow the definition of the ratio of rRNA to DNA. Generally, the faster the bacteria grow, the higher the ratio between rRNA and DNA. However, this indicates only the potential for rapid growth.

The advantages of PLFA technology lie in the ability to detect the incorporation of label from precursors into specific molecules. This is particularly powerful when used with nonradioactive ^{13}C and detection by capillary gas chromatography/mass spectrometry (GC/MS).

The nutritional status of biofilms or microbial consortia can be estimated by monitoring the proportions of specific endogenous storage compounds relative to the cellular biomass. The nutritional status of microeukaryotes (algae, fungi, or protozoa) in biofilms can be monitored by measuring the ratio of triglyceride glycerol to the cellular biomass (Gehron and White 1982).

Certain bacteria form endogenous lipid poly beta-hydroxyalkanoate (PHA) under conditions in which the organisms can accumulate carbon but have insufficient total nutrients to allow growth with cell division. Conditions in which PHA is accumulating and PLFA are not being formed represent unbalanced growth. In unbalanced growth the organisms accumulate PHA but are unable to form new membranes or divide. PLFA formation represents new membrane formation with cell division. The ratio of incorporation of labeled acetate into PHA/PLFA is an exquisitely sensitive indicator of nutritional status. Some metabolites, including some organic acids, are metabolized by bacteria only during conditions of unbalanced growth.

Starvation in microbes stimulates the secretion of exopolymers by microbiota. These may be critical to attachment to substrata. Generally, bacteria possess negatively charged exopolymers containing uronic acids: methods for their detection based on GC/MS with reduction with sodium borodeuteride of activated esters prior to hydrolysis have been utilized (Fazio et al. 1982). A much easier assay of exopolymer formation utilizes the ratios between the carbon–oxygen stretch ($\sim 1150\ cm^{-1}$ of carbohydrates) and the amide II ($\sim 1550\ cm^{-1}$ of the proteins in the bacteria) determined by diffuse reflectance with the Fourier transforming infrared spectrometer (DRIFT). With this system it is also possible to detect high concentrations of PHA by the carbonyl absorbance ($\sim 1730\ cm^{-1}$ [Nichols et al. 1985]). With the DRIFT microscope these ratios can be determined in areas as small as 20 μm in diameter.

The microbial component analysis utilizing capillary GC/MS involves the isolation and characterization of specific molecules by MS. If the soils and sediments in which the distribution and activities of the microbes are to be examined are exposed to ^{13}C-labeled precursor such as acetate prior to the analysis, then the enrichment of ^{13}C in specific biomarkers can be used to establish the activity of specific bacteria. The mass-labeled precursors are nonradioactive, have specific activities approaching 100%, and can be efficiently detected using the selective ion mode in mass spectroscopy. Their high specific activity makes possible the assay of critical reactions using substrate concentrations in the biofilms that are just above the natural levels. This is not possible with radioactive precursors. It is also possible to utilize ^{15}N to follow nitrogen metabolism by specific microbes. Improvements in analytical techniques have increased the sensitivity of this analysis. Utilizing a chiral derivative and fused silica capillary GLC with chemical ionization and negative ion detection of selected ions, it proved possible to detect 8 pg (90 femtomoles) of D-alanine from the bacterial cell wall (the equivalent of about 1000 bacteria the size of *E. coli*) (Tunlid et al. 1985). In this analysis it proved possible to detect reproducibly a 1% enrichment of ^{15}N in the ^{14}N–D-alanine.

The phospholipid signature technique based on MS offers a further tantalizing possibility. When lyophilized bacteria or biofilms are subjected to bombardment by ions or atoms of modest intensity, the naturally charged phospholipids are among the first molecules to be desorbed from the microbial films. Because of their charge, they are readily analyzed by MS. The technology is rapidly being developed to allow focused beams to sputter phospholipids from the biofilms directly. The sputtered lipids can be readily identified by MS. Phospholipids readily lose specific polar head groups (such as M^- and M-141 loss of ethanolamine representing phosphatidylethanolamine). If the sensitivity can be increased, it should be possible to image, at least crudely, the lipid distribution and to detect the areas of activity in biofilms previously exposed to ^{13}C (prior to lyophilization).

Validation

The isolation of specific organisms or groups of organisms for signature biomarker analysis, and the detection of these organisms in microbial consortia under conditions where their growth is induced, provide a validation of the signature biomarker methodology. This and other validation criteria for the biomarker methodology have been reviewed (White 1988).

Microbial Acid Formation

The actual formation of organic acids by living biofilm bacteria can be detected. Both direct and indirect nondestructive methods can be employed. Direct methods involve microelectrodes that respond to pH and have now been made with active surfaces of between 20 and 200 μm in diameter, which have sufficient stability to be useful. These electrodes can be placed into the biofilm and the pH monitored at specific places in the biofilm. The electrodes can be combined with oxygen electrodes (reported with diameters of 5 μm) (Revsbech and Jorgensen 1986). Microbial biofilms have been shown to be able to generate highly localized pHs as low as 5.2 in some areas and as high as 9.2 at the metal–biofilm surface using glass microelectrodes in growth systems with neutral pH in the bulk phase (Little et al. 1989). This area of research is rapidly developing. Other chemically modified ion-selective microelectrodes that respond directly to pH have been tried but stability problems have been rumored. Optrodes as terminal components of fiberoptic cables in which dyes respond to pH changes are currently under development.

A second direct method of acid formation detection involves the use of "reporter" molecules whose infrared spectrum shifts with changes in pH, which are detected in the evanescent wave on crystals exposed in the attenuated total reflectance (ATR) cells of the FT/IR. In this system, a biofilm is formed on the crystal in a flow cell connected to a continuous culture system. The microbes in the biofilm are then fed with an appropriate substrate to allow acid generation, which can be detected in the shifts of the reporter molecules.

Indirect methods are based on the changes in the ionized double layer of metal working electrodes on which microbial biofilms are grown. These systems also use a continuous culture system with flow rates so that the major microbial community in the apparatus is attached to surfaces. The electrochemical properties of the metal are affected by the propensity of the biofilm to elaborate organic acids. In these systems, the bacteria lower the pH by secreting acids, primarily acetate, with traces of propionate and butyrate. Other common interactions involve oxygen utilization, chelation of metals, and concentration of anions at the metal surface. These changes result in the local breakdown of the metal passivation layers with subsequent changes in the polarization resistance.

There are two electrochemical methods of detection that can be utilized to detect changes in the passivation that result in corrosion. The first method involves passively monitoring electrochemical events. Measuring the open-cell corrosion potential (OCP) between working stainless steel and reference electrodes has established that breakdown of passivation in local areas (pitting) is associated with a 200 to 300 mV negative shift in OCP relative to the calomel electrode. This shift is associated with the activity of microbial biofilms actively secreting acid. Preliminary experiments utilizing ultrafast multimeters which monitor electrochemical noise (ECN) between working and reference electrodes show that microbial activity on surfaces produces marked shifts in amplitude when the frequency spectrum of the noise is factored. When factored, spontaneous pulses of about 1 mV show a decrease in noise at higher frequencies compared to lower frequencies. These shifts between "pink" and "white" noise occurring at about 90 Hz appear to correlate with microbial activity and may be associated with local breaches in passivation films on the metal.

The second type of electrochemical monitoring involves the current responses when a small potential is applied to the working electrode. It is important to know that the currents or potentials applied to the biofilms are of short duration and low in magnitude, so no detectable damage to the electrode surface or to the cell metabolism is induced. The standard linear polarization analysis that employed linear potential sweeps of hundreds of millivolts from the open-cell potential irreversibly damaged the biofilms. Two methods, small amplitude cyclic voltametry (SACV) and electrochemical impedance spectroscopy (EIS), have shown utility in on-line nondestructive monitoring of biofilm microbial activity (Dowling et al. 1988, 1989).

SACV involves the application of a sawtooth (triangular) signal that is sufficiently small so that the change of the electrode potential remains in the linear range. This technique allows the accurate determination of the polarization resistance and provides as well some indication of the capacitance of the system. The polarization resistance is related to the reciprocal of corrosion rate. Generally, sweeps involve a maximum of 1 μA over a period of 64 s. This is slow enough to provide a pseudo-equilibrium that overcomes problems of steady state. The capacitive component of the system is indicated by the hysteresis. Both the capacitive contribution to the hysteresis and the problems of nonsteady state decrease at slower sweep rates. The limits on the slowness of the sweep rates are usually related to the stability of the potentiostat. In such a system the activity of acid-generating bacteria induced an increase in SACV of 3.7 Kohm cm^2 compared to 18 Kohm cm^2 for the uninoculated control.

The second transient response method involves EIS. EIS differs from SACV in that a small sinusoidal potential is applied to a working electrode over 5 decades of frequencies and the phase shift of the resultant sinusoidal current is measured. The results are usually plotted using a Nyquist diagram of Z'' (imaginary impedance) versus Z' (real impedance). This analysis gives more information about the system than the SACV measurement. However, the analysis can take several hours and the equipment tends to be more expensive. EIS subjects the microbially influenced corrosion (MIC) biofilm to greater perturbations than SACV and could lead to artifacts originating in the electrochemical measurement itself. Evidence that EIS does not significantly affect a living microbial biofilm comes from the fact that sequential EIS analyses show no differences. The small current resulting from the impression of a potential equivalent to the corrosion potential does not change during the analysis. Both of these observations indicate that the corrosion system including the biofilm is unchanged during the EIS analysis. Laboratory experiments can be used to demonstrate readily the effectiveness of EIS in the detection of microbial activity. EIS provides an additional advantage as it not only gives nondestructive estimates of the corrosion potential but also provides indications of local inhomogeneities in the corrosion process that become more pronounced as the sweep frequency decreases. The complication results from the influence of the local processes on the average corrosion rates which in turn are related to microbial biofilm activities. The use of these devices in monitoring microbially influenced corrosion has been reviewed (Dowling et al. 1988, 1989).

These electrochemical devices have the advantage of being nondestructive and can be utilized on-line to monitor changes in biofilm activities. The problem is that any process affecting the Helmholtz double layer can induce a response, and microbial organic acid production or bacterial growth are major but not exclusive processes monitored by these systems.

Independent on-line monitoring of the rate of bacterial colonization may be possible utilizing a piezoelectric Quartz Crystal Microbalance. In this device a transverse AT-cut quartz crystal is expected to detect changes in the viscoelastic properties of the surface exposed to the solution as the bacteria attach and colonize the surface. These changes are detected nondestructively as shifts in the resonant frequency of the crystal. This device can be combined with the electrochemical monitors described above to increase the specificity.

The specificity of electrochemical devices is being enhanced by the application of polymer coatings that contain enzymes or reactants which react only with specific analytes in the biofilm or solution. The membrane components in turn react with the electrochemical devices. If the electrodes themselves can be made sufficiently small (on the scale of microns), then many of the problems of inhomogeneity and imperfect mixing that plague electrochemistry will disappear. Microelectrodes also have the advantage of providing much greater spatial resolution.

Role of Microbes in Organic Acid Composition and Distribution

This paper has addressed the techniques by which microbial consortia in biofilms on soil, sediment, or suspended particles can be monitored. The biomass and community structure determinations can show who is there. If coupled to measures of labeled precursors incorporated, these techniques can also show the other critical consideration of not only who is there but who is metabolically active. The nutritional status can also be monitored with these techniques and can be important in processes where the organisms derive no direct benefit from specific metabolism yet in the presence of suboptimal growth conditions are able to modify chemical structures in the environment. This so-called co-metabolism has been demonstrated for microbial consortia in which aerobic methane or propane oxidation at suboptimal rates can degrade halogenated hydrocarbons (Ringelberg et al. 1988).

The role of specific microbial consortia in modifying the composition of humic organic acids could be examined in some direct experiments. Humic organic acid materials labeled with ^{13}C could be created by growth if plants were incubated with ^{13}C-carbon dioxide, such as cypress seedlings with $^{13}CO_2$. This material could be characterized as described in the preceding sections. If the labeled humic organic acids were placed in a flow-through microcosm where they were exposed to a microbial biofilm, the effects of shifts in microbial activity and community structure could be correlated with time to changes in the structures of the humic organic acids. Once specific microbial activities were associated with specific changes in the amount and composition of humic organic acids in the microcosm experiments, then insight into how the microbial metabolism affects the spatial and temporal distribution of these complex organics could be developed.

CONCLUSIONS

Other papers in this volume address novel methods by which the acids themselves can be detected, and these must be utilized coordinately with the techniques for monitoring the distribution, community structure, nutritional status, and metabolic activities of the microbes. The role of microbes in the temporal and spatial distribution of both biodegradation and possibly biosynthesis of the humic organic acids can be better defined.

Acknowledgements. This work would not have been possible without the dedicated work of the colleagues who formed this laboratory and continue to generate the insight and hypotheses. Particularly we wish to acknowledge the essential functions of Norah Rogers, who for 13 years has managed the laboratory at Florida State University with exemplary care and responsibility. This work has been supported by grants N00014-87-K00012, N00014-83-K-0275, and N00014-88-K-0489 from the Office of Naval Research; OCE-80-19757 and DPP-86-12348 from the National Science Foundation; and DE-FG05-88ER60643 from the Office of Health and Environmental Research, Department of Energy. The FT/IR was purchased with grant N00014-83-G-01066 from the Department of Defense, University Research Instrumentation Program, Office of Naval Research, and the VG Trio-3 GC/MS/MS was acquired with funds from the University of Tennessee, Grant ARO 24187-LS-RI, Department of Defense, University Instrumentation Program, Army Research Office, 2-4-01018 DEG-Lab Equipment from the Department of Education (to G.S. Sayler), and DE-F605-8ER75375 from the Department of Energy, University Research Instrumentation Program. The generous gift of the Hewlett-Packard HP-1000 RTE-6/VM data system for the HP5996A GC/MS system greatly facilitated the research reported here.

REFERENCES

Balkwill, D.L., F.R. Leach, J.T. Wilson, J.F. McNabb, and D.C. White. 1988. Equivalence of microbial biomass measures based on membrane lipid and cell wall components, adenosine triphosphate, and direct cell counts in subsurface sediments. *Microb. Ecol.* 16:73–84.

Dowling, N.J.E., J. Guezennec, M.L. Lemoine, A. Tunlid, and D.C. White. 1988. Corrosion analysis of carbon steels affected by aerobic and anaerobic bacteria in mono- and coculture using AC impedance and DC techniques. *Corrosion* 44:869–874.

Dowling, N.J.E., E.E. Stansbury, D.C. White, S.W. Borenstein, and J.C. Danko. 1989. On-line electrochemical monitoring of microbially induced corrosion. In: Microbial Corrosion: 1988 Workshop Proceedings, ed. G. Licena, pp. 1, 5–19. Research Project 8000-26, Electric Power Research Institute, Palo Alto, CA.

Fazio, S.A., J. Uhlinger, J.H. Parker, and D.C. White. 1982. Estimations of uronic acids as quantitative measures of extracellular polysaccharide and cell wall polymers from environmental samples. *Appl. Env. Microbiol.* 43:1151–1159.

Gehron, M.J., and D.C. White. 1982. Quantitative determination of the nutritional status of detrital microbiota and the grazing fauna by triglyceride glycerol analysis. *J. Exp. Mar. Biol. Ecol.* 64:145–158.

Guckert, J.B., C.B. Antworth, P.D. Nichols, and D.C. White. 1985. Phospholipid, ester-linked fatty acid profiles as reproducible assays for changes in prokaryotic community structure of estuarine sediments. *FEMS Microbiol. Ecol.* 31:147–158.

Guckert, J.B., M.A. Hood, and D.C. White. 1986. Phospholipid, ester-linked fatty acid profile changes during nutrient deprivation of *Vibrio cholerae*: increases in the *trans/cis* ratio and proportions of cyclopropyl fatty acids. *Appl. Env. Microbiol.* 52:794–801.

Jain, R.K., R.S. Burlage, and G.S. Sayler. 1988. Methods for detecting recombinant DNA in the environment. *CRC Crit. Rev. Biotech.* 8:33–84.

Little, B., R. Ray, P. Wagner, Z. Lewandowski, W.G. Characklis, J. Jacobus, and F. Mansfeld. 1989. Electrochemical reactions on stainless steels covered with biofilms. In: Corrosion Research Symposium 1989, ed. T.P. Ford and T.M. Devine, pp. 30–32. Houston, TX: National Association of Corrosion Engineers.

Mycke, B., F. Narjes, and W. Michaelis. 1987. Bacteriohopanetetrol from chemical degradation of an oil shale kerogen. *Nature* 326:179–181.

Nichols, P.D., J.B. Guckert, and D.C. White. 1986. Determination of monounsaturated fatty acid double-bond position and geometry for microbial monocultures and complex consortia by capillary GC-MS of their dimethyl disulphide adducts. *J. Microbiol. Meth* 5:49–55.

Nichols, P.D., J.M. Henson, J.B. Guckert, D.E. Nivens, and D.C. White. 1985. Fourier transform-infrared spectroscopic methods for microbial ecology: analysis of bacteria, bacteria-polymer mixtures and biofilms. *J. Microbiol. Meth.* 4:79–94.

Odham, G., A. Tunlid, G. Westerdahl, G. Larsson, J.B. Guckert, and D.C. White. 1985. Determination of microbial fatty acid profiles at femtomolar levels in human urine and the initial marine microfouling community by capillary gas chromatography-chemical ionization mass spectrometry with negative ion detection. *J. Microbiol. Meth.* 3:331–344.

Ogram, A.V., and G.S. Sayler. 1988. The use of gene probes in the rapid analysis of natural microbial communities. *J. Ind. Microbiol.* 3:281–292.

Olsen, G.J., D.J. Lane, S.J. Giovanni, N.R. Pace, and D.A. Stahl. 1986. Microbial ecology and evolution: a ribosomal approach. *Ann. Rev. Microbiol.* 40:337–365.

Ratledge, C., and S.G. Wilkinson. 1988. Microbial Lipids, vol. 1. New York: Academic.

Revsbech, N.P., and B.B. Jorgensen. 1986. Microelectrodes: their use in microbial ecology. *Adv. Microb. Ecol.* 9:293–352.

Ringelberg, D.B., J.D. Davis, G.A. Smith, P.D. Nichols, J.B. Nickels, J.M. Hensen, J.T. Wilson, M. Yates, D.H. Kampbell, H.W. Reed, T.T. Stockdale, and D.C. White. 1988. Validation of signature polar lipid fatty acid biomarkers for alkane-utilizing bacteria in soils and subsurface aquifer materials. *FEMS Microbiol. Ecol.* 62:39–50.

Smith, G.A., J.S. Nickels, B.D. Kerger, J.D. Davis, S.P. Collins, J.T. Wilson, J.F. McNabb, and D.C. White. 1986. Quantitative characterization of microbial biomass and community structure in subsurface material: prokaryotic consortium responsive to organic contamination. *Can. J. Microbiol.* 32:104–111.

Stahl, D.A., B. Flesher, H.R. Mansfeld, and L. Montgomery. 1988. Use of phylogenetically based hybridization probes for studies of ruminal microbial ecology. *Appl. Env. Microbiol.* 54:1079–1084.

Tunlid, A., G. Odham, R.H. Findlay, and D.C. White. 1985. Precision and sensitivity in the measurement of ^{15}N enrichment in D-alanine from bacterial cell walls using positive/negative ion mass spectrometry. *J. Microbiol. Meth.* 3:237–245.

White, D.C. 1983. Analysis of microorganisms in terms of quantity and activity in natural environments. *Symp. Soc. Gen. Microbiol.* 34:37–66.

White, D.C. 1988. Validation of quantitative analysis for microbial biomass, community structure, and metabolic activity. *Adv. Limnol.* 31:1–18.

White, D.C., R.J. Bobbie, J.S. Herron, J.D. King, and S.J. Morrison. 1979. Biochemical measurements of microbial mass and activity from environmental samples. In: Native Aquatic Bacteria: Enumeration, Activity and Ecology. *Soc. Test. Mater. Sp. Publ.* 695:69–81.

White, D.C., J.D. Davis, J.S. Nickels, J.D. King, and R.J Bobbie. 1979. Determination of the sedimentary microbial biomass by extractable lipid phosphate. *Oecologia* 40:51–62.

White, D.C., and R.H. Findlay. 1988. Biochemical markers of predation effects on the biomass, community structure, nutritional status and metabolic activity. *Hydrobiol.* 159:119–132.

Organic Acids in Aquatic Ecosystems
eds. E.M. Perdue and E.T. Gjessing, pp. 43–63
John Wiley & Sons Ltd

Compositional Indicators of Organic Acid Sources and Reactions in Natural Environments

J.I. Hedges

School of Oceanography, WB-10
University of Washington
Seattle, WA 98195, U.S.A.

Abstract. Both bulk chemical measurements (e.g., elemental, isotopic, and spectral analyses) and molecular tracers (biomarkers) can be used to determine the sources and reaction histories of organic acids in natural environments. This paper reviews selected characterization methods of these two types and critically evaluates their strengths and weaknesses. Finally, research directions are suggested which might lead to more accurate and sensitive procedures for determining the sources and pathways of organic acids in natural environments.

INTRODUCTION

Throughout the following discussion, the term "organic acid" will be used in reference to naturally occurring organic molecules of any size that are characterized by the presence of functional groups capable of doning protons in the natural pH range of approximately 4–9. Molecules in this category range from formic acid to high molecular weight humic substances and carry a variety of acidic substituents including carboxyl, phenolic, and nitrogen functionalities. Organic acids occur within almost all natural environments and in solid, colloidal, dissolved, and gaseous forms. Although molecules of this type may have a wide range of sources, properties, and histories, all share the characteristics that they can affect the pH of natural waters and form complexes with metal ions. Low molecular weight organic acids are often metabolically important substrates whereas humic polymers are one of the major forms of organic matter in soils, sediments, and natural waters and take part in a wide variety of geochemical processes (Thurman 1985; Frimmel and Christman 1988).

The sources and reactions of organic acids are fundamental considerations for investigations of many natural environments and the processes that occur within them. For several reasons, however, the origins and environmental histories of organic acids can be difficult to delineate. A major contributing factor to this uncertainty is that almost all natural organic acids occur in complex mixtures where the components may have different forms, ages, and transport mechanisms. In addition, small organic acids typically have simple structures and ubiquitous distributions which hinder definitive source assignments. Although individual humic molecules may be structurally unique, their large size, heterogeneity, and extreme diagenetic alteration lead to at least comparable characterization problems. In spite of these challenges, many organic acids carry structural and isotopic clues of their origins and pathways that can be read and deciphered if the appropriate methods are knowingly applied.

Almost all chemical methods for delineating the sources and reactions of organic acids can be categorized as measurements of either bulk or molecular properties. As will be discussed in the next two sections, these two methods of characterization exhibit sharply contrasting strengths and weaknesses (Norwood 1988) and yield fundamentally different types of geochemical information.

BULK CHEMICAL CHARACTERIZATIONS

Determinations of bulk chemical or isotopic properties have the characteristic that a property common to the constituents of a mixture is determined by a broadly descriptive analytical method. Examples of such approaches are determinations of isotopic, elemental, or functional group compositions. Characterizations by such methods have a number of intrinsic advantages. First, the analytical methods are comparatively comprehensive in that all the sample is usually analyzed, often for a quantitatively important constituent. Second, because they are extensive, bulk properties often can be interrelated (e.g., total acidity and oxygen content) or used together to advantage in balances of some type (e.g., determination of O by difference in elemental CHN analyses). Third, analyses of bulk properties often are readily accomplished and nondestructive (e.g., spectral analysis). Finally, bulk methods are ideally suited for the characterization of heterogeneous organic polymers, such as humic substances, which defy comprehensive structural analysis but may exhibit definitive compositional trends.

Bulk characterizations also have at least two basic weaknesses. One of these is that because the bulk properties are shared among many components of complex environmental mixtures, it is often difficult to detect and quantify a specific class of constituents (e.g., terrigenous humic substances) by a given characteristic. This is because an individual class of constituents usually exhibits a range of values for a given property against which the effect of admixing a second type of organic matter must be assessed (Fig. 1). Thus, the uncertainty in characterizing individual mixture components proportionately reduces the sensitivity with which additional components can be detected. This problem is often exacerbated by the second drawback that only limited information can be drawn from many bulk characterizations. For example CHN analyses directly provide only three features of an organic mixture and most stable isotope analyses result in only one datum per sample.

Selected common bulk analyses that provide source or reaction history information will be reviewed in the following section. In this treatment, "source" will represent the biochemical, biological, or geographic origin of the organic acid. These different levels of source distinction are often linked, for example when an organism characteristic of a particular physical environment produces a unique biochemical or bulk compositional property. Likewise, "reaction history" will include any chemical changes that occur in the environment following the death of the organism in which the organic acid (or its precursor) was biosynthesized.

Elemental Analyses

Elemental composition is a useful and often overlooked method for delineating both the sources and reactions of organic acids. Although potentially applicable to small simple molecules, this technique is particularly well suited for humic substances and other structurally complex polymers. The general approach involves the direct analysis of the weight percentages of C, H, N, S, and O (the latter can be analyzed directly by pyrolysis) which together comprise 99% or more of most organic materials.

The results of elemental analyses are often presented in a van Krevelen plot of atomic H/C versus atomic O/C (e.g., Fig. 2), or some derivative presentation such as N/C versus O/C. Algebraic solutions for specific mixture components also can be obtained, if a limited number of endmembers of known composition are present. This method is useful for source analysis because many major biochemical classes have characteristic elemental compositions and distinct abundance patterns in different kinds of living organisms or geographic regions. For example, woody vascular plants are confined essentially to land and contain high levels of the aromatic polymers, lignin and tannin, which usually impart a characteristic low H/C to terrestrially derived humic substances (Fig. 2). In contrast, plankton have lipids and proteins among their major components and typically contribute toward aquatic humic substances with high H/C and N/C ratios. Polysaccharides are among the most oxygen-rich of all biochemicals (Fig. 2) and impart this characteristic to derivative organic acids, including uronic acids. Mixtures

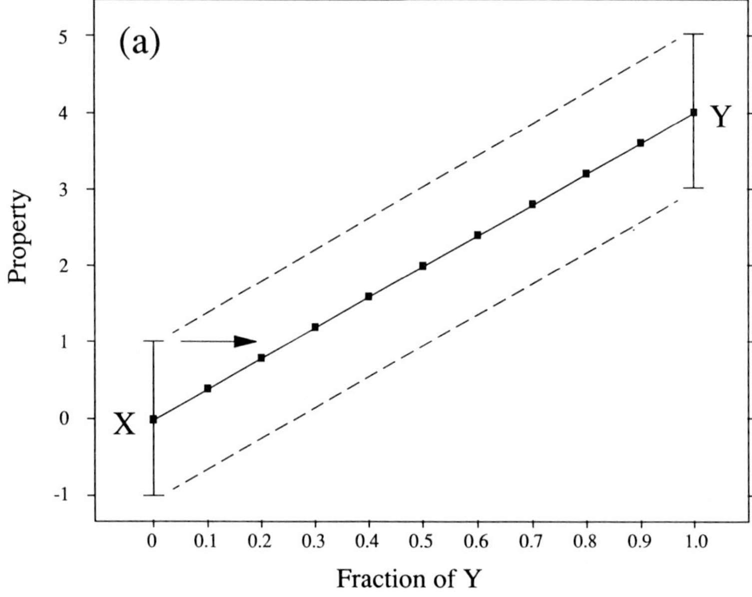
(a)
5
4
3
2
1
0
-1
Property
X
Y
0
0.1
0.2
0.3
0.4
0.5
0.6
0.7
0.8
0.9
1.0
Fraction of Y

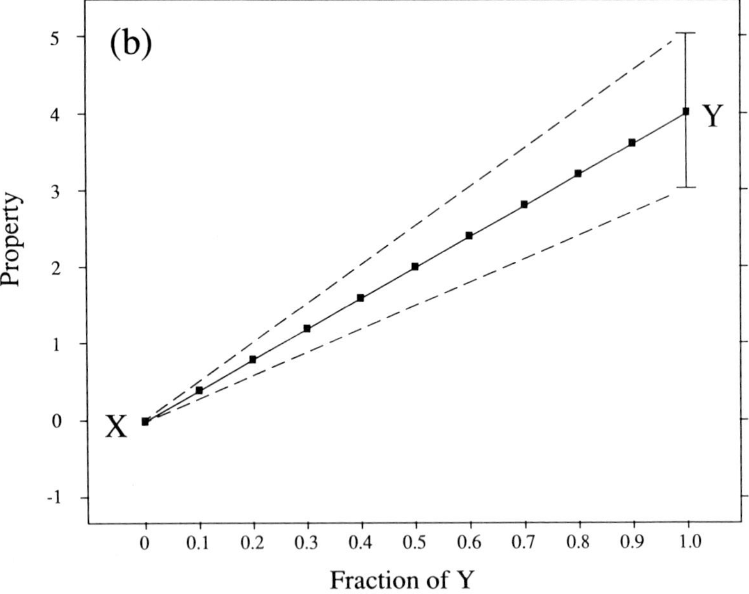
(b)
5
4
3
2
1
0
-1
Property
X
Y
0
0.1
0.2
0.3
0.4
0.5
0.6
0.7
0.8
0.9
1.0
Fraction of Y

of the above biochemicals in individual organisms (or environmental samples) must lie within the compositional range of the components, unless their elemental compositions are greatly altered by diagenesis (see below).

The power of elemental analyses and van Krevelen-type plots for determinations of both the source and reaction histories of humic substances has been demonstrated by Reuter and Perdue (1984). They point out that in van Krevelen plots the compositional trajectories of organic materials undergoing diagenesis often are diagnostic of the reaction that is occurring. For example, selective loss or gain of a specific carbon-containing moiety from a humic material will shift the H/C and O/C of the reaction product along a line which runs between the two composition points for the initial material and its lost component (Fig. 2). Uptake or removal of oxygen causes only horizontal displacements from the original compositional point whereas gains or losses of water (hydration and dehydration) always shift composition points along a line with a slope of 2.

Given these relationships, either the position of an individual composition point or a trend among a suite of related samples can provide a reaction history record. Thus a humic substance with low H/C whose composition point lies to the right (high oxygen) side of the biochemical mixing region (point X in Fig. 2) could be formed diagenetically from the oxidation of a lignin- or tannin-rich terrigenous precursor. Similarly, a trend of slope approximately 2 along a line between the polysaccharide and lignin endmembers (broken line in Fig. 2) could result either from (*a*) preferential carbohydrate loss from a woody tissue or (*b*) dehydration of a polysaccharide-rich material. Visser (1983) demonstrated how size fractions obtained from individual humic polymers follow consistent trends in van Krevelen plots that apparently reflect the diagenetic history of the sample.

Although useful, elemental compositional analyses are also subject to a variety of sample preparation and measurement artifacts. For example, the high sulfur contents often measured for humic substances from reducing environments may be in part an artifact of basic extractions carried out in the presence of reduced inorganic sulfur compounds (Francois 1987).

Fig. 1—Schematic representation of the comparative lower bounds for the quantification of Y when (a) X and Y share a common property with equal natural variability in both endmembers and (b) the measured property is unique to Y. In case (a) it is impossible to confidently quantify less than about 20% of Y (arrow) by measurements of the common property, due to the natural variability in X (brackets). In case (b), however, where the property is only expressed by Y, the potential exists to detect and measure extremely small concentrations of Y ($< 1\%$) in binary mixtures with X. The accuracy of the low-level measurement of Y ultimately will depend on the sensitivity and precision of the analytical method for the unique property and on contributions of background interferences from other properties (such as coeluting compounds in chromatographic analyses).

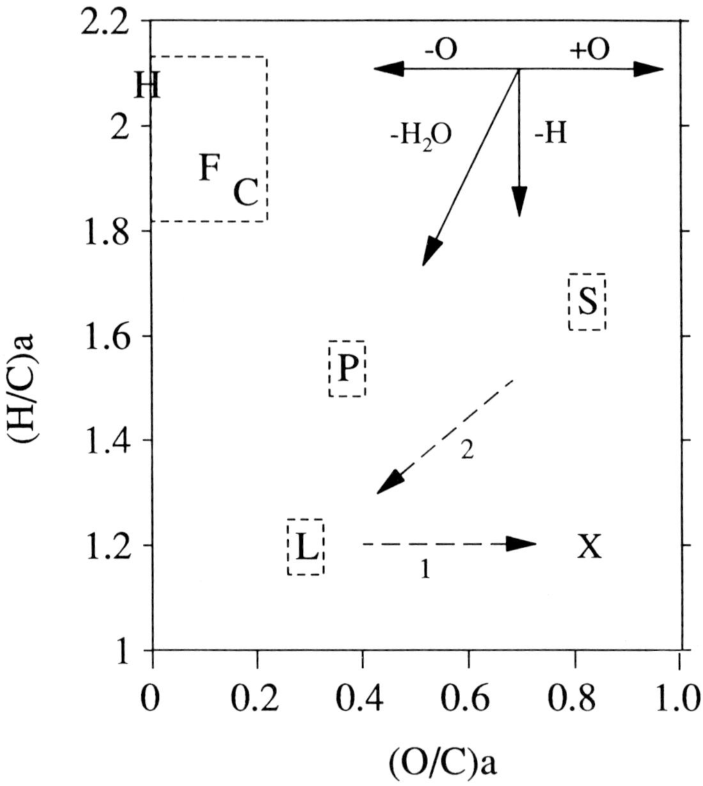

Fig. 2—A van Krevelen plot illustrating the compositional positions of some major biochemical components of living organisms and the displacements ("trajectories") that can be expected for different types of environmental reactions. Abbreviations: (H/C)a, atomic hydrogen to carbon ratio; (O/C)a, atomic oxygen to carbon ratio; H, hydrocarbon; F, fat; C, cutin; S, polysaccharide; P, protein; and L, lignin. The joined arrows at the top of the plot represent the directions of compositional changes that result, at any point on the plot, when an organic substance loses oxygen (−O) or is oxidized (+O), dehydrated ($-H_2O$), or dehydrogenated (−H). Toward the bottom of the plot, trajectory 1 represents the compositional change to be expected when lignin is oxidized to form a humic acid (X) derived from this source only. Trajectory 2 could result either from the preferential loss of polysaccharide from wood (which is about 1/4 lignin and 3/4 polysaccharide) or, less likely because the slope does not match closely, from the dehydration of polysaccharide alone. The elemental compositions of all biochemicals except protein were calculated from commonly available literature data (e.g., Sarkanen and Ludwig 1971; Holloway 1982). The protein value is directly from Reuter and Perdue (1984) which was the source and inspiration for the treatments illustrated in this figure.

Likewise, analyses of oxygen as the difference between total sample mass and measured C, H, N, and ash contents are intrinsically imprecise and subject to error from multiple sources such as the presence of molecular water, sulfur, and mineral phases that change mass on combustion. Finally, most "dry" humic substance isolates contain some adsorbed water that is difficult to remove or quantify on small samples and leads to erroneously high H/C ratios (see Frimmel, this volume).

Stable Isotopes

Stable isotope compositions are among the most comprehensive and least diagenetically sensitive parameters for source designations among organic acids of all types (Fritz and Fontes 1980). Stable carbon isotope compositions ($^{13}C/^{12}C$), the most commonly applied nuclear-level characterizations, are particularly representative because organic compounds contain about 50% by weight carbon, essentially all of which is combusted to CO_2 and analyzed by ratio mass spectrometry. Because carbon forms the structural framework of organic molecules and is not exchangeable (as potentially are O and H), this element is particularly resistant to diagenetic alteration and carries one of the most permanent of all source signatures.

The resolving power of stable carbon isotopic measurements, however, is limited to only a few biological sources. The most ^{13}C-rich ("heavy") major source is C_4 vascular land plants, such as *Spartina*, corn, and many grasses, which fix CO_2 by the Hatch and Slack pathway and have $\delta^{13}C$ values in the range of $-10‰$ to $-14‰$ (Fry and Sherr 1984). In contrast, ordinary vascular land plants, which fix CO_2 by the C_3 pathway, have $\delta^{13}C$ values in the range of $-23‰$ to $-30‰$. Marine plankton have intermediate $\delta^{13}C$ values ($-18‰$ to $-24‰$) whereas freshwater plankton have an isotopic signature similar to that of C_3 land plants. The previous values are all for total tissue within which proteins, carbohydrates, lignins, and lipids typically become increasingly depleted in ^{13}C ("lighter").

Most applications of stable carbon isotope analysis for source distinctions among organic acids have been for humic polymers, for which definitive structural characterizations are difficult. A classic example of this approach was the demonstration by Nissenbaum and Kaplan (1972) that humic and fulvic acids from ocean sediments have stable carbon isotopic compositions that closely resemble those of local marine plankton and are thus likely to have an *in situ* origin. Similar isotopic measurements have demonstrated a primarily indigenous origin of dissolved humic substances in seawater (Stuermer and Harvey 1974), a finding supported by the low lignin content of these mixtures (Meyers-Schulte and Hedges 1986).

Stable isotope distributions of nitrogen ($^{15}N/^{14}N$), hydrogen ($^{2}H/^{1}H$), oxygen ($^{18}O/^{16}O$), and sulfur ($^{34}S/^{32}S$) also bear potential information about the sources of organic acids in nature (Fritz and Fontes 1980). Isotopic differences among all four elements can be useful indicators of marine versus terrestrial origin. Sulfur isotopes also have the potential to distinguish biologically fixed S of sulfate origin from sulfide-derived S that is incorporated as an artifact of humic isolation procedures. Both oxygen and hydrogen stable isotopes are fractionated during the hydrologic cycle. Their increasing inland depletion in precipitation is reflected by local vegetation and offers a unique means of distinguishing among the different possible terrestrial source regions of organic acids (Fritz and Fontes 1980). For example, organic acids of identical chemical composition from low and high elevations within an individual drainage basin should be discernable by the lower ($^{2}H/^{1}H$) and ($^{18}O/^{16}O$) of molecules from the more upland source region.

Radiocarbon

In addition to conventional dating methods, ^{14}C analyses provide a means of delineating both the sources and recent dynamics of organic acids in natural environments. Such applications usually are based on the near doubling of the ^{14}C content of atmospheric CO_2 during the early 1960s as a result of atmospheric nuclear testing and the subsequent decrease in $^{14}CO_2$ as this radiocarbon was mixed into seawater and the biosphere (Fig. 3). All subsequently biosynthesized terrestrial organic matter has been labeled with characteristically high concentrations of this "bomb" ^{14}C, whose radioactive decay rate is extremely slow (half-life $\approx$ 5700 yr). Thus, "excess" radiocarbon in any organic acid indicates that at least a portion of the material (if it is a mixture) was photosynthesized since the onset of active atmospheric nuclear testing in the early 1950s. Organic acids which were biosynthesized subaerially during the 1960s carry uniquely high ^{14}C concentrations versus those that were made at other times on land, or at any time in the sea where ^{14}C excesses have consistently been much smaller (Fig. 3).

The dynamics of organic acids within a finite terrestrial reservoir, such as a soil profile, lake, or drainage basin, can be inferred from the average extent of radiocarbon enrichment over prebomb levels. For such estimates the minimal fraction of bomb carbon in the sample can be determined from the percentage of atmospheric bomb CO_2 that would have to be added (at the local historic maximal ^{14}C) to a mixture that otherwise contained only modern, prebomb carbon (ca. 1954). Division of the maximal number of years since active atmospheric testing (presently about 35) by the previously described minimal fraction of bomb carbon yields a maximal average residence time for organic carbon in the pool of question. Calculations of this type were used to determine a maximal residence time of less than 150 years between photosynthetic fixation and riverine export of carbon within

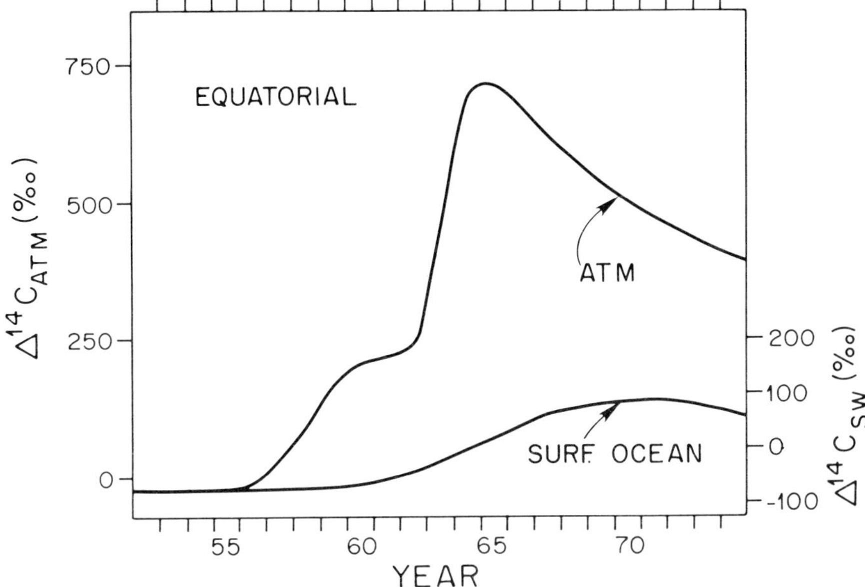

Fig. 3—A plot of the radiocarbon contents ($\Delta^{14}C$) of atmospheric CO_2 and dissolved inorganic carbon dissolved in surface ocean water of equatorial regions between 1940 and the present. The sharp increase in the ^{14}C content of the atmosphere and ocean in the early 1960s was the result of atmospheric testing of thermonuclear devices. The spike of "bomb" ^{14}C has decreased since the ban on testing due to mixing of the excess radiocarbon into the ocean and biosphere. This plot is after Broecker and Peng (1982, p. 389) and is based on data cited in that reference.

dissolved humic substancs from the Amazon River drainage basin (Hedges et al. 1986).

Spectral Analyses

Spectral measurements have been one of the most commonly applied methods for characterizing organic acids from natural environments, especially humic substances. The methods employed include IR, ESR, UV, fluorescence, and NMR spectroscopy of different types. Although all these techniques can yield useful information on the sources of organic acids, ^{13}C NMR analyses of solid samples by the cross-polarization, magic angle spinning (CP/MAS) method have been especially rewarding (see review by Wilson 1987). This technique allows absorbance measurement at up to 15 bands which correspond to different chemical environments around carbon atoms within organic molecules (Fig. 4). CP/MAS ^{13}C NMR spectra can be

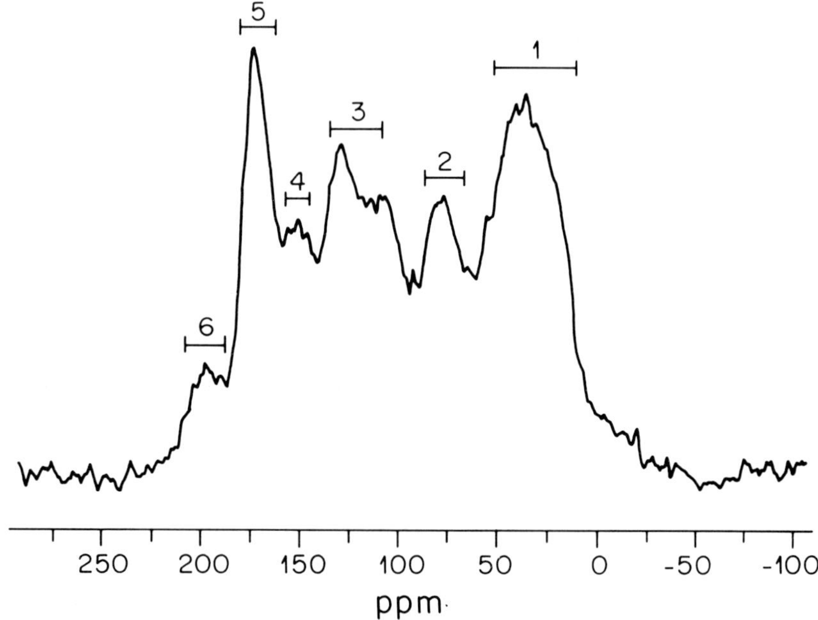

Fig. 4—A CP/MAS ^{13}C NMR spectrum of dissolved fulvic acid isolated from the Rio Negro by adsorption of an acidified solution onto XAD-8 resin (Ertel et al. 1986). The spectrum was obtained by Patrick Hatcher (U.S.G.S., Reston, VA) under conditions similar to those described by Hedges et al. (1985). Assignments of the numbered major absorbance bands are as follows: 1, alkyl carbon; 2, carbon bonded singularly to one O or N; 3, aromatic or doubly bonded nonaromatic carbon bonded to C or H; 4, aromatic carbon substituted by O or N; 5, carboxyl carbon; 6, carbonyl carbon (see Wilson 1987).

obtained nondestructively on milligram size samples and (under proper operating conditions) reveal most carbon types with comparable efficiencies. Because the measurement does not necessitate dissolution of the sample, spectra can be obtained without chemical fractionations that usually attend physical separations. The overall result of the various advantages is that CP/MAS ^{13}C NMR analysis provides a detailed and representative structural characterization that is consistent with, and complementary to, corresponding elemental and functional group analyses (Wilson et al. 1987). Although ^{1}H NMR can yield supplementary compositional information (Wilson et al. 1987), especially concerning different types of aliphatic structures, this method presently requires dissolution of the sample and is not as widely employed as CP/MAS ^{13}C NMR for the characterization of humic substances.

The key to interpreting organic acid sources based on ^{13}C NMR spectral data is to relate individual absorbance peaks (or groups of peaks) to specific biochemicals. For example, lignin structures give characteristic absorbances for methoxyl (55 ppm) and aromatic carbons (120–150 ppm) which provide clear evidence both for the presence of vascular plant and terrigenous carbon (Fig. 4). The oxygen-substituted carbons of carbohydrates adsorb at four distinctive bands in the 65–105 ppm range. Finally, lipids give predominantly low field (<25 ppm) absorbances due to methyl and methylene carbons (Fig. 4). Biochemical distributions indicated by ^{13}C NMR also can be used to gain information about the reaction histories and environments of organic acids. An example of such an application is the demonstration by Orem and Hatcher (1987) that dissolved organic matter (DOM) from the pore waters of anaerobic sediments contains high carbohydrate levels whereas in aerobic sediments the microbial degradation of lignin releases aromatic structural units to pore water DOM.

MOLECULAR BIOMARKER CHARACTERIZATIONS

In general, biomarker methods for characterizing organic acids depend on the selective measurement of a specific structure within a sample. In most cases, these direct measurements necessitate the separation and detection of relatively small molecules (<1,000 mass units) of known structure. Common examples would be the extraction and chromatographic separation of small organic acids which preexist in a sample or are released from the degradation of large acidic polymers such as humic substances. By definition, the biomarker must be structurally related to a specific biological or geographic source.

Biomarker characterizations also have intrinsic strengths and weaknesses. The first strength is that due to the great diversity of different organic molecules in living organisms, biomarkers often allow finer distinctions among potential organic sources than are possible with measurements of bulk chemical properties. In addition, these distinctions sometimes can be carried out with sensitivity orders of magnitude greater than is possible with bulk analyses. This increased sensitivity results from the fact that a unique biomarker often can be quantified against an effective analytical background of zero, as opposed to a shared bulk property which can never be quantified with an uncertainty less than the variability in the coexisting components (Fig. 1).

Biomarker applications, however, have a number of associated drawbacks. The first of these is that information on the distribution of natural products in specific organisms is very incomplete. For example, individual sterols which were once considered to be unique to vascular land plants are now known to occur in marine algae as well (Volkman 1986) and a highly aliphatic biopolymer of unusual chemical stability has only recently been discovered in the cuticles of higher plants (e.g., Nip et al. 1986). As a result of such spotty chemotaxonomic information, biomarkers sometimes are obtained from undocumented sources which can confuse interpretations of geochemical relationships and processes. A second drawback is that most biomarkers occur only as trace components of living organisms and environmental mixtures and thus do not directly affect commonly measured bulk properties. The last, and by far the most troublesome, problem is that biomarker applications are sensitive to the effects of diagenesis, which is the alteration of biochemicals in nature following the death of the source organism. This shortcoming can be particularly acute in biomarker applications to humic substances, which themselves are thought to be the products of extensive degradation of natural products.

Although a wide variety of organic acids have been directly extracted from natural samples or produced from them by degradative reactions (Thurman 1985), only a fraction of these have unique sources and therefore qualify as biomarkers. Because individual biomarkers often can be quantified in different types of samples by a wide variety of methods, it is least confusing to review the topic of source discrimination from the perspective of compound type, rather than by study environment or analytical method. The following is an overview of several classes of biomarkers which are themselves acids or carry information about the sources of acidic precursors such as humic substances.

Lipid Biomarkers

Fatty acids are aliphatic lipids containing one carboxyl group per molecule along with eleven or more other carbons. Their predominant sources are animal fats (esters of glycerol) or plant waxes (esters of long-chain alcohols). Fatty acids are nonvolatile, sparingly soluble in water, and have a high affinity for adsorption by particles. They can be directly extracted with nonpolar solvents from soils, sediments, and natural waters or released degradatively from humic substances and other polymers. Although lower molecular weight aliphatic acids are also common in nature (Thurman 1985), they carry limited structural information about their sources.

A variety of structural features of fatty acids bear source information. One of the most fundamental of these is molecular size. Although almost all types of organisms synthesize linear fatty acid molecules containing 12–22 carbons, vascular land plants are the primary source of straight-chain fatty acids of higher molecular weight (Fig. 5a). These plant wax fatty acids occur in suites characterized by a strong predominance of homologs having even numbers of carbon atoms per molecule. A second diagnostic feature of fatty acid structures is the degree of branching in the carbon backbone of the molecule. Plants primarily produce straight-chain structures whereas bacteria typically synthesize fatty acids that have methyl branches on the first (or second) carbon from the tail (uncarboxylated) end of the molecule (Fig. 5f), or cyclopropyl structures (Fig. 5h).

Other distinctive features are the placement, number, and geometries of carbon double bonds in fatty acids. Vascular plant fatty acids usually contain no more than three double bonds, which are all unconjugated (more than one carbon between) and of *cis* geometry (Fig. 5c). In contrast some phytoplankton, such as diatoms and dinoflagellates, biosynthesize fatty acids with 4 to 5 carbon double bonds (Fig. 5e). Bacteria seldom form more than one double bond per molecule and are the only important direct source of *trans* fatty acids (Fig. 5g). Bacterial fatty acids also sometimes have their single carbon double bond at an unusual position versus its usual position in most other organisms (nine linkages in from the carboxyl group). Many of these characteristic patterns are especially pronounced in phosphorus-containing polar lipids, which can be selectively extracted from their nonpolar counterparts and then analyzed by conventional means (Guckert et al. 1985; White, this volume).

Many examples of the application of fatty acids (from nonpolar and polar lipids) as source indicators can be found in the geochemical literature (e.g., Parrish 1988). These compounds have been identified in both free and esterified forms in water and soil samples. Fatty acids also become bound over time into humic substances and polymeric organic materials from which they can only be released by hydrolysis or other harsh chemical treatments (e.g., Nishimura and Baker 1987).

Resin acids are also synthesized by vascular plants and are thus unique to terrestrial sources. These diterpenoid acids of which abietic acid (Fig. 5d) is representative, are characterized by 20 carbons in branched, polycyclic structures. Such substances are particularly informative because they also exhibit characteristically different structural modifications among different taxa of higher plants. Resin acids have been directly extracted from soils and identified in pelagic marine sediments where they record the input of the remains of land-dwelling resinous plants (Simoneit 1977).

Centrally hydroxylated C_{16} and C_{18} fatty acids (Fig. 5b) also are unambiguous indicators of land plant sources. These molecules comprise complex natural polyesters known as cutin and suberin which coat the leaves and roots, respectively, of many vascular plants (Holloway 1982). Cutin and suberin polyesters are not soluble in nonpolar solvents; the component acid molecules must be directly released by either basic hydrolysis or reduction with $LiAlH_4$ or $LiAlD_4$. Although cutin- and suberin-derived hydroxyacids often vary considerably among plants in molecular weight, number and position of hydroxyl groups, and degree of unsaturation, unique chemotaxonomic relationships are not as yet evident (Holloway 1982). Hydroxyacids have been applied as tracers of vascular plant remains in lacustrine and marine sediments, but do not appear to be sufficiently stable to persist over geologic time (Cardoso and Eglinton 1983).

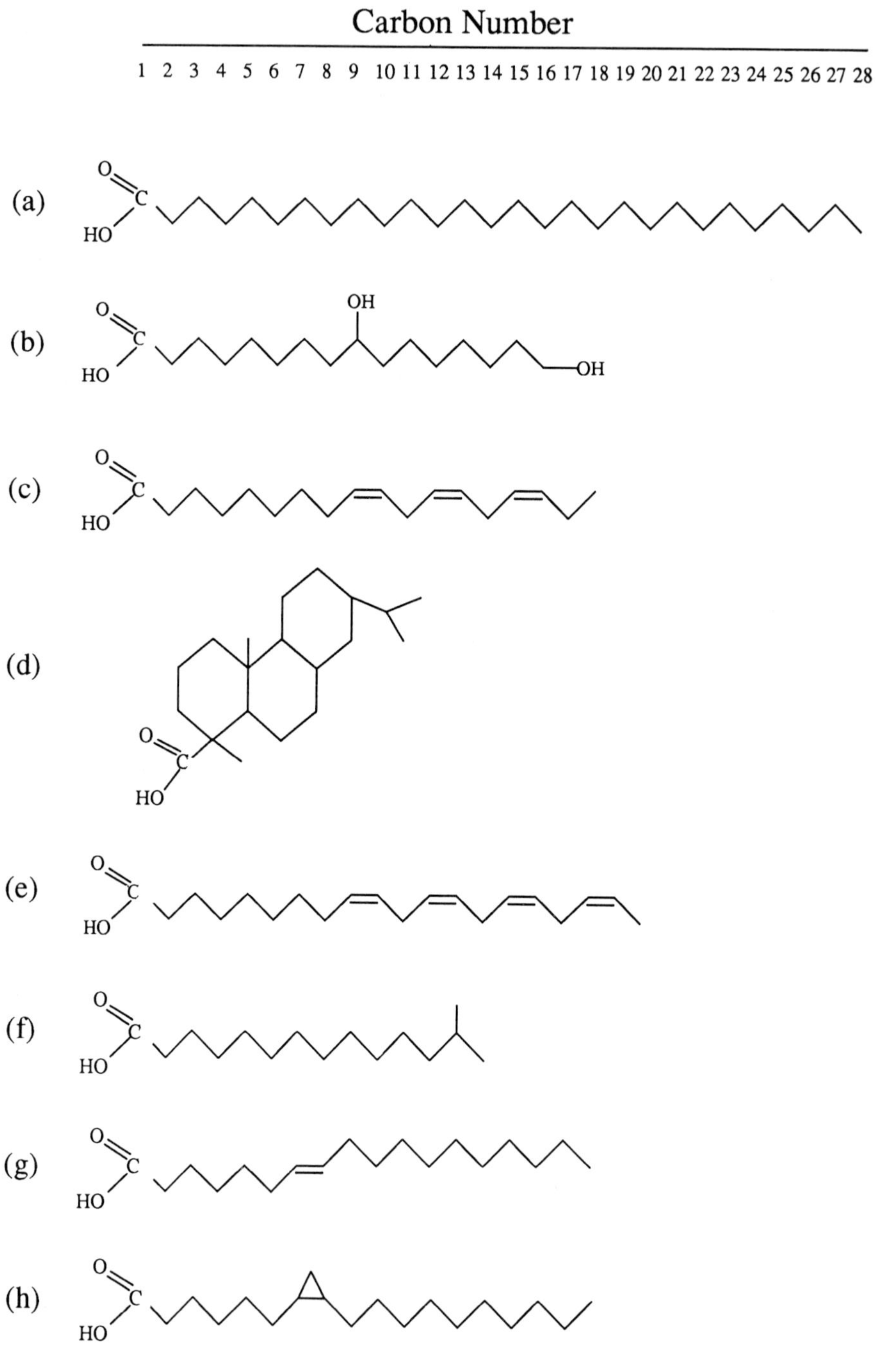

Carbon Number

1 2 3 4 5 6 7 8 9 10 11 12 13 14 15 16 17 18 19 20 21 22 23 24 25 26 27 28

(a)

(b)

(c)

(d)

(e)

(f)

(g)

(h)

Lignin Composition

Lignin compounds have proven to be useful biomarkers for both the vascular plant and terrestrial provenances of organic acids. Lignins are structurally complex, high molecular weight phenolic polymers that are unique to the support structures of vascular plants (Sarkanen and Ludwig 1971). These polymers are intrinsically acidic and, upon microbial degradation, release characteristic phenols and phenolic acids which occur in soils, sediments, and natural waters both as simple compounds and as structural units within humic substances (Thurman 1985).

Lignin compounds have several charactersistics, other than a unique vascular plant source, that lend advantages as biomarkers. First, these highly cross-linked polymers are among the most refractory of all plant components and thus tend to persist in chemically recognizable form into the later stages of organic diagenesis. This trait enables them to serve as source indicators for humic substances, within which many of the other major structural precursors can no longer be identified. Second, the phenolic building blocks of lignin polymers differ in clear taxonomic patterns among various broad types of vascular plants and their tissues. Angiosperms can be distinguished from gymnosperms by the presence of doubly methoxylated (syringyl) phenol building blocks whereas nonwoody vascular plant tissues are characterized by the presence of cinnamyl acids (such as *p*-coumaric and ferulic acid) with unsaturated three-carbon side chains. Finally, lignin structural units become systematically enriched in acidic structural units in response to white-rot fungal degradation (Hedges et al. 1988), thereby providing a "built-in" indication of the state of preservation of the remnant lignin.

A variety of methods can be used to analyze for lignin in environmental samples. Most of these analyses involve degradation of lignin polymers to simple phenols which retain the methoxylation and side chain structures that characterize different vascular plant sources. One of the most common

---

Fig. 5—Chemical structures of eight different lipid molecules which carry features characteristic of different biological sources. The various sources and the related diagnostic structural features of their tracer molecules ("biomarkers") are as follows. *Vascular land plants* characteristically contain: (a) high molecular weight ($> C_{22}$), linear fatty acids with a predominance of homologs having an even number of carbon atoms per molecule, (b) centrally hydroxylated C_{16} and C_{18} fatty acids of cutin origin, (c) shorter chain ($< C_{20}$), polyunsaturated fatty acids with all *cis* double bonds at common, unconjugated positions, and (d) polycyclic, branched diterpenoid resin acids such as abietic acid. *Phytoplankton* such as diatoms also contain (e) very unsaturated (>3 C = C) C_{20} and C_{22} linear acids with all *cis* bonds. *Bacteria* contain low molecular weight lipids which often incorporate unusual structural features such as: (f) methyl branching at the ultimate (iso form) or penultimate (anteiso form) carbon, (g) *trans* carbon double bonds, often at unusual positions, and (h) mid-chain cyclopropyl groups (see text).

techniques is basic cupric oxide (or nitrobenzene) oxidation at elevated temperature (170°C) followed by liquid or gas chromatographic analyses of the resulting phenol mixture. Although somewhat tedious, such procedures are sensitive, reproducible, and directly applicable to a wide variety of sample types including humic substances. This method also benefits from an extensive background of data for fresh plant tissues from the geochemical and wood products literature (Thurman 1985). Pyrolysis is an equally popular method for lignin characterization which, although not quantitative or applicable to organic-poor materials, can rapidly provide a detailed compositional fingerprint from microgram-size samples (Saiz-Jimenez and de Leeuw 1986). Lignin compositions can also be characterized by permanganate oxidation and reductive degradation techniques (see review by Francois 1989).

Simple phenols and phenolic acids of lignin origin have been reported in river and groundwater (Thurman 1985). Most organic acid characterizations, however, have been reported for humic substances. For example, humic acids dissolved in river water have been shown to contain lignin structural units that reflect local vegetation, although the coexisting fulvic acids do not (Ertel et al. 1984). In general, humic substances dissolved in river water yield high ratios of acidic to aldehydic phenol CuO reaction products that indicate a high degree of microbial oxidation (Ertel et al. 1986). The observation that acid/aldehyde ratios are invariably higher for fulvic acids than the associated humic acids suggests that the lignin in fulvic acids is diagenetically more altered and thus is not the precursor of lignin in humic acids (Ertel and Hedges 1984). Small yields of lignin-derived phenols have also been obtained from DOM in seawater, indicating that a small fraction of this acidic material is also terrigenous (Meyers-Schulte and Hedges 1986).

Biomarker Representativeness

At this point it is appropriate to address how well biomarkers in humic substances represent the sources and histories of the bulk material. The nonspecific procedures by which humic substances are isolated (and defined) result in the recovery of a complex mixture of dissimilar molecules that may have different provenances and degrees of diagenetic alteration. For example, it has been demonstrated that lignin, carbohydrate, and nitrogen-containing structural units can be directly extracted from sedimentary plant debris and recovered along with more altered substances in humic and fulvic acid isolates (Stevenson 1982; Ertel and Hedges 1985). Although these biochemicals can be rationalized away as artifacts unassociated with the "core" of "true" humic substances, they are in fact almost impossible to remove without drastic chemical treatment and may be incorporated no differently than other major humic components. Nevertheless, the hazard exists that biomarkers may be biased toward more recent (or labile) sources of natural humic mixtures.

POSSIBLE FUTURE RESEARCH DIRECTIONS

Biomarkers provide unique advantages as indicators of the sources of organic acids in natural environments. Several lines of research, however, could lead toward a more efficient use of this great potential.

Identify More Bulk Chemical and Biomarker Tracers

Although progress has been made toward more comprehensive characterizations of humic substances for source and reaction history analysis, a broader range of indicators is needed. Among bulk chemical characterizations, one obvious means of advancement is to routinely make better use of elemental data. CHN analyses are not difficult or expensive and should be more widely published as a part of the routine description of humic samples. Although stable carbon isotope methods for bulk samples have most likely been developed to near their full diagnostic potential, stable nitrogen, hydrogen, and oxygen isotope characterizations have not. The latter two "water isotopes" have particular appeal as discriminators of regional sources in terrestrial environments and can be applied both to small and large organic acids. Finally, the recent development of accelerator mass spectrometer (AMS) methods for measuring ^{14}C in milligram size samples of organic materials has provided the unique opportunity to gain information on both the recent sources and dynamics of a wide range of natural organic acids.

Because biomarkers can give biased perspectives on organic acid sources, they are best used in conjunction with each other or with bulk chemical characterizations. Along these lines, determination of stable isotope compositions on individual types of biochemicals affords an especially powerful marriage of biomarker and isotopic techniques. Although most applications to date of this technique have been for fossil organic molecules (e.g., Engel and Macko 1986; Hayes et al. 1987), molecular-level isotopic analysis can also provide independent information on the biological or regional sources of biomarkers in modern aquatic environments. Measurements of this type will be greatly facilitated by the recent commercial availability of gas chromatograph/ratio mass spectrometers that can rapidly analyze small amounts of different individual components within complex organic mixtures.

The identification of new biomarkers is also necessary for more accurate source distinctions among natural organic acids. For example, additional tracers are needed for the sources of volatile low molecular weight (<12

carbons) acids which occur as dissolved components of rain and are widely distributed in other natural waters (Thurman 1985). Hydrophilic biomarkers, which would have very different physical properties and environmental pathways from conventional lipid biomarkers (e.g., fatty acids), would also be very useful. With such tracers it should be feasible to explore the origins and fates of the highly polar acidic polymers that comprise a major fraction of DOM in river and seawater. These substances remain poorly characterized because they are not easily adsorbed onto XAD resins (Thurman 1985). Highly soluble low molecular weight biomarkers with a low affinity for adsorption onto particles might also be excellent tracers for the surface sources and flow pathways of ground and stream waters. Examples of polar biochemicals that might serve in this function would include uronic acids, many of which have unique bacterial or higher plant sources (Keene and Lindberg 1983), as well as cyclitol methyl ethers, which are highly water soluble simple carbohydrates that often occur in characteristic patterns within vascular plant tissues (Anderson 1972).

Define the Effects of Isolation Procedures

A problem of all organic acids, and humic substances in particular, is that their measured characteristics are often dependent upon the isolation procedure. For example, the fatty acid content of humic isolates is affected by sample preextraction with nonpolar solvents, and the measured nitrogen levels depend on whether or not NH_4OH is used in the workup procedure (Francois 1989). To meaningfully compare results for different sample types and laboratories, it is necessary to investigate the effect of variations in the isolation procedure on the measured compositions. Definition of this variability is especially important for biomarkers, which are a minor, and often weakly associated, component of bulk organic matter and humic substances and may be easily added or lost as a result of subtle change in isolation procedures. An important step toward this goal can be taken by the comparative study of readily available reference materials, such as the International Humic Substances Society standard samples (Thurman et al. 1988), for which the various isolation procedures are well defined. Mass balances are also useful in identifying procedural artifacts and for judging how representative a given biomarker might be of the bulk material with which it is associated.

Seek Out Diagenetic Indicators

Because of the extreme alteration of organic acids in most environmental mixtures, bulk chemical and biomarker measurements can often lead to misinterpretations about their sources and pathways. For example, the absence of a diagenetically labile tracer (or associated bulk characteristic) does not necessarily mean that all other acids of a common origin are also not present. On the other hand, a particularly refractory biomarker such as lignin may overrepresent land plant sources. To assess such biases accurately, it is necessary to know the relative reactivities of the biomarkers as well as the extent of diagenetic alteration that the sample material has undergone. At present few biomarkers other than lignin are recognized to also carry diagenetic information, and only limited means exist to gauge the diagenetic state of natural organic mixtures.

Systematic comparisons of organic matter degradation in natural and laboratory settings will be necessary to determine the relative reactivity patterns that must be known for better source assignments. These studies will also help to identify biomarkers of intermediate reactivity that presently are needed to study natural processes, such as river–floodplain exchange and stormwater routing through small watersheds, that occur on time scales of hours to days. Establishing such reactivity sequences for a variety of coexisting biomarkers is also a necessary step toward the ultimate goal of assessing the diagenetic history and nutritional potential of natural organic acid mixtures.

Acknowledgements. The preparation of this manuscript was supported by National Science Foundation Grant OCE-8716481. This paper benefitted from reviews by H. De Haan and M. Peterson and from comments by many other participants of this Dahlem Workshop.

REFERENCES

Anderson, L. 1972. The cyclitols. In: The Carbohydrates: Chemistry and Biochemistry, ed. W. Pigman and D. Horton, vol. 1A, 2nd ed., pp. 519–579. New York: Academic.

Broecker, W.S., and T.-H. Peng. 1982. Tracers in the Sea. Pallisades, NY: Lamont-Doherty Geological Observatory.

Cardoso, J.N., and G. Eglinton. 1983. The use of hydroxyacids as geochemical indicators. *Geochim. Cosmo. Acta* 47:723–730.

Engel, M.H., and S.A. Macko. 1986. Stable isotope evaluation of the origins of amino acids in fossils. *Nature* 323:531–533.

Ertel, J.R., and J.I. Hedges. 1984. The lignin component of humic substances: distribution among soil and sedimentary humic, fulvic, and base-insoluble fractions. *Geochim. Cosmo. Acta* 48:2065–2074.

Ertel, J.R., and J.I. Hedges. 1985. Sources of sedimentary humic substances: vascular plant debris. *Geochim. Cosmo. Acta* 49:2097–2107.

Ertel, J.R., J.I. Hedges, A.H. Devol, R.R. Richey, and M.N. Ribeiro. 1986. Dissolved humic substances of the Amazon River system. *Limnol. Ocean.* 31:739–754.

Ertel, J.R., J.I. Hedges, and E.M. Perdue. 1984. Lignin signature of aquatic humic substances. *Science* 223:485–487.

Francois, R. 1987. A study of the extraction conditions of sedimentary humic acids to estimate their true *in situ* sulfur content. *Limnol. Ocean.* 32:964–972.

Francois, R. 1989. Sedimentary humic substances: structure, genesis, and properties. *Rev. Aquat. Sci.*, in press.

Frimmel, F.H., and R.F. Christman, eds. 1988. Humic Substances and Their Role in the Environment. Dahlem Konferenzen. Chichester: Wiley.

Fritz, P., and J.C. Fontes, eds. 1980. Handbook of Environmental Isotope Geochemistry, vol. 1. New York: Elsevier.

Fry, B., and E.B. Sherr. 1984. $\delta^{13}C$ measurements as indicators of carbon flow in marine and freshwater ecosystems. *Contr. Mar. Sci.* 27:13–47.

Guckert, J.B., C.P. Antworth, P.D. Nichols, and D.C. White. 1985. Phospholipid, ester-linked fatty acid profiles as reproducible assays for changes in prokaryotic community structure of estuarine sediments. *FEMS Microbiol. Ecol.* 31:147–158.

Hayes, J.M., R. Takigiku, R. Ocampo, H.J. Callot, and P. Albrecht. 1987. Isotopic compositions and probable origins of organic molecules in the Eocene Messel shale. *Nature* 329:48–51.

Hedges, J.I., R.A. Blanchette, K. Weliky, and A.H. Devol. 1988. Effects of fungal degradation on the CuO oxidation products of lignin: a controlled laboratory study. *Geochim. Cosmo. Acta* 52:2717–2726.

Hedges, J.I., G.L. Cowie, and J.R. Ertel. 1985. Degradation of carbohydrates and lignins in buried woods. *Geochim. Cosmo. Acta* 49:701–711.

Hedges, J.I., J.R. Ertel, P.D. Quay, P.M. Grootes, J.E. Richey, A.H. Devol, G.W. Farwell, F.W. Schmidt, and E. Salati. 1986. Organic carbon-14 in the Amazon River system. *Science* 231:1129–1131.

Holloway, P.J. 1982. The chemical constitution of plant cutins. In: The Plant Cuticle, ed. D.F. Cutler, K.L. Alvin, and C.E. Price. *Linn. Soc. Symp. Ser.* 10:45–85. London: Academic.

Keene, L., and B. Lindburg. 1983. Bacterial polysaccharides. In: The Polysaccharides, ed. G.O. Aspinall, vol. 2, pp. 287–363. New York: Academic.

Meyers-Schulte, K.J., and J.I. Hedges. 1986. Molecular evidence for a terrestrial component of organic matter dissolved in ocean water. *Nature* 321:61–63.

Nip, M., E.W. Tegelaar, H. Brinkhuis, J.W. de Leeuw, P.A. Schenck, and P.J. Holloway. 1986. Analysis of modern and fossil plant cuticles by Curie point Py-GC and Curie point Py-GC-MS: recognition of a new, highly aliphatic and resistant biopolymer. *Org. Geochem.* 10:769–778.

Nishimura, M., and E.W. Baker. 1987. Compositional similarities of non-solvent extractable fatty acids from recent marine sediments deposited in differing environments. *Geochim. Cosmo. Acta* 51:1365–1378.

Nissenbaum, A., and I.R. Kaplan. 1972. Chemical and isotopic evidence for the *in situ* origin of marine humic substances. *Limnol. Ocean.* 17:570–581.

Norwood, D.L. 1988. Critical comparison of structural implications from degradative and nondegradative approaches. In: Humic Substances and Their Role in the Environment, ed. F.H. Frimmel and R.F. Christman, pp. 133–148. Dahlem Konferenzen. Chichester: Wiley.

Orem, W.H., and P.G. Hatcher. 1987. Solid-state ^{13}C NMR studies of dissolved organic matter in pore waters from different depositional environments. *Org. Geochem.* 11:73–82.

Parrish, C.C. 1988. Dissolved and particulate marine lipid classes: a review. *Marine Chem.* 23:17–40.

Reuter, J.H., and E.M. Perdue. 1984. A chemical structural model of early diagenesis of sedimentary humus/proto-kerogens. *Mitt. Geol.-Paläont. Inst. Univ. Hamburg* 56:249–262.

Saiz-Jimenez, C., and J.W. de Leeuw. 1986. Chemical characterization of soil organic matter fractions by analytical pyrolysis-gas chromatography-mass spectrometry. *J. Anal. Appl. Pyr.* 9:99–119.

Sarkanen, K.V., and C.H. Ludwig. 1971. Lignins. New York: Wiley.

Simoneit, B.R.T. 1977. Diterpenoid compounds and other lipids in deep-sea sediments and their geochemical significance. *Geochim. Cosmo. Acta* 11:463–476.

Stevenson, F.J. 1982. Humus Chemistry (Genesis, Composition, Reactions). New York: Wiley.

Stuermer, D.H., and G.R. Harvey. 1974. Humic substances from seawater. *Nature* 250:480–481.

Thurman, E.M. 1985. Organic Geochemistry of Natural Waters. Boston: W. Junk.

Thurman, E.M., et al. 1988. Isolation of soil and aquatic humic substances. In: Humic Substances and Their Role in the Environment, ed. F.H. Frimmel and R.F. Christman, pp. 31–43. Dahlem Konferenzen. Chichester: Wiley.

Visser, S.A. 1983. Application of van Krevelen's graphical-statistical method for the study of aquatic humic material. *Env. Sci. Tech.* 17:412–417.

Volkman, J.K. 1986. A review of sterol markers for marine and terrigenous organic matter. *Org. Geochem.* 9:83–99.

Wilson, M.A. 1987. N.M.R. Techniques and Applications in Geochemistry and Soil Chemistry. New York: Pergamon.

Wilson, M.A., A.M. Vassallo, E.M. Perdue, and J.H. Reuter. 1987. Compositional and solid-state nuclear magnetic resonance study of humic and fulvic acid fractions of soil organic matter. *Analyt. Chem.* 59:551–558.

Organic Acids in Aquatic Ecosystems
eds. E.M. Perdue and E.T. Gjessing, pp. 65–73
John Wiley & Sons Ltd

Biochemical Methods for the Analysis of Organic Acids in Water

H.G. Gassen

Institut für Biochemie
Technische Hochschule Darmstadt
6100 Darmstadt, F.R. Germany

Abstract. Humic substances in water are a nonstoichiometric mixture of high molecular weight compounds that resists any detailed structured analysis. However, methods such as immunological analysis, enzyme assays, two-dimensional separations, and gene technology have recently advanced to a level of sophistication that makes them potentially useful for the analysis of these natural compounds. The principles of these methods are outlined, and examples of successful application are cited from the literature. Future prospects in the field are discussed using two-dimensional separation techniques and genetic enzyme modifications as examples.

INTRODUCTION

Biochemistry and molecular biology have advanced tremendously during the last ten years. This is true not only for new scientific achievements but also for the development of new preparative and analytical methods (Obst 1986). The latter have contributed much to progress in biology, in particular to the correlation of structure and function. The most outstanding achievements have been fast DNA-sequencing methods and the use of monoclonal antibodies for the quantitative detection of a variety of products (Gassen et al. 1987). Although the more classic methods such as chemical/physical methods (NMR, GC/MS) and enzymatic diagnosis have advanced as well, the analytical problems connected especially with environmental pollution have grown exponentially. The number of anthropogenic substances that may be toxic to living beings exceeds 100,000 (Marthaler et al. 1989). Quite often their concentrations in soil, water, or air are below the nanomolar range and thus below the level of conventional detection. From the viewpoint of an analyst, contaminants in water are much easier to detect than in soil or air. Thus aquatic ecosystems may be an ideal environment for the development of new biochemical methods with respect to trace analysis.

This chapter reviews, in unspecialized terms, the use of enzymatic, immunological, and genetic methods for the determination of trace contaminants in water.

THE USE OF ENZYMATIC METHODS AS A SCREENING TEST FOR POLLUTANTS

Any type of low molecular weight organic substance can be analyzed enzymatically if an enzyme is involved in either its generation or its degradation. Well-known examples are the mono- and dicarboxylic acids such as acetic and maleic acids. These compounds can be oxidized to carbon dioxide in either one step or in a series of reactions (Bergmeyer 1974). Most often, either nicotinamide-adenine-dinucleotide (NAD) or flavin-adenine-dinucleotide (FAD) is the coenzyme that is reduced by the process. Because the molar extinction coefficient of NAD is $A_{340} = 6.2 \times 10^3$ l $\times$ mol^{-1} cm and an A_{340} difference of 0.01 can easily be detected, it is possible to measure a substrate concentration of 1–10 ng/ml easily. For practical purposes, however, one has to use an enzyme that is stable, has a high affinity (K_m) for the substrate, and has a high turnover number (k_2-value). Enzymatic analysis is much more complex for xenobiotic substances such as halogenated aromatic or aliphatic hydrocarbons (Amy et al. 1985). A number of bacterial enzymes can oxidatively split the carbon–halogen bond; however, their turnover number and substrate affinity are very low. Recently the "cholinesterase inhibition test" has been developed to monitor pesticides and herbicides that are phosphoesters (Alsen et al. 1982).

For the detection of organophosphorous compounds in water, a continuous column perfusion system with carrier-bound enzymes was developed (Alsen 1975). The substrate acetylthiocholine is broken down into acetic acid and thiocholine. Thiocholine in turn is reacted with Ellman's reagent, and the resulting thionitrobenzoate is determined photometrically. With paraoxon as an internal standard, a detection limit of 1–10 ng/ml can be reached. One of the disadvantages of this method is the sensitivity of the acetylthiocholine towards many substances occurring in drinking water. Water samples must, therefore, be extracted with methylene chloride, and the organic phase must be concentrated.

Acetylcholine esterase is a serine protease, so it has a very nucleophilic primary alcoholic functional group in its active center. For other serine esterases such as trypsin, chymotrypsin, and elastase, protein inhibitors are known that bind with a high specificity to the active center of these hydrolytic enzymes. Apparent association constants of 10^{-12} mol/l document the high affinities of the inhibitors for their target enzymes. Using the esterase-inhibitor pair, assays similar to the acetylcholine esterase assay for phosphoesters may be developed in the future.

The very critical point in all these new analytical systems will be the affinity of the toxic substance in its role as a substrate analogue for the corresponding enzyme, because of its low concentration in water. Proteins will become ideal scavengers for these substances and thus allow the multifold enrichment without substrate alteration.

Antibodies are the best-suited proteinaceous compounds for this purpose (Engvall 1980; Frimmel and Christman 1988; Habermehl 1985; Maggio 1980; Tijssen 1985; Vanderlaan et al. 1988). With the methods of gene technology, especially site-directed mutagenesis, one can exchange definite amino acids in the active center of the enzyme. This can be done in such a way that the catalytic activity of the enzyme is lost but its binding properties remain unchanged. Thus one can use the very same enzyme in modified form as an affinity reagent and in the unmodified form for the analytical determination of the substrate concentration.

THE USE OF MONOCLONAL ANTIBODIES FOR THE ANALYSIS OF TOXIC SUBSTANCES

Antibodies are symmetrical serum proteins with two sides that enable binding to target proteins (antigen). The antibody has a complex folded surface and binding is influenced by hydrophobic and ionic forces as well as geometric fit. The bulk of the antibody protein is not involved in the combining site and can provide a region for covalent attachment to marker molecules such as enzymes, radioactive markers, and fluorophores (those readers not familiar with the basic concepts of immunochemistry are referred to Vanderlaan et al. [1988] and Paul [1984]).

Immunoassays, because of their capacity to handle large numbers of samples in low-cost, rapid, and semiautomated procedures, have become the most common analytical methods in a clinical laboratory. The most frequent use of this method is the determination of metabolites, enzymes, viruses, and bacteria. From this group only the metabolites may be comparable to the low molecular weight substances occurring in high diversity in natural waters. In the last 20 years, methods have been worked out to generate antibodies which selectively bind low molecular weight molecules such as dinitrophenol (Erlanger 1980). Lately there has been a heightened interest in generating antibodies that will bind to environmentally significant chemicals for trace residue analysis (Vanderlaan et al. 1988).

Small molecules with a molecular mass of less than 1,000 (known in this discipline as haptens) do not induce an immune response by themselves. A hapten must be linked to a carrier molecule to form a hapten–protein conjugate that proves to be strongly antigenic in a warm-blooded animal.

The dioxin–protein conjugate may be used as an example to demonstrate the binding characteristics of the conjugate (Stanker et al. 1987). The antibody produced by lymphocytes primarily binds to the carrier protein or the linker between protein and hapten. Only a minority of antibodies specifically recognizes the hapten portion of the conjugate. Each lymphocyte produces identical antibodies with a unique binding site; therefore, the multiplicity of antibodies in the serum of the immunized animal reflects the multiplicity of lymphocytes. It is thus necessary to either purify the hapten-specific antibody from the serum or prepare monoclonal antibodies (Mabs). Purification of specific antibodies can be achieved by affinity column chromatography, but this is a very laborious task.

In 1975 Georg Köhler and Caesar Milstein (Milstein 1986) revolutionized the production of antibodies by developing a method for culturing the particular lymphocytes that secrete antibodies. Spleen cells of an immunized mouse are fused with myeloma tumor cells and the advantages of the two cell types are combined: the production of a specific antibody and the possibility to culture the cells continuously (immortalization). About 10^4 hybridomas are formed per fusion. Among these should be several cell clones that contain a specific Mab which recognizes only the hapten portion of the antigen. The laborious task is now to screen for the desired hybridoma cell. Once selected, the clones are expanded into the larger wells of a 24-well plate and then to bulk spinner flasks. Ultimately, stable uniform cells are obtained that secrete one antibody optimized towards recognition of the hapten. Basically one year is required to obtain a Mab in sufficient quantities; however, this antibody can then be produced in unlimited amounts. Mabs have the characteristics of a normal chemical reagent. They may be obtained with consistent quality; they can be standardized and incorporated into regulated methods. The Mab can be produced in large enough quantity for billions of assays.

Once the antibody has been produced, it must be incorporated into an immunoassay and the assay must be validated. There are numerous methods in immunology, and the best-suited for the determination of ppm quantities of low molecular weight compounds is the competition enzyme-linked immunoadsorbent assay (cELISA). The basic principle of competition immunoassays is that a limited amount of antibody will distribute between a free antigen in the sample and an immobilized antigen on a solid surface.

The final step in the assay is the detection of the antibody that remains bound to the solid phase. For cELISA this is done by detecting enzymatic activity associated with the antibody. The enzyme can either have been directly conjugated to the hapten-specific antibody used in the second screen or introduced via an enzyme-conjugated second antibody that binds to the antihapten antibody. A wide variety of enzymes are suitable (peroxidase, alkaline phosphatase, or urease) and there is a wide choice in substrates for these enzymes. Sensitive assays based on the detection of fluorescent, radiolabeled, luminescent, or electrically charged enzyme products have also been reported.

Immunological assays have been applied in the past to two major fields: pesticides and carcinogens or toxic chemicals (see Tables 1 and 2). With polyclonal antibodies, sensitivity is in the range of 50–100 ng/ml. However, there is intensive cross reactivity among structurally related compounds. The problem of specificity can be avoided by using Mabs; however, many of the Mabs have a reduced affinity towards their target substrate. A cELISA was used to determine about 1 ng/ml of dioxin in a very convenient assay.

Although many of the new assays are still experimental, they hold promise that passive monitoring probes could be engineered to monitor water or air continuously for the presence of specific chemicals. Coupling the specificity of immunological reactions with all the advantages of sensor techniques such as enzyme electrodes is one of the challenges for the future (Wilson 1984).

TABLE 1. Immunoassays for pesticides since 1980.

Chemical detected	Method	Detection limit (ng/ml)	Mab or antisera
Herbicides:			
Chloro	ELISA		antisera
2,4,5-T and 2,4-D	RIA	7	
Paraquat	ELISA	3.0	Mab, antisera
Tabutrym	ELISA	250	antisera
Atrazine	ELISA	0.32	antisera
Insecticides:			
Diflurbenzuron	ELISA	3.0	antisera
Parathion	RIA	50	antisera
Parmethrin	ELISA	17	Mab
Fungicides:			
Benomyl	RIA	2.0	antisera
Metataxyl	ELISA	0.5	antisera
Triadimeton	ELISA	2.4	antisera
Iprodione	ELISA	3.0	antisera

ELISA = enzyme-linked immunoadsorbent assay
RIA = radioimmunoassay
FIA = fluorescence immunoassay
EIA = enzyme immunoassay

Data taken from Vanderlaan et al. (1988).

TABLE 2. Some immunoassays for DNA adducts, carcinogene, and toxic chemicals.

Chemical detected	Method	Detection limit (ng/ml)	Mab or antisera
DNA adducts:			
Benzo(a)pyrene-DNA	ELISA	0.05	Mab
Ethylated and	ELISA	0.02	antisera
methylated bases	RIA		Mab
Aflatoxin B1-DNA	ELISA	6.0	Mab
1-aminopyrene-DNA	ELISA	0.05	Mab
Mutagene, carcinogene, and toxic chemicals:			
Aflatoxine	RIA	3.3	Mab
Ochratoxine	ELISA	—	antisera
Nitrofluoranthrenes	RIA	120	Mab
Dibenzofurans and	RIA	1	antisera
dibenzodioxins	ELISA	1	Mab
PCBs	RIA	1	antisera
Benzidine metabolite	RIA	0.25	antisera
4-Aminobiphenyl-metabolite	RIA	2.5	antisera
3,3′-dichlorobenzidine	ELISA	500	antisera
Aminoimidazoaza-arenes	ELISA	1.0	Mab
Diethylstilbestrol	RIA	0.1	antisera

Data taken from Vanderlaan et al. (1988).

PROSPECTS OF A DETAILED ANALYSIS OF HUMIC MATERIALS

When nonspecialists such as myself are confronted with the field of humic substances, they readily realize that very little is known with respect to a detailed molecular structure of the corresponding high molecular weight compounds (Frimmel and Christman 1988). First, everything that can stem from the decomposition of the organic materials existing in bacteria, plants, and animals is potentially present. New classes of substances may be formed during humification, including intermediate metabolites. Soil-specific complexes between organic matter and ion exchange-type inorganic substances may be formed. Quite a number of products generated by mineralization (such as NH_3, H_2S, and CH_4) are gaseous and still quite reactive. Many of the metal oxides (e.g., Fe_2O_3) may act as catalysts and accelerate otherwise slow reactions. The same holds true for the heat that is generated during the decomposition process.

Initially, high molecular weight compounds stemming mainly from the cell walls are hydrolyzed or oxidized by nonenzymatic means. Starches will form sugar molecules, proteins are transformed into amino acids, and chlorophyll yields aromatic systems.

Worms and other animals will mix the generated compounds into a heterogeneous soup. If a biochemist dwells on how to analyze this complex mixture, he will at once ask the basic question of his profession: can we find and identify enzymes as well as the corresponding reactions which they catalyze? Either all reactions forming new polymer compounds are nonenzymatic—then he is out of business—or they are enzymatically catalyzed and he can set himself to work. The first candidates to look for are the oxidoreductases and all hydrolytic enzymes such as amylases, proteases, and lipases. Many of these enzymes are quite stable under soil conditions, especially in the presence of excess substrate or product. One quite simple experiment would be to spike an aqueous extract of humus with either NADH or NAD. Because most oxidoreductases use these compounds as coenzymes, the presence of such enzymes can be checked. Under partially anaerobic conditions and in the presence of heavy metals, a number of enzymatic reactions presumably do not go to completion. Thus aldehydes may be formed from primary alcohol groups and imino compounds from either aromatic or aliphatic amines. These intermediates could form Schiff's bases and thus lead to higher molecular weight compounds. Especially prone to this type of reaction are the amino acid side chains of lysine and serine.

Many bacteria and fungi that live in a high variety and concentration in soil may secrete enzymes in high concentrations. This is well documented with cell wall degrading filamentous fungi like *Aspergillus* (Dean and Timberlake 1989). One of the fungi called *Tritirachium album* produces a very active protease, namely proteinase K. This enzyme can digest even feathers to free amino acids, which are, in turn, internalized by the fungus and used for the synthesis of high molecular weight compounds (Gunkel and Gassen 1989).

Macromolecules with a molecular mass above 10,000, which display polar or hydrogen bond-forming groups, may aggregate in aqueous solutions. The degree of aggregation depends on the chemical nature of the compound, its actual conformation, its concentration, and the pH and ionic strength of the solution. Because this type of aggregation does not involve covalent bond formation (exception disulfide bonds and Schiff's bases), a dynamic equilibrium exists between the monomers and the sum of all possible aggregates.

Whenever one attempts to separate a mixture of such closely related compounds (e.g., proteins or carbohydrates), one finds a "smear" because of the inhomogeneity in the molecular masses. Especially for the separation of membrane proteins, which have an intrinsic tendency to stick together, the so-called denaturing conditions to improve separation have been perfected. The classic solvents are 7 mol/l urea, 4 mol/l HCL, 2–3 mol/l guanidiniumhydrochloride, 5–10% dimethylformamide or dimethylsulfoxide. Recently detergents such as sodium dodecylsulfate (SDS) or nonionic detergents such as Tween80 have been used. The application of these denaturing reagents results in sharp peaks, small bands, or focused spots in one- or two-dimensional separations using either chromatography or electrophoresis. With proteins, for example, up to 3000 compounds can be separated on a 20 cm × 20 cm polyacrylamide plate. The use of SDS in the solvent shows the further advantage that all proteins, because of the binding of excess detergent, become polyanions regardless of their pK_a values. Thus separation in an electric field is based only on the differences in molecular masses.

Perhaps this method, which has proved to be successful in protein chemistry, could be applied to the separation of humic substances as well. However, even if humic substances can be separated on a two-dimensional screen, it will be very difficult to find a staining reagent which reacts with all compounds to make them visible under normal or UV light.

Ten years ago one could only analyze for the presence of certain classes of enzymes. There was no possibility of isolating and characterizing the protein in question. With the molecular cloning methods at hand, the situation has changed completely. With about 100 picomoles of protein, one can determine the sequence of 30–40 amino acids from the amino terminus by Edman degradation. This information allows the construction of an oligonucleotide probe to screen either for the so-called complementary DNA (cDNA), which has been synthesized enzymatically as a complement of the nucleotide sequence of the messenger RNA (mRNA), or for the genomic DNA in a library. The identified structural gene may be introduced into a producing organism such as *E. coli* or *Aspergillus niger*. With the enzymes in hand, one can either produce antibodies or look in detail into the type of reaction which is catalyzed by the enzyme in soil (Obst 1983).

Biochemical techniques have improved tremendously in recent years. Perhaps they can be of some help in identifying some of the substances forming the humic acids.

REFERENCES

Alsen, C. 1975. Studies on acetylcholinesterase and acetyl cholinesterase covalently bound to polymaleinic anhydride. *Biophys. Biochem. Acta* 377:297–302.

Alsen, C., O. Christensen, and H. Kruse. 1982. Nachweis von Acetylcholinesterase-Hemmstoffen in Trink- und Oberflächenwasser. In: Vom Wasser, ed. Fachgruppe Wasserchemie in der Gesellschaft Deutscher Chemiker, Band 58, pp. 1–12. Weinheim, F.R. Germany: VCH.

Amy, P.S., J.W. Schulke, L.M. Frazier, and R.J. Seidler. 1985. Characterisation of aquatic bacteria and cloning of genes specifying partial degradation of 2.4-dichlorophenoxyacetic acid. *Appl. Env. Microbiol.* 49:1237–1245.

Bergmeyer, H.U. 1974. Methoden der enzymatischen Analyse. Weinheim, F.R. Germany: VCH.
Dean, R.A., and W.E. Timberlake. 1989. Production of cell-wall degrading enzymes by *Aspergillus nidulans*: a model system for fungal pathogenesis of plants and regulation of the *Aspergillus nidulans* pectate lyase gene (pel A). *The Plant Cell* 1:265–284.
Engvall, E. 1980. Enzyme immunoassay ELISA and EMIT. In: Methods Enzymology, ed. S.P. Colowick and N.O. Kaplan, vol. 70A, pp. 419–439. New York: Academic.
Erlanger, B.F. 1980. The preparation of antigenic hapten-carrier conjugates: a survey. *Meth. Enzymol.* 70:85–104.
Frimmel, F.H., and R.F. Christman, eds. 1988. Humic Substances and Their Role in the Environment. Dahlem Konferenzen. Chichester: Wiley.
Gassen, H.G., A. Martin, and S. Bertram. 1987. Gentechnik. Stuttgart: G. Fischer.
Gunkel, F.A., and H.G. Gassen. 1989. Proteinase K from *Tritirachium album* limber, *Eur. J. Biochem* B179:185–194.
Habermehl, K.O. 1985. Rapid Methods and Automation in Microbiology and Immunology. New York: Springer.
Maggio, E.T. 1980. Enzyme Immunoassay. Boca Raton, FL: CRC.
Marthaler, R., H. Gebhardt, and M. Linnenbach. 1989. Gewässerversauerung. *Biol. Uns. Zeit* 1:22–24.
Milstein, C. 1986. From antibody structure to immunological diversification of immune response. *Science* 231:1261–1268.
Obst, U. 1983. Biochemische Methoden zur Untersuchung von Stoffwechselaktivitäten in Sediment und Schlamm. In: Vom Wasser, ed. Fachgruppe Wasserchemie in der Gesellschaft Deutscher Chemiker, Band 61, pp. 187–198. Weinheim, F.R. Germany: VCH.
Obst, U. 1986. Die Relevanz biochemischer Methoden für die Wasserpraxis. In: Vom Wasser, ed. Fachgruppe Wasserchemie in der Gesellschaft Deutscher Chemiker, Band 67, pp. 1–10. Weinheim, F.R. Germany: VCH.
Paul, W.E. 1984. Fundamental Immunology. New York: Raven.
Stanker, L.H., B. Watkins, N. Rogers, and M. Vanderlaan. 1987. Monoclonal antibodies for dioxin: antibody characterization and assay development. *Toxicology* 45:229–243.
Tijssen, P. 1985. Practice and theory of enzyme immunoassay. In: Laboratory Techniques in Biochemistry and Molecular Biology, ed. T.S. Work and E. Work, vol. 15, p. 549. Amsterdam: Elsevier.
Vanderlaan, M., B.E. Watkins, and L. Stanker. 1988. Environmental monitoring by immunoassay. *Env. Sci. Tech.* 22:247.
Wilson, J.R. 1984. Instrumentation for process measurement and control in the bioprocess industry. *Trends Anal. Chem.* 3:223–229.

Standing, left to right:
John Hedges, André Braun, David White, Michael Klug, Fritz Frimmel, Günter Gassen, Russell Christman, Walter Giger

Seated, left to right:
Alexander Nehrkorn, Bill Glaze, Gudrun Abbt-Braun, Michael Thurman

Organic Acids in Aquatic Ecosystems
eds. E.M. Perdue and E.T. Gjessing, pp. 75–95
John Wiley & Sons Ltd

Group Report
What Is the Composition of Organic Acids in Aquatic Systems and How Are They Characterized?

W.H. Glaze, Rapporteur
G. Abbt-Braun
A.M. Braun
R.F. Christman
F.H. Frimmel
H.G. Gassen
W. Giger
J.I. Hedges
M.J. Klug
A.H. Nehrkorn
E.M. Thurman
D.C. White

INTRODUCTION

Despite decades of research, dissolved organic material in aquatic environments remains poorly characterized at the molecular level. Indeed, many scientists believe that there is no good purpose served by pressing efforts to obtain a molecular formula for the organic acids of natural waters; rather, they maintain that it will be more productive to characterize the material only so far as we need to in order to identify its major roles in ecological and geochemical processes. In this view, the degree of characterization should fit the intended purpose. In some cases, bulk property measurements such as elemental or isotopic composition will be sufficient; in other cases, more molecular specificity is needed, for example, the oxidation state of sulfur or the number of carboxylate units per gram of carbon. The alternative view is that we should not be satisfied until the molecular nature of aquatic humic material is fully characterized, or at least until we have a molecular model that is consistent with all the known properties of the material.

During this workshop we did not focus on this debate, although it arose from time to time. Rather, the overall goal of the workshop was to identify the major processes that cause spatial and temporal variability in properties and concentrations of aquatic acids, which make up the overwhelming bulk of aquatic organic matter. The specific purpose of our group, therefore, was to examine what level of characterization is needed in order for scientists to reach this overall goal, and to identify the most promising tools available for that characterization. While the pursuit of this purpose may eventually lead to a more thorough elucidation of the structure of aquatic organic material, that is not the primary goal.

The Genesis and Fate of Aquatic Organic Acids

Several excellent reviews are available on this subject and it will not be discussed here in depth (Schnitzer and Kahn 1972; Gjessing 1976; Aiken 1985; Thurman 1985; Wilson 1987; Frimmel and Christman 1988). It is useful, however, to recognize that many unanswered questions remain about the genesis of the organic material found in natural waters and the processes that affect the material over time as it passes into different parts of the aquatic environment.

It is generally recognized that aquatic organic matter is a complex mixture of organic acids (OAs) with small amounts of neutral and basic components more or less tightly bound to the acidic molecules. One may use the term "organic acids" qualitatively, i.e., to designate the collection of organic acids that make up this complex mixture—phenols, carboxylates, etc. Alternatively, one may attempt to describe the mixture quantitatively as the sum of the analytical concentrations of these acids. There is no method presently available for the direct determination of a parameter "total organic acids," and the fraction of aquatic organic matter represented by the analytical sum of the individually determined OAs may be obtained only by calculation. Moreover, it is recognized that the term "acid" can mean a Brönsted (proton-donating) acid, or more generally, an electron acceptor (Lewis acid). Aquatic organic matter undoubtedly contains Brönsted and non-Brönsted acids, the latter including hydrogen-bonding moieties such as the carbohydrates. These may play important roles in determining the properties of "organic acids." They are, however, not the principal focus of the present work.

Much of the OA content of natural freshwaters is presumably of terrestrial origin, solubilized by microbiological processes that occur in soils of woodlands and wetlands. Also, man may affect the nature of organic matter found in natural waters, especially through the discharge of domestic and industrial wastewaters, including substances that are not readily biodegraded. As is described by other groups of this workshop, it is not yet clear to what extent the OAs in an aquatic system reflect their natural or anthropogenic sources, what time scales are involved in their transformations, and what processes occur in the aquatic environment that affect their properties. It is clear, however, that processes do occur in the aquatic environment that affect the nature of the OA assemblage. These processes include:

microbiologically mediated metabolism including mineralization, abiotic oxidation or photochemical processes, adsorption, precipitation, and repolymerization. The result is an "aging" process wherein the composition and properties of the OA mixture change with time. Some of the relevant processes are probably very fast, e.g., microbial mineralization of very reactive species such as unbound fatty acids. Others are of intermediate rates and some are extremely slow. Figure 1 shows a plot (modified from Meybeck) depicting the types of OAs that make up the spectrum of reactivities and the concentrations that one finds in natural waters. In the upper left-hand quadrant of the graph are the recalcitrant "aquatic humic substances" (Thurman et al. 1988) found ubiquitously in natural waters. In the lower right-hand quadrant are the very reactive OAs, such as free fatty acids, that generally are found in natural waters (individually) at very low concentrations, except perhaps near primary sources. Such a plot is instructive in that it suggests several research questions that formed the basis of our group's task:

1. To what extent does such a plot accurately reflect the composition of waters of various types (surface, ground, sea, wetland, etc.)?
2. For a given water, what changes occur in the composition of OAs as time and transport proceed?

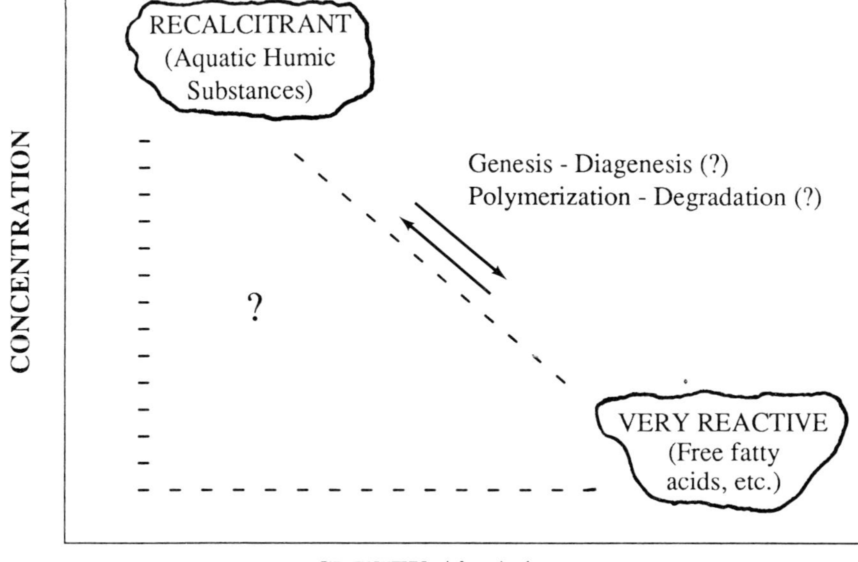

Fig. 1—Representation of the spectrum of concentrations and reactivities of aquatic organic acids (suggested by Meybeck at this workshop).

3. Does the aquatic humic substances (recalcitrant) component differ from one water to another? Does it metabolize at all? Are abiotic processes responsible for its "reactivity?" Do abiotic and biotic processes combine to remove it eventually from the system?
4. To what extent does "reactivity" on the abscissa mean biological or abiotic chemical reactivity, or abiotic physical removal? What are the relative magnitudes of each of these processes in different environments?
5. What are the types of OAs that lie within the triangular (shaded) area? How do they reflect the source of the OA pool? How do their molecular structures (functional groups, *cis/trans* ratios, etc.) reflect their reactivity? What roles do they play, if any, in ecological and geochemical cycles?
6. To what extent is there interconversion between reactive chemicals and less reactive humic substances? Is the formation of recalcitrant substances a polymerization process or the result of a slow degradation of a recalcitrant biopolymer precursor, or both?

These are precisely the types of questions raised by ecologists and geochemists, and to obtain answers to them, one must be able to characterize the OAs in a natural system. Ideally, this characterization should accurately portray the system under study *although it is not necessary that the system be completely characterized*. Much can be learned about the nature of the OA pool, its genesis, and the processes that occur within it, by selective characterization. One such approach is the use of chemical tracers, including those of biochemical origin ("biomarkers"). With such tracers one hopefully could determine the contribution of each source term to the OA mixture at various times and places within a system (Hedges, this volume).

As examples of this approach, Ertel and Hedges (1984) have used relatively stable phenols and phenolic acids of lignin origin to determine that lignin residues in fulvic acid are more altered (and thus more diagenetically "downstream") than lignin in humic acids, and Meyers-Schulte and Hedges (1986) have shown that a small fraction of OAs in seawater are of terrestrial origin. Biomarkers, or more generically, "chemical tracers," are needed to help expand our understanding of the genesis of OAs and how OAs are transformed diagenetically. Workshop participants identified several specific research questions in these areas: how do we distinguish aerobic and anaerobic microbiological communities as the source of OAs? How do we distinguish riparian from upland vegetation as sources? What are the contributions of specific higher plant communities such as vascular plants? What is the contribution of peat bogs to stream OAs? While it is beyond the scope of this report to list the biomarkers that may be used to examine these and similar research questions, readers are referred to works by White et al. (1979).

Bulk Parameters of Aquatic Organic Acids

Bulk or "group" parameters such as dissolved organic carbon (DOC) or nitrogen, color, UV absorbance, etc., have also been used to track OAs in aquatic systems. However, the value of these bulk parameters is limited because they contain so little information that is specific to a given source or process. What is needed are bulk parameters that have such specificity, but which are not too difficult or costly to be used by scientists who are not specialists in analytical chemistry. Parameters of this type are needed for several purposes:

1. To measure the sources of OAs in the given system, including the differentiation of anthropogenic from natural sources and different natural sources from one another (e.g., atmospheric from terrestrial sources).
2. To determine the bulk potential of OA mixtures as substrates for biodegradation. Of particular interest are OAs with intermediate reactivities that might support microbial respiration on the time scales that are involved when waters from flood plains mix into river mainstems or through soil horizons into groundwater.
3. To act as conservative tracers for OAs for transport studies.
4. To measure overall abiotic reactivity and removal via specific mechanisms, e.g., oxidation via OH radicals, hydrolysis, photooxidation, adsorption, etc.

Further Characterization of Refractory Aquatic Humic Substances

For a variety of reasons there continues to be an interest in the recalcitrant aquatic humic substances (AHS) that make up a large fraction of the organic carbon in aquatic systems. These are complex, colored OAs with relatively high molecular weights (up to several thousand daltons), high metal complexing capability, and low bioreactivity (Thurman et al. 1988; Aiken 1985; Thurman 1985). Because of their ubiquity and importance in geochemical processes, scientists continue to pursue questions about AHS. These include:

1. What is the nature of the moieties that are responsible for acidity and metal complexation of AHS? How do these vary from source to source and in time?
2. Do AHS act as sources of energy and carbon for microorganisms, and if so, what is the mechanism of these processes? Do such processes

represent significant vectors for the removal of carbon from the OA pool?
3. Are AHS being modified/removed by abiotic chemical or photochemical processes, and if so, how do these processes compare in magnitude with biological processes? Are the AHS adsorbed on particulate matter when these processes occur?

To answer these and related questions one needs tools to determine either the relevant structural component of the AHS or a parameter which is an appropriate surrogate for the process involved. Thus, there appears to be the need for further research on the recalcitrant AHS found in natural waters, as well as further study on the remainder of the OA pool. In the following two sections we shall discuss conventional and alternative approaches to these types of studies. Later sections will summarize our recommendations and conclusions as well as emphasize the need for interdisciplinary cooperation in the study of aquatic ecosystems.

CONVENTIONAL METHODS FOR THE CHARACTERIZATION OF ORGANIC ACIDS: NEW DEVELOPMENTS

In this section, we will not attempt to review in detail the methods used for the characterization of organic acids (see Frimmel, this volume; Frimmel and Christman 1988). Rather, we shall point out improvements that have occurred recently in this area, new methods that have special promise, and research that is needed in order for ecologists, geochemists, and other scientists to have available the tools needed to reach the goal of this workshop. We have separated the section into parts on isolation methods, fractionation methods, and spectroscopic and chemical characterization methods.

Methods for the Extraction of Organic Acids from Water

For over a decade, the method of choice for the isolation of the AHS fraction of OAs in water has been the XAD-8 method (Aiken et al. 1979). Using this method in combination with cation- and anion-exchange resins, Leenheer (1984) published a method for the isolation of four fractions of OAs, bases, and neutral compounds. These methods, though laborious, require little special technical skill or expensive equipment. They should be considered by scientists for application in field and model ecosystem studies. However, there are uncertainties associated with the XAD isolation method, including losses of hydrophilic material, that require further research. Of special interest is the possible effect on OA structure through the high and low pH conditions to which the OA mix is exposed in the method. Recently, other solid phase adsorbents have become available that may be more suitable for isolation of the OAs (derivatized silica cartridges, etc.). These adsorbents offer the promise of ease of use, low cost, and (perhaps) more specificity. Further research into the use of these adsorbents is needed.

Membranes may also be used to isolate organic matter from water (Leenheer 1984; Taylor et al. 1987; Weber 1988). Reverse osmosis and ultrafiltration membranes may be attractive alternatives to XAD resins, especially if the latter are shown to affect the nature of the OA mixture because of the high and low pH conditions involved. However, there are problems associated with the use of membranes for the isolation of low molecular weight OAs. For example, losses are certain to occur due to passage of low molecular weight OAs through membranes. Also, high molecular weight OAs and special substituted OAs may be lost via adsorption onto the membranes. Comprehensive recovery studies are needed before membranes can be used confidently for isolation of OAs.

Freeze drying (lyophilization) is another conventional method for the isolation of OAs from water. Although laborious, lyophilization may affect the structure of OAs less than other isolation methods, although volatile compounds may be lost in the process. After lyophilization one may use extraction or derivatization procedures to isolate a particular substance or class of substances from the sample. A new extraction method of interest is supercritical fluid extraction (SCFE), in which the solvent is a gas, such as carbon dioxide, above its critical point (Davies et al. 1988). Supercritical fluids behave neither as gases nor as liquids; e.g., the solubility of a substance in a supercritical fluid is strongly dependent on the pressure of the system. Thus, one can use specific pressure conditions to extract a specific compound from a lyophilized mixture of OAs. SCFE is a relatively new technique that must be investigated more thoroughly but it may be very useful, especially in combination with supercritical fluid chromatography (see below) for the characterization of complex OA mixtures.

Fractionation Methods

Fractionation in this section is taken to mean the separation of an OA mixture into component acids, either with full resolution (which is usually impractical) or with partial resolution. As noted in the introduction, many studies may not require full resolution of the mixture. In such a case (for example, the isolation of a specific OA), a customized isolation/fractionation procedure may be developed. Of the methods available now, gas chromatography (GC) has the highest potential resolving power. Combined with mass spectrometry, GC continues to be a powerful method for separation, identification, and quantification of OAs. However, OAs generally must be derivatized beforehand to achieve maximum resolution and sensitivity (Kawamura and Kaplan 1987). While this step is often laborious and can be accompanied by loss of analyte, introduction of artifacts, etc., it can increase the sensitivity of the determination if the derivative is detectable by very sensitive detectors such as the electron capture detector or negative chemical ionization mass spectroscopy.

Methods based on liquid chromatography or electrophoresis are alternatives that generally do not require prior derivatization of the OA. Reverse-phase liquid chromatography (RP/HPLC) is a well-developed separations method for OAs. The method is especially useful for organic acids with UV-absorbing chromophores, because one of the most sensitive detectors available is the UV-absorption detector. With some OAs that are intensely fluorescent, even higher sensitivities are possible with RP/HPLC fluorescence systems. These two systems should be considered especially when polar biochemicals are being monitored during ecological studies. Important developments taking place in these fields include: (*a*) the marketing of new, high resolution ion-exchange columns for HPLC separation of OAs, (*b*) multidimensional chromatography systems, and (*c*) the development of high resolution capillary-focused electrophoresis (CFEP).

It should be noted, however, that conventional HPLC and electrophoresis methods do not generally lend themselves to the identification of unknown OAs in a complex mixture; they are most useful for analysis of target OAs and for separation of fractions which may then be subjected to identification methods, such as the spectroscopic methods discussed below. Recently, HPLC and CFEP have been combined with mass spectrometry to overcome this deficiency, but systems for routine use are not likely to be available for some time. Nevertheless, these systems may be very useful for the study of complex OA mixtures in order to identify potential biomarkers or to elucidate the source of OAs in a given ecosystem.

Supercritical fluid chromatography (SCFC) is a new chromatographic system developed recently that may have a large impact on the characterization of OAs. SCFC combines high resolution (comparable to capillary GC) with the ability to handle compounds of very high molecular weight or low volatility (Davies et al. 1988). The method, which is related to supercritical extraction (SCFE) discussed above, utilizes a supercritical fluid (such as carbon dioxide at pressures and temperatures above its critical point) as the mobile phase and differential solubility as a function of pressure gradient as the separations parameter. Although at the present SCFC is not well suited for separation of ionic substances, it may be modified toward that purpose in the future. Also, it may be immediately useful for the study of methylated AHS or derivatized low molecular weight OAs (see below.

Comment on Isolation and Fractionation of Aquatic Organic Matter

The isolation and fractionation of aquatic organic matter are often essential if one wishes to characterize the material or use it for controlled experimentation. This is necessary because many experimental methods are not sensitive enough to measure properties of OAs at natural concentration levels, the natural OA mixture is judged to be too complex to characterize in bulk, or perhaps because one is interested only in a particular fraction of the material. While isolation and fractionation methods are certainly justified in many cases, there is the danger that these methods will introduce biases into the measurements that are not readily apparent. Isolation and fractionation procedures are seldom perfectly efficient (Aiken 1988). One commonly loses a fraction of the material or the material is transformed in some way by the procedure. Investigators, particularly those who are not expert in separations processes, should be aware that such losses or transformations occur and may affect the outcome of an experiment. Research is needed to determine to what extent this may be the case for various isolation/fractionation procedures *as applied to specific research purposes*. Thus, a particular isolation/fractionation method may prove to be quite unbiased for one type of study (e.g., metal complexation) but quite unacceptable for another (e.g., bioavailability). Collaborative research between separations specialists and investigators who wish to study the properties of OAs is needed to address this issue.

Spectroscopic Characterization Methods

Scientists have used several spectroscopic methods to study AHS samples. Of these, nuclear magnetic resonance (NMR) spectroscopy, mass spectroscopy (MS), infrared spectrophotometry (IR), and UV/visible spectrophotometry have been the most revealing. Recent developments in NMR and MS methods have been reviewed (Wilson 1987; Frimmel, this volume). Of particular importance to the OA field are new developments in solid phase NMR (Norwood 1988) and new MS systems (Bellar and Budde 1988) which can handle HPLC eluates directly (i.e., without manual collection of fractions). Combined with the new, high resolution HPLC columns, these methods may prove to be of considerable value in elucidating the structures of OAs of intermediate molecular weight. HPLC combined with isotope ratio mass spectrometry may be particularly useful for ecological and geochemical research that attempts to elucidate the source of OAs.

Nonspecialists in the field of mass spectrometry should not be discouraged from using this method because of its apparent degree of sophistication. MS-based methods, while elegant, are becoming routine in the monitoring of food, clinical, and environmental samples (Scully et al. 1987; Swineford and Belisle 1989). GC/MS methods are amenable to use with laboratory robots (autosamplers, injectors, and data stations) which are also easy to use. Low-cost quadrupole analyzers (mass selective detectors) are also available. Thus, rapid, routine methods may be developed that involve isolation, derivatization (or derivatization then isolation), and GC/MS analysis. Compared to older methods, these are highly specific for a given OA, precise, and amenable to studies that generate large numbers of samples. (See also *Functional group analysis and derivatization methods*, below).

Fourier transform infrared spectroscopy (FTIR) is gaining credibility as a tool for identification and analysis of environmental samples. FTIR can now be combined with GC or in series with a GC/MS system (GC/FTIR/MS) to yield an extremely powerful combination (Gurka and Pyle 1988). FTIR cannot be conveniently coupled directly to HPLC systems, but HPLC fractions can be collected onto KBr discs which can then be scanned by FTIR. In general, FTIR gives information that is complementary to MS data, but in some cases FTIR is possible when no mass spectral data can be obtained. This is the case for medium and high molecular weight OAs; thus, FTIR may become a tool of increasing value in the study of these materials.

Chemical Characterization Methods

Bulk chemical and isotopic properties. Few new methodological developments have occurred in these fields during the past few years. However, bulk chemical and isotopic methods continue to be valuable tools for ecologists and geochemists. Isotopic methods are especially valuable in that they give information on the age of OA samples and in some cases may be useful in determining their origin and history. Isotopic ratios of several elements carry important information relative to the OA pool, especially those of carbon, sulfur, and nitrogen. Recent advances in the determination of ^{14}C levels by accelerator mass spectrometry, requiring only small amounts of sample, are very promising. Bulk isotopic measurements are subject to problems of interpretation if the sample consists of inputs from several undifferentiated sources. Hence, separations methods are equally as important in isotopic studies as in the use of chemical parameters, but are seldom employed. More applications of combined separations and isotopic measurement methods may yield useful information on OAs in natural systems. Also, more studies need to combine isotopic studies with measurements of chemical tracers to provide confirmatory information, when appropriate.

There also continues to be the need to evaluate and standardize the methods used for the measurement of DOC and total organic carbon.

Recent studies in marine systems suggest that current carbon analyzers based on wet chemical oxidation techniques may give values that are substantially lower than those obtained with high temperature platinum catalyzed oxidation. This needs to be verified and also tested in freshwater systems.

Average molecular weight and size measurements are bulk properties that may also be useful to ecologists and geochemists. These methods include ultracentrifugation (Schnitzer and Kahn 1972), small angle light scattering (Wershaw et al. 1967), field flow fractionation (Caldwell 1988), and size exclusion chromatography (SEC) (Bombaugh 1984; Provder 1980). Of the methods available, those which involve new columns for aqueous SEC appear to fit the requirements for field and ecosystem studies most satisfactorily. To be most useful, however, SEC should be coupled to a total carbon detector (Gloor et al. 1981). Such detectors are not commonly available from HPLC suppliers though they have been used in the past. SEC systems with UV detectors are also useful for ecological studies because most OAs in aquatic samples have some UV absorbance. These systems (SEC/UV) are relatively low in cost, have proven reliability, and supply a level of information which may make them extremely useful and cost-effective for field studies. It is also true that new bulk parameters will expedite field and ecosystem studies. Workshop participants identified the following as parameters of potential usefulness in such studies:

— a rapid method for biodegradability
— degree of aromaticity
— critical micellular concentration
— hydrophobicity
— C, N, S, and P content of DOC fractions
— average molecular surface area
— percentage organic N compounds.

In these and other cases that may arise, it is important that new methods, if developed, be carefully validated and field tested.

Functional group analysis and derivatization methods. Functional group analysis of the AHS mixture in natural waters has been used for many years to establish estimates of the number of carbohydrate, carboxylate, phenol, etc., units in these polymers. These methods may also be used to determine levels of such units in more hydrophilic portions of the OA mixture. However, in this case, one may obtain substantially more information, including levels of specific biomarkers, if these functional group tests are used in conjunction with separations methods.

Thus, it is possible to combine tests for specific functional units with chromatographic or electrophoretic separation in order to obtain information about the levels of specific members of a class of compounds (e.g., hydroxy acids). Alternately, one may obtain the total concentration of members of that class either by summing the concentrations of individually measured compounds or by using a batch method (without separation).

Functional group tests are generally one of two types: (*a*) those wherein the test species causes a response in another chemical system, or (*b*) those wherein the test species is converted to a derivative that is then detected. Both of these approaches are useful and appropriate for field and ecosystem studies of OAs. However, as we shall reiterate in the following section, parameters are needed that accurately reflect the nature of the OA environment and its changes. It is most likely that this will be accomplished through the use of differentiated chemical tests (for specific species) rather than tests that measure large groups of compounds from a given class. For that reason, more attention should be given to the development of simple, reliable chemical derivatization methods in combination with high resolution chromatographic separations for OAs of different types. Methods are available for the common fatty acids and for selected other classes of mixed-functional OAs. However, further work in this area is needed for other OAs, especially those that are identified as potential biomarkers.

Microscale methods and special techniques. Ecologists would benefit immensely from the development of methods that measure the properties of aquatic OAs *in situ*, i.e., without disturbing the system. Moreover, microscale methods are needed which may be applied in spatially limited environments (soil pores, leaf tissue, etc.). Few such methods are available other than pH, redox, and other electrometric measurements; however, recent advances in fiber optics technology combined with more sensitive spectrometers may offer methods which fit these criteria. Further research in this area may be extremely useful.

Also, there are recent developments of portable analyzers that may be of interest to ecologists and geochemists. These include capillary GC and GC/MS systems that are purportedly rugged and sensitive. At present the advantages of these systems for field studies of OAs seem minimal, primarily because they cannot be used directly on water samples (i.e., prior separation and/or derivatization is needed). Future developments in this area may overcome this problem, especially if field HPLC systems become available.

ALTERNATIVE APPROACHES TO THE CHARACTERIZATION OF ORGANIC ACIDS

As the goal of this workshop implies, scientists are now asking more and more demanding questions about the roles of OAs in the environment. To answer these questions analytical methods are needed that are concomitantly more elegant, i.e., sensitive, reliable, accurate, convenient, and "cost-effective." The parameters which these methods measure should reflect the (preferably subtle) changes in the environment from which the sample was taken. Increasingly, it is apparent that the available bulk parameters will never provide detailed enough information to elucidate the ecological roles and processes of OAs. To provide this information, more detailed chemical information is needed, i.e., the nature of the OA mixture must be elucidated. However, as noted in the introduction, it may not be necessary to elucidate the entire OA mixture in a given ecological setting. Indeed, given the lack of success over the past six decades, it is unlikely that the recalcitrant AHS mixture will ever be fully elucidated at the molecular level. Perhaps it is a mixture whose composition is different at each place and time on this planet and, therefore, cannot even be considered to have constant physical and chemical properties. With this as a possibility, it becomes imperative to rethink the goals of chemical characterization of OAs. Perhaps it will be sufficient to characterize the AHS mixture only partially and to give more attention to the other constituents of the natural OA mix, compounds that may be in trace quantities but whose levels may reveal substantial information about the ecological sources and history of OAs. Finally, it may be possible to characterize the OA mixture in an ecologically more relevant fashion by picking parameters and methods that are drawn from the environment itself.

To elaborate upon this new approach to the characterization of OAs, we will discuss two examples: (*a*) the techniques derived from biochemistry and molecular biology for the characterization of OAs, and (*b*) the use of information drawn from the ecosystem itself, especially the microbiological communities thereof, to guide the characterization process.

Characterization Techniques Derived from Biochemistry and Molecular Biology

Gassen (this volume) has summarized two types of analytical methods that may be useful in the characterization of OAs in complex mixtures: (*a*) enzymatic methods and (*b*) immunoassays. In the former, one utilizes enzymes of known specificity to search for specific substrates in natural systems. The degree to which this process may be useful in characterizing OA mixtures is not known, but in principle the method could be very useful if enzymes are available which are specific for compounds (OAs) of particular ecological relevance. Thus, if a particular OA is discovered that is a biomarker of a given process, following that biomarker may be possible by an enzymatic assay. Such assays are often highly sensitive and reasonably specific, though the latter must be proven. Moreover, the same enzyme may be synthesized using gene technology (Gassen, this volume) with a modified active center so that it can not function as an enzyme but still binds to the substrate. Thus, it may be possible to use the inactive enzyme for enrichment of the desired analyte and the active enzyme for its analysis.

Environmental applications of immunoassay methods have been reviewed by Vanderlaan et al. (1988). Recently, immunoassays have become increasingly more specific, applicable to lower molecular weight compounds (haptens), more sensitive (50–100 pg), and extremely rapid and less costly compared to other chemical assay methods. Immunoassays are now common analytical methods in clinical laboratories.

In principle, immunoassays can be applied to any compound of interest as long as the compound can be linked to a carrier molecule or is large enough to cause a response itself. Hence, OAs in principle could be characterized individually, as classes, or as components of macromolecules. Unfortunately, the process of making monoclonal antibodies (Mabs) is not inexpensive, and there are some problems with cross-reactivity (or alternately loss of sensitivity) that must be worked out by further research. Nonetheless, it appears that the application of immunoassays for the study of natural environments will accelerate immensely in the near future. The choice of what systems to study, and therefore which antibodies to clone, must be made carefully and wisely.

Enzyme and immunoassay techniques will certainly be applicable to the study of OAs of relatively low molecular weight—species that behave as chemically distinct species. It remains to be determined how useful these methods will be in following small molecules that are "bound" to the AHS core or in following the core itself. Will a bound amino acid or fatty acid react with its respective enzyme or antibody? Will this not depend on the type of "boundness?" Can enzyme or immunoassay techniques be used to follow high molecular weight OAs from different terrestrial sources or through diagenetic processes? These are research questions that must be addressed before these new techniques can be fully exploited in the study of aquatic OAs.

Making Use of the Ecosystem in the Characterization of Organic Acids

A common approach to the study of natural systems has been to apply chemical and spectroscopic methods to analyze the system, to develop parameters that measure some property of the system, and to use those parameters to monitor the system. This approach has shown that natural organic material is largely made up of acidic substances which, up to now, appear to have no regular (repeating) molecular structure, either in terms of the identifiable components of the mixture or the molecular structure of the most prevalent part of the mixture, the AHS "core."

Of increasing interest to ecologists and geochemists is not the structure of this material but its function. Thus, it is appropriate to return to the ecosystem and search for new parameters that will be more relevant to the function of the OA mixture than those now available. Two approaches to the task are suggested: (*a*) the search for enzymes that exist within the system and the determination of their target substrates, and (*b*) the development of immunoassays for gross (unidentified and perhaps composited) organic material; such assays would be used to track the material in its pathways through the environment. Both of these approaches are still highly experimental, but may in time yield much valuable information about systems that contain OAs.

Analysis of the Microbial Community and Microbiological Enzyme Ensembles in an Ecosystem Containing Organic Acids

In this approach, one develops an understanding of the OAs in an ecosystem through an analysis of the microbial communities and microbiological enzyme ensembles which exist in the ecosystem. For example, the microbial community structure involved in the bioprocessing of OAs may be defined with phospholipid biomarkers and nucleic acid probes against 16S RNA's (White et al., this volume). Aerobic versus anaerobic processes can be defined by comparing the incorporation of ^{13}C-acetate into specific phospholipid ester-linked fatty acids (PLFA) characteristic of aerobic bacteria, such as the short-branched PLFA of pseudomonads compared to the ether-linked alkyl glycerols of methane-forming bacteria or the specific-branched PLFA of sulfate-reducing bacteria. The potential fates of OAs in contact with aerobic microbes can be quite different from those whose primary bioprocessing involves anaerobic terminal electron acceptors. Aerobic microbes can utilize highly reactive oxygen-derived components to degrade polyaromatic OAs, and these organisms are considerably less likely to form short-chain OAs. Anaerobic systems using oxidized sulfur as a terminal electron acceptor apparently have a much less versatile bioprocessing potential than the methanogenic consortia.

Characterization of the microbial communities will allow the genes for specific enzymes (or the enzymes themselves) to be probed, and will ultimately allow the enzymes to be synthesized by genetic engineering techniques (Gassen, this volume).

Using this technique it may be possible to identify key OAs that are biomarkers of a given localized environment and to track those biomarkers as the OA mixture ages or leaves that environment (White et al. 1979). Alternately, one may take the information about the biomarker to the chemistry laboratory, synthesize sizeable quantities of the biomarker with appropriate isotopic labels, and conduct transport and fate studies of the compound in field or laboratory experiments. Also, one may develop cheap immunoassays or chemical assays for the biomarker that will be used in subsequent studies.

Immunoassays for uncharacterized organic acids. It is possible, in principle, to develop immunoassays for molecularly uncharacterized substances. Thus, one may postulate the use of immunoassay methods to track high molecular weight AHS through the aquatic environment. There are several assumptions involved in such projections, including the postulate that AHS would yield a useful set of antibodies (see above). However, the prospect appears to be reasonable enough so that experiments should be tried.

In addition, the use of monoclonal and polyclonal antibodies and enzymatic techniques may assist in characterizing the AHS molecules themselves. Thus, if Mabs or OA-biomarker-specific enzymes are available, one may probe the AHS complex with these tools. It may be possible then to determine details of the AHS macromolecules that have not been obtained through classical approaches.

In summary, new approaches to the characterization of OAs through the application of modern principles of biochemistry and molecular biology may be extremely useful in providing structure and functional information about OAs. Approaches such as those discussed above should be pursued.

RESEARCH RECOMMENDATIONS AND CONCLUSIONS

Isolation Methods

1. Research is needed to explore the possible effects of the high and low pH conditions to which OAs are exposed in the XAD isolation procedure.
2. Because of their low cost and ease of use, prepacked cartridges of various adsorbents should be evaluated for isolation of OAs from aqueous samples.
3. Membrane methods appear to be attractive methods for the isolation of OAs but should be evaluated carefully, especially with respect to possible losses of very low and very high molecular weight OAs.

Fractionation Methods

1. GC continues to be the separations method of highest resolving power for low molecular weight organic compounds. Despite the fact the OAs

generally require derivatization prior to GC analysis, scientists who wish to study OAs in the environment should consider this method (with appropriate derivatization).
2. Further research is needed in the development of HPLC and electrophoretic methods for the analysis of OAs, including the evaluation of mass spectrometric as well as low-cost, field-adaptable detectors.
3. SCFC may be immediately useful for the study of derivatized OAs including AHS and eventually for the study of free OAs.

Spectroscopic Characterization Methods

1. Further use of new NMR methods for the study of solid samples may yield new information on the structure of AHS, especially if combined with good fractionation procedures.
2. Recently developed HPLC/MS methods should be applied to the study of OA mixtures. Of special interest is the potential for using HPLC separations methods in combination with tandem isotope ratio mass spectrometry.
3. GC/MS methods are now amenable to routine use with autosamplers, injectors, and data analysis for rapid turnover of samples; thus, they should be seriously considered for field and ecosystem studies that generate large amounts of data.
4. FTIR and surface-enhanced Raman spectroscopy should be used to study fractions of OAs especially of moderate to high molecular weights in order to derive information which is complementary to that obtained by mass spectrometry.

Chemical Characterization Methods

1. More applications of combined separations and isotopic measurements are needed to exploit fully the power of isotopic methods in determining the properties of OAs.
2. Also, more studies are needed that combine isotopic studies with measurements of chemical tracers.
3. Given the fact that methods for the measurement of DOC and total organic carbon may give low values on natural water samples, DOC methods should be reevaluated and standardized so that DOC measurements will be comparable from one study to another.

4. Size exclusion HPLC systems probably can be adapted easily for field and routine studies and may provide a useful parameter for the characterization of bulk OAs.
5. Reliable methods are needed for other bulk parameters such as a rapid method for biodegradability, degree of aromaticity, critical micellular concentration, hydrophobicity, C, N, S, and P content of DOC fractions, average molecular surface area, and percentage organic N compounds.
6. Chemical derivatization methods in combination with high resolution chromatographic separations are needed, especially for OAs of moderate molecular weight and with polyfunctionality.
7. Methods are needed that measure the properties of OAs without disturbing the ecological setting. Microscale methods such as those utilizing fiber optics and portable GC analyzers are promising possibilities.

New Approaches to the Characterization of Organic Acids

1. Enzymatic analysis systems should be applied to the study of aquatic systems containing OAs.
2. Also, immunoassay procedures should be applied to the same purpose, especially for biomarker compounds and for the AHS complex as a whole. The latter may prove to be a useful bulk parameter which is characteristic of OA mixtures from a given environment.
3. The characterization (and production by genetic synthesis methods) of enzymes and 16S RNAs of microbes from natural environments may be used to determine key OAs that are biomarkers of aquatic ecosystems or sources of OAs.
4. Immunoassay and enzyme analytical techniques may be useful to characterize the AHS mixture by probing the macromolecules for specific chemical moieties or combinations of moieties.

Other Observations and Conclusions

Several other aspects of the problem of OA characterization were discussed by our group that are not included in the preceding pages. These include the following subjects, some of which are not new to the humic acid scientific community:

1. What is the molecular weight distribution of OAs in actual (natural) environments as opposed to isolated (and therefore fractionated) samples? How could such a property be determined?

2. What is the best procedure to use to obtain large amounts of humic material (OAs) for experimental purposes?
3. What artifacts are introduced by the use of membranes with charged surfaces?
4. What derivatizing agents are best suited for polyfunctional OAs, or are other (biochemically based methods) inherently more suitable?
5. Are residual OAs in groundwaters substantially different (in structure and properties) than the OAs that make up the majority of DOC in surface waters?
6. What approaches should be used to characterize "dissolved" vs. total OAs, to differentiate abiotic from biotic processes, and to measure both processes in unfractionated (natural) samples?
7. How does a reconstituted OA mixture (the recombined fractions) compare in properties with the original OA mixture?

THE NEED FOR INTERDISCIPLINARY TEAMWORK IN THE STUDY OF ORGANIC ACIDS

It is apparent from the preceding discussion that the characterization of OAs from aquatic environments has long since ceased to be the domain of analytical chemists; indeed, it was never such a proprietary domain. Now, however, the need for teamwork is clearer than ever as we move more actively into the study of the processes in which OAs are involved. Such studies must involve teams of ecologists, microbiologists, analytical chemists, and sometimes biochemists and molecular biologists. In addition to their principal roles, ecologists and microbiologists must assist analytical chemists in the identification of problems for which assays are needed; they must also provide insight into the biology and biochemistry of these systems. Likewise, analytical chemists and biochemists must assist in the development of useful assays and the transfer of these methods to the user community. They must be sensitive to the needs for methods that not only meet scientific criteria, but are also convenient, minimally disturbing to the environment, and amenable to automation. Also, team members must be sensitive to the different traditions and purposes of their colleagues' work and be able to appreciate these differences and work together towards common research goals.

Finally, our group briefly discussed a grim possibility: is it possible that within the OA ensemble in natural systems there is important information regarding global change that is going undetected because of our inability to characterize this ensemble? Such a possibility brings clearly into focus the need for effective research on the nature and behavior of aquatic OAs, research that can only be effective if it involves interdisciplinary approaches to sound science.

REFERENCES

Aiken, G.R. 1985. Isolation and concentration techniques for aquatic humic substances. In: Humic Substances in Soil, Sediment and Water, ed. G.R. Aiken, D.M. McKnight, R.L. Wershaw, and P. MacCarthy, pp. 363–385. New York: Wiley.

Aiken, G.R. 1988. A critical evaluation of the use of macroporous resins for the isolation of aquatic humic substances. In: Humic Substances and Their Role in the Environment, ed. F.H. Frimmel and R.F. Christman, pp. 15–28. Dahlem Konferenzen. Chichester: Wiley.

Aiken, G.R., E.M. Thurman, R.L. Malcolm, and H.F. Walton. 1979. Comparison of XAD macroporous resins for the concentration of fulvic acid from aqueous solution. *Analyt. Chem.* 51:1799–1803.

Bellar, T.A., and W.L. Budde. 1988. Determination of nonvolatile organic compounds in aqueous environmental samples using liquid chromatography/mass spectrometry. *Analyt. Chem.* 60:2076–2083.

Bombaugh, K.J. 1984. The use of HPLC for water analysis. In: Organic Species, ed. R.A. Minear and L.H. Keith. Vol. 3 of Water Analysis, pp. 317–379. Orlando, FL: Academic.

Caldwell, K.D. 1988. Field-flow fractionation. *Analyt. Chem.* 60:959A–971A.

Davies, I.L., M.W. Raynor, J.P. Kithinji, K.D. Bartle, P.T. Williams, and G.E. Andrews. 1988. Interfacing LC, SFE, GC and SFC. *Analyt. Chem.* 60:683A–702A.

Ertel, J.R., and J.I. Hedges. 1984. The lignin component of humic substances: distribution among soil and sedimentary humic, fulvic, and base-insoluble fractions. *Geochim. Cosmo. Acta* 48:2025–2074.

Frimmel, F.H., and R.F. Christman, eds. 1988. Humic Substances and Their Role in the Environment. Dahlem Konferenzen. Chichester: Wiley.

Gjessing, E.T. 1976. Physical and Chemical Characteristics of Aquatic Humus. Ann Arbor, MI: Ann Arbor Science.

Gloor, R., H. Leidner, K. Wuhrman, and T. Fleischmann. 1981. Exclusion chromatography with carbon detection: a tool for further characterization of dissolved organic carbon. *Water Res.* 15:457–463.

Gurka, D.F., and S.M. Pyle. 1988. Qualitative and quantitative environmental analysis by capillary column gas chromatography/lightpipe Fourier transform infrared spectrometry. *Env. Sci. Tech.* 22:963–968.

Kawamura, K., and I.R. Kaplan. 1987. Motor exhaust emissions as a primary source for dicarboxylic acids in Los Angeles ambient air. *Env. Sci. Tech.* 21:105–110.

Leenheer, J.A. 1984. Concentration, partitioning and isolation techniques. In: Organic Species, ed. R.A. Minear and L.H. Keith. Vol. 3 of Water Analysis, pp. 83–166. Orlando, FL: Academic.

Meyers-Schulte, K.J., and J.I. Hedges. 1986. Molecular evidence for a terrestrial component of organic matter dissolved in sea water. *Nature* 321:61–63.

Norwood, D.L. 1988. Critical comparison of structural implications from degradative and nondegradative approaches. In: Humic Substances and Their Role in the Environment, ed. F.H. Frimmel and R.F. Christman, pp. 133–148. Dahlem Konferenzen. Chichester: Wiley.

Provder, T. 1980. Size Exclusion Chromatography (GPC). *ACS Symp. Ser.* 138. Washington, D.C.: American Chemical Society.

Schnitzer, M., and S.U. Kahn. 1972. Humic Substances in the Environment. New York: Dekker.

Scully, F.E., Jr., G.D. Howell, H.H. Penn, K. Mazina, and J.D. Johnson. 1987. Small molecular weight organic amine nitrogen compounds in treated water. *Env. Sci. Tech.* 22:1186–1190.

Swineford, D.M., and A.A. Belisle. 1989. Analysis of trifluralin, methyl paraoxon, methyl parathion, fenvalerate and 2,4-D dimethylamine in pond water using solid-phase extraction. *Env. Tox. Chem.* 8:465–468.

Taylor, J.S., D.M. Thompson, and J.K. Carswell. 1987. Applying membrane processes to groundwater sources for trihalomethane precursor control. *J. Am. Wat. Wks. Assn.* 79:72–82.

Thurman, E.M. 1985. Organic Geochemistry of Natural Waters. Dordrecht: Nijhoff/ Junk.

Thurman, E.M., et al. 1988. Isolation of soil and aquatic humic substances. In: Humic Substances and Their Role in the Environment, ed. F.H. Frimmel and R.F. Christman, pp. 31-43. Dahlem Konferenzen. Chichester: Wiley.

Vanderlaan, M., B.E. Watkins, and L. Stanker. 1988. Environmental monitoring by immunoassay. *Env. Sci. Tech.* 22:247–254.

Weber, J.H. 1988. Binding and transport of metals by humic materials. In: Humic Substances and Their Role in the Environment, ed. F.H. Frimmel and R.F. Christman, pp. 165–178. Dahlem Konferenzen. Chichester: Wiley.

Wershaw, R.L., P.J. Burcar, C.L. Sutula, and B.J. Wiginton. 1967. *Science* 157:1429–1431.

White, D.C., R.J. Bobbie, J.S. Herron, J.D. King, and S.J. Morrison. 1979. Biochemical measurements of microbial mass and activity from environmental samples. In: Native Aquatic Bacteria: Enumeration, Activity, and Ecology. *Am. Soc. Test. Mater.* 695:69–81.

Wilson, M.A. 1987. NMR Techniques and Applications in Geochemistry and Soil Chemistry. Oxford: Pergamon.

Organic Acids in Aquatic Ecosystems
eds. E.M. Perdue and E.T. Gjessing, pp. 97–109
John Wiley & Sons Ltd

Carboxyl Acidity of Aquatic Organic Matter: Possible Systematic Errors Introduced by XAD Extraction

M.S. Shuman

Department of Environmental Sciences and Engineering
University of North Carolina
Chapel Hill, NC 27599–7400, U.S.A.

Abstract. Literature values of carboxyl acidity are compared and found to be exceedingly uniform for XAD extracts of aquatic dissolved organic matter (DOM). Although the data base for drawing any conclusion is incomplete, there is some evidence indicating that these values are lower than those obtained from anion exchange resin extracts or from samples that have not undergone an extraction step. It is suggested that XAD resins select a uniform fraction of the DOM, which gives the illusion of uniform chemical properties, and that the acidic fraction that is not extracted, the so-called "hydrophilic acid" fraction, carries important and differentiating information that is ignored. Relying principally on XAD extracts as surrogates for investigations of native aquatic DOM chemistry may lead to serious errors in modeling DOM for acidity or metal complexation, and may bias our understanding of how the chemical properties of DOM vary over time and space. Recommendations are made for research to study whole water samples and the "hydrophilic acid" fraction to test these hypotheses.

INTRODUCTION

Ask anyone working with aquatic organic matter to tell you the value of its carboxyl acidity and the answer will be 10 microequivalents per milligram of organic carbon (μeq/mg C). This is a felicitous value, one which is easy to remember; it is also an average value taken from a voluminous research literature that has been made possible because of the introduction of XAD macroporous resin extraction techniques. These techniques are easy to use for processing large volumes of water to obtain gram quantities of research materials. They also introduce an operational definition to the material that is ultimately studied, i.e., this material is not the original dissolved organic matter (DOM) but a certain fraction of it defined as XAD-extracted material. This fraction may or may not accurately represent the chemical properties of the original DOM in the sample. One indication that it may not be representative comes from the observation that XAD resins extract only about 45–70% of the total dissolved organic carbon (DOC). Another comes from Cu binding studies. XAD extracts obtained from a number of different sources are all described by a single model, whereas binding to DOM in samples that have not undergone any extraction or concentration procedure cannot be so described (Cabaniss and Shuman 1988). This uniformity of binding character and the low and variable percent extraction together suggest that XAD selects from the total DOM (for XAD resins, the selectivity is based on molecular size or solubility) a fraction that has *uniform* chemical properties but which does not likely have the *same* chemical properties as unfractionated DOM. In this way, XAD extracts could reflect the same information from all DOM sources, and the information which distinguishes samples could be lost in the DOM that is not extracted and discarded.

Although the operational nature of XAD-extracted DOM is recognized and discussed (Aiken 1988), no systematic study has been attempted to reveal how our understanding of the chemistry of DOM may be biased when our investigations are based on XAD extracts. With regard to carboxyl acidity, there are no studies to establish that XAD extracts accurately represent the carboxyl acidity of native DOM. What is required is evidence that both XAD extracts of a sample and the unextracted sample give the same value for carboxyl acidity. Because this evidence is not available from any one study, this chapter reviews the literature to look for uniformity among the carboxyl acidity values of various XAD extracts, of whole water samples, and of extracts obtained using sorbents other than XAD. This comparison is complicated by the fact that two methods are used to estimate carboxyl acidity: direct titration and reaction with calcium acetate. The calcium acetate method can give higher acidity values than the direct titration method (Perdue 1979), and these considerations are discussed briefly at the beginning of the chapter. Comparison is also frustrated by the paucity of data for samples other than XAD extracts, a situation that can only be remedied by careful collection of data under identical conditions on whole water samples and on extracts and effluents of various sorbents, including XAD. Collection of these data is one of the principal recommendations of this chapter, made in concert with suggestions about how it should be done.

DEFINITIONS OF CARBOXYL ACIDITY

XAD extracts of DOC dissolved in distilled water equilibrate at about pH 3–4. Titration with standard base to pH 7 indicates 9.0–12.0 μeq/mg C and an average mass action quotient, pK_a, in the range 3–5. Infrared spectroscopy and ^{13}C NMR data suggest that this acidity arises principally from carboxylic acids. Titration continued to about pH 11 neutralizes additional weak acids (about 2–3 μeq/mg C, pK_a 8–11) that are thought to be phenols and alcohols.

When the mass action quotient, $K_a=[H][A]/[HA]$, is calculated at any point along the titration curve, its value is not constant, as is expected of a simple organic acid such as acetic acid, but decreases with increasing neutralization. This suggests the presence of a mixture of acidic groups with different acid strengths. Attempts to describe the chemical behavior of this mixture and its neutralization curve by using a small number of nonidentical, noninteracting acidic functional groups, by using a Gaussian or undetermined pK_a distribution of such groups, or by assuming the presence of coulombic interactions between identical groups where the pK_a varies with degree of neutralization, have all been about equally successful, even though each model begins with entirely different assumptions about physical reality. The endpoint observed in the neutralization curve apparently reflects a change in buffer capacity of this complex mixture of acids, not a true equivalence point of any individual group or ensemble. Nevertheless, the amount of base used to titrate the sample to pH 7–9 is routinely used to obtain a value for what is called the "carboxyl acidity." Clearly this is an operational definition because the endpoint is not a true equivalence point and the titratable protons may not all come from carboxylic acids but may include other weak acids that are wholly or partially neutralized at the pH where the titration is terminated.

Another operational definition of carboxyl acidity that is widely used is based on reaction of the organic matter with excess Ca-acetate to form acetic acid, which is then titrated with NaOH to pH 9.8. Rather high concentrations of organic matter are usually necessary for this technique (50–100 mg for 20 ml of reagent solution is recommended). Both this and the direct titration technique are thoroughly critiqued by Perdue (1985), who also includes recommendations for obtaining consistent values. Variability can be quite low if these recommendations are followed (about 10% relative error).

$Ca(OAc)_2$ AND DIRECT TITRATION METHODS

Although the direct potentiometric titration method and the $Ca(OAc)_2$ method both give an operational definition of carboxyl acidity, consistent and comparable results are desirable. The $Ca(OAc)_2$ method requires certain precautions if the values it gives are to be compared to the values obtained by the direct titration method. $Ca(OAc)_2$ reacts with the more acidic functional groups on the organic matter (those with pK_a less than approximately 7) to form acetic acid (HOAc) which is titrated with standard base to pH 9.8. From the discussion above, it is evident that acids remain in solution which are too weak to react with $Ca(OAc)_2$ but which will nevertheless be titrated with the strong base used to titrate the HOAc. Ultrafiltration with Amicon UM-2 after $Ca(OAc)_2$ reaction and before titration is suggested by Perdue (1985) to reduce the amount of titratable organic material. If ultrafiltration is carried out before titration, the estimate of carboxyl acidity is comparable to the direct titration method. If filtration is done with Whatman #2, as recommended by Schnitzer and Khan (1972), carboxyl acidity values are about 25–30% higher than those obtained with direct titration to pH 7. Data in the literature are nearly all the result of Whatman #2 filtration so that values higher than direct titration values are expected.

Perdue (1985) also recommends NaOAc instead of $Ca(OAc)_2$ to avoid Ca complexation of weakly acidic functional groups such as phenolic acids. Complexation releases protons which would be counted as carboxyl acidity. In addition, because the organic matter–$Ca(OAc)_2$ mixture is poorly buffered (the equilibrium pH is typically 6–7, well outside the buffering zone of acetic acid–acetate solutions), the equilibrium pH will decrease and thus the extent of HOAc formation will increase with the concentration of organic matter. Therefore, carboxyl acidity per gram of organic matter appears to increase with organic matter concentration, especially when ultrafiltration, which eliminates other concentration effects, is used. Direct titrations do not seem to show a concentration effect (Dempsey and O'Melia 1983; Oliver et al. 1983).

Direct titrations are carried out to an endpoint or to some fixed pH, usually somewhere between pH 7 and pH 7.6. This pH range occurs where the apparent carboxyl content is relatively invariant with pH so that pH selection is not particularly critical. For a typical titration, the estimate of carboxyl acidity at pH 7.6 is about 5% higher than that at pH 7. Thus, differences among estimates by direct titration methods are small and consistent. The variability among estimates by the $Ca(OAc)_2$ method is also small, but their values are systematically and consistently higher than direct titration values. It is therefore convenient and sensible to separate these two analytical methods to inspect them for uniformity.

EXTRACTION/CONCENTRATION METHODS

Organic materials are commonly extracted from soils or water samples to separate them from the background matrix and to concentrate them to levels where instrumental and wet chemical methods can be used. Various bases (e.g., NaOH, $Na_2P_2O_7$, Na_2CO_3) are employed for extracting organic matter from soils, but sorption chromatography is more common for extracting them from fresh and oceanic waters because large volumes can be processed conveniently. The most popular sorbents are macroporous resins, Amberlite XAD nonionic resins, and Duolite A–7, a weak base anion exchange resin. Other anion exchange resins and activated carbon have also been used. Sometimes sorbents are used in combination, e.g., XAD resins are often used with a cation or an anion exchange resin for the purpose of desalting the sample or extracting different fractions of the total organic material. The eluate and its concentration or its pH can also be varied for fractionation.

The use of macroporous resins has been reviewed extensively (Aiken 1985, 1988, and references therein). A typical method for extracting with XAD-8, the most popular sorbent, is to acidify the filtered sample to pH 2.0, pass it through a column of the resin, elute with 0.1 N NaOH, acidify and concentrate further on a smaller XAD-8 column, precipitate the humic acid portion at pH 1, desalt the fulvic acid portion by another pass through the column, protonate on a cation exchange column, and freeze dry or use within a short time.

Extraction schemes such as this have become almost a standard workup procedure for obtaining material to study acidity or metal complexation. The chemical character of this material will vary among different isolation schemes, as has been pointed out by Aiken (1988). Each isolation scheme gives its own operational definition to the extract that is studied. An important question is whether the chemical properties being studied with the extract accurately reflect the chemical properties of the DOM as a whole. If they do not, effort is not only wasted, but our view of the chemistry of DOM in the environment will be distorted, and our predictive models of acidity and metal binding will be useless.

THE DATA

Extraction Efficiency Data

The efficiency of extracting organic carbon by XAD or other sorbents varies widely from method to method and from sample to sample. Table 1 lists some examples. Data listed in this table are recoveries of DOC from actual samples, and are not the recoveries that are sometimes also reported based on readsorption of previously extracted material. The data in the table indicate that marine organic material is poorly sorbed on carbon or XAD and that the most widely used sorbent, XAD-8, consistently averages about 50% extraction efficiency for all samples reported. Efficiencies for XAD-2

TABLE 1. Percent organic matter extracted.

Sorbent	%DOC Extracted/Sample	Reference
XAD-2	69–78% Lakes, rivers	Mantoura and Riley (1975)
	4–5% Oceanic surface	Stuermer and Harvey (1977)
	22% Deep ocean	Stuermer and Harvey (1977)
XAD-7	70% Satilla River	Perdue et al. (1980)
	30–40% Oyster River	Weber and Wilson (1975)
XAD-8	45% Suwannee River	Aiken (1985)
	44–60% Various sources	
	38–58% Lakes, rivers	Collins et al. (1986)
DEAE	86–89% Suwannee River	Miles et al. (1983)
CARBON	14–45% Seawater	Kerr and Quinn (1975)

and XAD-7 are based on just a few available data, but these suggest efficiencies quite a bit less than 100%. Suwannee River organic material is reported to be 45% extracted by XAD-8 and 86–89% extracted by diethylaminoethylcellulose (DEAE). Although these data were not collected on the same sample of the Suwannee, nor by the same research group, the reported efficiency for DEAE is significantly higher than those reported for other sorbents.

Carboxyl Acidity Data

Table 2 summarizes the carboxyl acidity of samples of aquatic organic material determined by the $Ca(OAc)_2$ method. None of these data are the result of a procedure where the sample is ultrafiltered before titration as recommended by Perdue, so it is assumed that they are higher values than might have been obtained if the direct titration method had been employed. Table 3 lists the carboxyl acidity of samples determined by direct potentiometric titration. These two methods are grouped separately to compare absolute values independent of the positive bias introduced by the $Ca(OAc)_2$ method.

TABLE 2. Carboxyl acidity by $Ca(OAc)_2$ method.

Carboxyl μeq/mg C	Sample	Sorbent	Reference
20.7	Hatchett Creek	DEAE	Miles et al. (1983)
19.0	Newport River	None	Gaskill (1978)
18.2 (ave)	Satilla River	None	Beck et al. (1974)
17.4 (<25,000)	Aquatic FA	XAD-2	Visser (1982)
13.2 (>25,000)			
13.3	Oyster River	XAD-2	Weber and Wilson (1975)

TABLE 3. Carboxyl acidity by potentiometric titration.

Carboxyl μeq/mg C	Sample	Sorbent	Reference
		XAD Extracted Samples	
11.7	Suwannee River FA	XAD-8	Thurman and Malcolm (1983)
10.7–11.6	Lake Drummond	XAD-7	Dempsey and O'Melia (1983)
10.9+/−0.5	8 Rivers/streams	XAD-2 & 8	Oliver et al. (1983)
10.0–10.7	Lake FA	XAD-8	McKnight et al. (1982)
9.3	Bersbo	XAD-8	Ephraim et al. (1989)
9.0	Satilla River	XAD-7	Perdue et al. (1980)
8.0–12.2	Lake, rivers	XAD-8	Collins et al. (1986)
6.4	Aquifer	XAD-8	Collins et al. (1986)
		Other Samples	
16.0	Newport River	None	Gaskill (1978)
13.5	Biscayne Aquifer	Duolite A-7	Thurman and Malcolm (1983)
10.3	3 rivers, mixed	Carbon	Huizenger and Kester (1979)
9.0–11.6	Various oceanic	Carbon	Huizenger and Kester (1979)

What is striking about these data is the uniformity of values in the XAD-extracted material summarized in Table 3, with an average value of about 10.5 μeq/mg C and a relative standard deviation of about 11%. This is remarkable and almost too good to be true from an analyst's standpoint if consideration is given to the many steps involved in the extraction/titration procedure, the wide geographical distribution of the samples, and the large number of research laboratories involved. Although there are no data available to confirm it, it is difficult to imagine that the variability seen in this table could be much larger than the variability in the combined extraction/titration procedure itself.

A comparison of carboxyl acidity of a sample extracted by two different techniques has been reported for only one sample. Biscayne groundwater was extracted with XAD-8 and with Diamond Shamrock Duolite A-7, a phenol-formaldehyde weak-base anion exchange macroporous resin, and the carboxyl acidity values are given in Table 3. The acidity of 13.5 μeq/mg C for the Duolite extract is higher by about twice the standard deviation of what is assumed to be the variability in the rest of the data in the table and may indicate a significant difference. It is interesting, anyway, that this value and the value obtained on an unextracted sample (Newport River, 16.0

μeq/mg C) are larger than any of the XAD extract values. There are several hypotheses that could explain these differences, and they are discussed below, but one among them is that XAD resins selectively extract a fraction which has lower carboxyl acidity than the unextracted organic matter.

The data in Table 2 reflect the higher absolute values expected of the $Ca(OAc)_2$ method. In common with Table 3, samples extracted with an XAD resin appear to give smaller values. If the extract is separated into two molecular size groups by ultrafiltration before carboxyl acidity is determined, the smaller size fraction (<25,000 MW) gives larger acidity values than the larger size fraction (>25,000 MW).

DISCUSSION OF ERRORS

For the following discussion it is assumed that for any given sample of aquatic DOM, there exists a correct value for the operationally defined carboxyl acidity. If estimation of this value is viewed as a two-step process (an extraction step followed by an analytical step), errors can be conveniently broken out for each step. One can imagine both positive and negative errors for the extraction step. The selectivity of the sorbent may produce an extract containing either higher or lower acidity than the native DOM, depending on what basis the sorbed fraction is selected. Positive errors may arise from the carboxyl acidity production that takes place by ester hydrolysis during alkali elution of DOM from the sorbent. During the analytical step, Al or Fe present in the sample may undergo hydrolysis and produce positive errors. Nonsterile solutions in the presence of oxygen will also produce acids via bacterially mediated oxidation reactions. These errors are discussed in more detail in the following sections.

Extraction Step

In the chromatographic fractionation scheme introduced by Leenheer (1981), an XAD-8 column is followed in series by a strong cation exchange resin in H form (Bio-Rad AG-MP-50), and then by a weak anion exchange resin (Duolite A-7) in free base form. The portion of the DOM that sorbs onto the XAD-8 is called "hydrophobic," which is most of the colored materials or "humic substances" of the sample (this "hydrophobic"/"hydrophilic" distinction is somewhat confusing, because to be dissolved, the organic matter must be quite hydrophilic overall). This "hydrophobic" portion is typically about 50% of the DOM. The material that passes the XAD resin is called "hydrophilic." The "hydrophilic acids," which are sorbed onto the Duolite A-7 and eluted with 3N NH_4OH, account for about 20–50% of the DOM. 0.1 N NaOH is used to elute "hydrophobic acids" from the XAD resin. It is this fraction that is normally assayed for carboxyl acidity.

In general, adsorption strengths on XAD resins are linearly related to average molecular weights (Mantoura and Riley 1975) and inversely related to solubility (Thurman et al. 1977). It is the larger sized, less soluble materials which are retained by the resin. Carboxyl acidity, on the other hand, follows a reverse trend; acidities are inversely related to molecular size and directly related to solubilities. The smaller, more soluble aquatic fulvic acids have acidities about 40% higher than aquatic humic acids (Oliver et al. 1983). Satilla River DOM fractionated by Sephadex G-50 gel permeation chromatography shows this same trend. The smallest sized molecules, those retarded by the gel, have carboxyl acidities over twice that of the largest sized molecules, those excluded from the gel (Beck et al. 1974).

It can be argued that estimates of DOM carboxyl acidity based on analysis of XAD extracts are low because the "hydrophilic acid" fraction is ignored. This fraction, up to half the DOM, has not yet been analyzed for carboxylic acidity, although its IR and ^{13}C NMR spectra (Leenheer 1981; Aiken 1988) show carboxyl bands comparable to the "hydrophobic acid" fraction. If the trends of carboxyl acidity with molecular size are valid, then the "hydrophilic acid" fraction, because of its small size, is expected to have a higher acidity than the XAD-extracted fraction.

The alkali elution of sorbents is often mentioned as a possible source of positive error due to acid production via ester hydrolysis. However, Greger and Powell (1987) observed no difference in carboxyl acidity after 16 hours reaction during alkali extraction of XAD-8, and only 6–9% increase after alkali treatment of a pyrophosphate leached peat fulvic acid for 72 hours. This error is well within the assumed analytical error of the extraction/analytical procedure.

Analytical Step

The presence of Al or Fe in the sample can introduce a positive error due to hydrolysis. Studies with soil organic matter indicate that removal of these metals will also greatly reduce hysteresis observed in acid-base titration curves of humic acids (Flaig et al. 1975). This hysteresis has been related to the rate-limited dissociation of the metals (Dempsey and O'Melia 1983).

One of the advantages of using XAD resins is that they sorb relatively large, neutral organic molecules and not inorganic salts. The pH of the sorption step, usually around pH 2, should not favor formation of Al and Fe complexes. Even so, the metal content of the extract can be a several mole percent of the carboxyl acidity (Dempsey and O'Melia 1983) and affect titration of the extract.

The metal content of whole water samples or their concentrates are of more concern. For example, the Fe concentration of the Newport River sample of Table 2 is about 25% of the estimated carboxyl acidity and will account for a portion of it. The Satilla River concentrates of Table 2 are calculated from the original data to be about the same percent Fe. Beck et al. (1974) noted that desalting with a strong cation exchange resin in the H form removed only about 1/3 of the Fe, and they speculated that the remainder was complexed to neutral organic molecules or present as colloidal Fe.

The exact extent to which Fe or Al introduces a positive error into the estimation of carboxyl acidity is difficult to estimate. From studies done with soils, it is not clear if the errors are associated with complexed or colloidal Fe and whether they are time dependent.

CONCLUSIONS

Carboxyl acidity values of XAD extracts are surprisingly uniform among aquatic samples. Factors such as geographical location, sample type (lake, river, groundwater, marine), DOC concentration, or chemical composition of the sample seem to have little effect on a value that is always about 10 μeq/mg C. In fact it is so uniform that it is questionable whether it ever needs to be measured again for its extremely low information value. The same cannot be said of the 20–50% of the DOC that passes the XAD column. It has high information value because it has not been studied to any great degree and yet may contain all the geographical, sample concentration/composition variability missing from the extract.

XAD resins select a fraction comprising about 40–60% of the aquatic sample DOM that not only has the same carboxyl acidity, sample to sample, but also has the same copper binding properties. This would be serendipitous if the point of studying XAD extracts was to illuminate their own chemical properties, but this is not usually the case. The point is to investigate the properties of DOM, for which the extract is a convenient substitute. Models based on XAD extracts do not describe copper binding to native DOM in unextracted samples. There have been no similar attempts to validate proton binding models or even to compare carboxyl acidity values.

A review of the literature revealed that, in general, XAD extracts give lower values for carboxyl acidity than whole water samples or weak anion exchange resin extracts. This conclusion is based on pitifully few samples, but one might reason from the selectivity of the XAD resin that this is plausible. The acid fraction of the DOM that passes the resin is low molecular weight and has high carboxyl acidity. The sorbed fraction thus has lower carboxyl acidity than the DOM in the sample. By similar reasoning, anion exchange resin extracts should have acidity closer to DOM because the anion exchange resin extracts a higher percent of the DOM than does XAD.

XAD extracts are used for studies of metal binding, acid-base properties, metal or organic adsorption and transport, and many other chemical properties. The results of these studies, the models, the product distributions, etc., are used, many times without validation with the original DOM, to predict the behavior of DOM in the environment. If predictions are wrong, the mistakes could be costly. To possibly understand this, one simply needs to think about calculations to determine the contribution of DOM to the acidity of low alkalinity lakes which use faulty values for the carboxyl acidity of DOM.

RECOMMENDATIONS

The hypotheses of this chapter are (*a*) that XAD resins extract a uniform fraction of aquatic DOM and (*b*) that measurements of carboxyl acidity and perhaps other chemical properties of XAD extracts are biased, uniform, and devoid of information that would differentiate between samples. Research is recommended that would test these hypotheses by collecting data not only on the XAD extract, which is the current widespread practice, but also on the unextracted fraction and the sample untreated by the resin. Physical characteristics such as the molecular size distribution as well as the acid-base characteristics should be compared. Two specific lines of investigation are suggested.

1. Samples that have not undergone XAD extraction should be prepared for comparative studies. Combinations of reverse osmosis, dialysis, diafiltration, evaporation, etc., could be used judiciously to concentrate and desalt the DOM if necessary. Steps should be taken to minimize the concentrations of Al and Fe in these DOM concentrates.
2. XAD-treated samples should also be studied, but not just the "hydrophobic acid" fraction. The "hydrophilic acid" fraction by itself and in combination with the "hydrophobic acid" fraction should be investigated. Almost any information about this fraction is desirable, because up to now, most of it has been flushed down the drain, and with it perhaps everything that could tell us about how one DOM sample differs from another.

Acknowledgements. I thank D. Amaral, S. Cabaniss, and L. Sonnenberg for helpful suggestions.

REFERENCES

Aiken, G.R. 1985. Isolation and concentration techniques for aquatic humic substances. In: Humic Substances in Soil, Sediment, and Water, ed. G.R. Aiken,

D.M. McKnight, R.L. Wershaw, and P. MacCarthy, pp. 363–386. New York: Wiley.

Aiken, G.R. 1988. A critical evaluation of the use of macroporous resins for the isolation of aquatic humic subtances. In: Humic Substances and their Role in the Environment, ed. F.H. Frimmel and R.F. Christman, pp. 15–28. Dahlem Konferenzen. Chichester: Wiley.

Beck, K.C., J.H. Reuter, and E.M. Perdue. 1974. Organic and inorganic geochemistry of some coastal plain rivers of the southeastern United States. *Geochim. Cosmo. Acta* 38:341–364.

Cabaniss, S.E., and M.S. Shuman. 1988. Copper binding by dissolved organic matter. I. Suwannee River fulvic acid equilibria. II. Variation in type and source of organic matter. *Geochim. Cosmo. Acta* 52:185–200.

Collins, M.R., G.L. Amy, and C. Steelink. 1986. Molecular weight distribution, carboxylic acidity, and humic substances content of aquatic organic matter: implications for removal during water treatment. *Env. Sci. Tech.* 20:1028–1032.

Dempsey, B.A., and C.R. O'Melia. 1983. Proton and calcium complexation of four fulvic acid fractions. In: Aquatic and Terrestrial Humic Materials, ed. R.F. Christman and E.T. Gjessing, pp. 239–274. Ann Arbor: Ann Arbor Science.

Ephraim, J.H., J. Borem, C. Pettersson, I. Arsenie, and B. Allard. 1989. A novel description of the acid-base properties of an aquatic fulvic acid. *Env. Sci. Tech.* 23:356–362.

Flaig, W., H. Beutelspacher, and E. Rietz. 1975. Chemical composition and physical properties of humic substances. In: Organic Components. Vol. 1 of Soil Components, ed. J. Gieseking, pp. 1–211. New York: Springer.

Gaskill, A.G. 1978. Methods for the characterization of metal-humic interactions. MSPH thesis, Dept. of Environmental Sciences and Engineering, Univ. North Carolina, Chapel Hill.

Greger, J.E., and H.K.J. Powell. 1987. Effects of extraction procedures on fulvic acid properties. *Sci. Total Env.* 62:3–12.

Huizenga, D.L., and D.R. Kester. 1979. Protonation equilibria of marine dissolved organic matter. *Limnol. Ocean.* 24:145–150.

Kerr, R.A., and J.G. Quinn. 1975. Chemical studies on the dissolved organic matter in seawater isolation and fractionation. *Deep-Sea Res.* 22:107–116.

Leenheer, J.A. 1981. Comprehensive approach to preparative isolation and fractionation of dissolved organic carbon from natural water and wastewaters. *Env. Sci. Tech.* 15:578–587.

Mantoura, R.F.C., and J.P. Riley. 1975. The analytical concentration of humic substances from natural waters. *Anal. Chim. Acta* 76:97–106.

McKnight, D.M., W.E. Periera, C.L. Zeazan, and R.C. Wissmar. 1982. Characterization of dissolved organic materials in surface waters within the blast zone of Mt. St. Helens, Washington. *Org. Geochem.* 4:85–92.

Miles, C.J., J.R. Tuschall, Jr., and P.L. Brezonik. 1983. Isolation of aquatic humus with diethylaminoethylcellulose. *Analyt. Chem.* 55:410–411.

Oliver, B.G., E.M. Thurman, and R.L. Malcolm. 1983. The contribution of humic substances to the acidity of colored natural waters. *Geochim. Cosmo. Acta* 47:2031–2035.

Perdue, E.M. 1979. Solution thermochemistry of humic substances. Acid-base equilibria of river water humic substances. In: Chemical Modeling in Aquatic Systems, ed. E.H. Jenne. *ACS Symp. Ser.* 93:94–114. Washington, D.C.: Am. Chem. Soc.

Perdue, E.M. 1985. Acidic functional groups in humic substances. In: Humic

Substances in Soil, Sediment, and Water, ed. G.R. Aiken, D.M. McKnight, R.L. Wershaw, and P. MacCarthy, pp. 493–526. New York: Wiley.
Perdue, E.M., J.H. Reuter, and M. Ghosal. 1980. The operational nature of acidic functional group analyses and its impact on mathematical descriptions of acid-base equilibria in humic substances. *Geochem. Cosmo. Acta* 44:1841–1851.
Schnitzer, M., and S.V. Khan. 1972. Humic Substances in the Environment. New York: Marcel Dekker.
Stuermer, D.H., and G.R. Harvey. 1977. The isolation of humic substances and alcohol-soluble organic matter from seawater. *Deep-Sea Res.* 24:303–309.
Thurman, E.M., and R.L. Malcolm. 1983. Structural study of humic substances: new approaches and methods. In: Aquatic and Terrestrial Humic Materials, ed. R.F. Christman and E.T. Gjessing, pp. 1–24. Ann Arbor: Ann Arbor Science.
Thurman, E.M., R.L. Malcolm, and G.R. Aiken. 1978. Prediction of capacity factors for aqueous organic solutes adsorbed on a porous acrylic resin. *Analyt. Chem.* 50:775–779.
Visser, S.A. 1982. Acid functional group content of aquatic humic matter: its dependence upon origin, molecular weight and degree of humification of the material. *J. Env. Sci. Health* A17:767–788.
Weber, J.H., and S.A. Wilson. 1975. The isolation and characterization of fulvic acid and humic acid from river water. *Water Res.* 9:1079–1084.

Organic Acids in Aquatic Ecosystems
eds. E.M. Perdue and E.T. Gjessing, pp. 111–126
John Wiley & Sons Ltd

Modeling the Acid-base Chemistry of Organic Acids in Laboratory Experiments and in Freshwaters

E.M. Perdue

School of Geophysical Sciences
Georgia Institute of Technology
Atlanta, GA 30332, U.S.A.

Abstract. This paper briefly evaluates the sensitivity of experimental titration data to slight errors in pH measurement and to neglect of activity coefficients at finite ionic strengths. The common mathematical models of the acid-base properties of complex mixtures of organic acids in aquatic ecosystems (including polymer-based models, multi-site models, and continuous distribution models) are discussed and compared.

INTRODUCTION

Aquatic ecosystems contain an incredibly complex mixture of organic acids, including simple low molecular weight compounds such as acetic, oxalic, and lactic acids and more complex acidic substances. The simple compounds typically account for only a few percent of the dissolved organic carbon (DOC) and a correspondingly small fraction of the organic acidity. The bulk of the DOC and the organic acidity occurs in more complex substances that are commonly known as humic and fulvic acids. The current state of knowledge on the origin, structure, and properties of humic and fulvic acids has been the subject of several recent books (Aiken et al. 1985; Thurman 1985; Frimmel and Christman 1988). The first of these books contains a comprehensive review of the acid-base chemistry of humic substances by Perdue (1985).

The ability to measure and model the acid-base chemistry of organic acids in aquatic ecosystems is a prerequisite for identifying processes that cause their properties and concentrations to vary over time and space. This chapter will present an overview of the mathematical and chemical backgrounds of several models of the acid-base chemistry of aquatic organic acids, including an assessment of the similarities and differences between models.

BACKGROUND

In order to discuss the acid-base chemistry of a complex mixture of organic acids, taking into account effects of DOC concentration, ionic strength, etc., it is instructive to define a number of terms for titration of the ith proton binding site of the mixture and for the mixture as a whole. The necessary definitions are given in Table 1, where charges of ions (including H^+) have been omitted. The challenge facing scientists who investigate the role of organic acids in aquatic ecosystems is to develop concise descriptive and predictive models of mixtures that contain perhaps thousands of different proton binding sites, even though the properties defined in Table 1 for the ith proton binding site cannot generally be measured for a significant fraction of those sites.

BASIC DATA QUALITY

The collection and interpretation of accurate acid-base titration data is prerequisite to the development of a mathematical model, whose parameters

TABLE 1. Definition of terms for titration of a single proton binding site and for a complex mixture of binding sites.

General Definitions		
$[X]$	=	concentration of species X
γ_X	=	activity coefficient of species X
$\{X\}$	=	activity of species X, where $\{X\} = [X]\gamma_X$
C_B	=	concentration of added strong base (e.g., NaOH)
C_A	=	concentration of added strong acid (e.g., HCl)
Definitions for the ith Proton Binding Site		
HL_i	=	acidic form of ith site
L_i	=	basic form of ith site
z_i	=	charge of L_i
C_i	=	total concentration of ith site, where $C_i = [HL_i] + [L_i]$
K_i	=	equilibrium constant for ionization of ith site, where $K_i = \{H\}\{L_i\}/\{HL_i\}$
K_i^c	=	concentration quotient for ionization of ith site, where $K_i^c = [H][L_i]/[HL_i] = K_i\gamma_{HL_i}/\gamma_H \cdot \gamma_{L_i} = K_i/\Gamma_i$
Definitions for the Whole Mixture of Proton Binding Sites		
C_L	=	total concentration of sites, where $C_L = \Sigma C_i = \Sigma[HL_i] + \Sigma[L_i]$
$\overline{K}$	=	average concentration quotient, where $\overline{K} = [H]\Sigma[L_i]/\Sigma[HL_i] = \Sigma K_i^c[HL_i]/\Sigma[HL_i]$

are usually determined by fitting the model to experimental data. Although a more comprehensive discussion of experimental methodologies is beyond the scope of this review, some basic concepts should be presented.

Consider the titration of a mixture of organic acids, each of which can be characterized by the parameters given in Table 1. The charge balance equation at any point in the titration is:

$$C_B + [H] = C_A + [OH] + \Sigma z_i[L_i] . \quad (1)$$

Given that $K_w = \{H\}\{OH\}$ and that $[H] = \{H\}/\gamma_H$, and assuming that $\gamma_H = \gamma_{OH}$, the concentration of organic anion charge at a given pH is simply:

$$\Sigma z_i[L_i] = C_B - C_A + [\{H\} - K_w/\{H\}]/\gamma_H . \quad (2)$$

Errors in the calculated value of $\Sigma z_i[L_i]$ can arise from errors in C_B or C_A but arise more likely either from errors in pH measurements or from assuming that γ_H equals one. Errors in pH measurement can arise from temperature variations during an experiment, CO_2 contamination of high pH buffers, bacterial contamination of phthalate buffers, liquid junction potential differences between buffers and samples, or fouling of the glass electrode by components of the sample. Such errors are easily in the range of 0.01–0.03 pH units in complex environmental samples, even under the best conditions. Neglecting activity coefficients for the moment, the errors in $\Sigma z_i[L_i]$ (in μeq/l) caused by pH errors of 0.01–0.03 pH units are given in Table 2 at selected pH values. Note that $\Sigma z_i[L_i]$ is overestimated if pH is underestimated and vice versa.

Additional errors in $\Sigma z_i[L_i]$ can occur if γ_H is neglected in Eq. 2, i.e., if it is assumed to equal one. Because γ_H is less than one in dilute aqueous solutions, it is evident from Eq. 2 that assuming that $\gamma_H = 1.0$ will cause the calculated organic anion charge concentration $\Sigma z_i[L_i]$ to be underestimated at low pH and overestimated at high pH. The resulting error in $\Sigma z_i[L_i]$ is given in Table 3 at selected ionic strengths and pH values. Errors are largest at extremely low or high pH values and at high ionic strengths. Procedures for calculating γ_H at each point in a titration are outlined in the APPENDIX.

To place such errors in perspective, consider that DOC samples reportedly contain, on average, about 10 microequivalents of carboxylic acid functional groups [COOH] per milligram of organic carbon (10 μeq/mg C). Whenever acid-base titrations are attempted at low DOC concentrations (e.g., 10 mg/l;

TABLE 2. The error in organic anion charge due to pH errors.

Absolute pH Error	Absolute Value of Error in $\Sigma z_i[L_i]$ in μeq/l pH 2	pH 3	pH 5	pH 7	pH 9	pH 11	pH 12
0.01	230.3	23.0	0.2	0.0	0.2	23.0	230.3
0.02	460.5	46.0	0.5	0.0	0.5	46.0	460.5
0.03	690.8	69.1	0.7	0.0	0.7	69.1	690.8

$$(\text{Error in } \Sigma z_i[L_i]) = -(\ln(10)/\gamma_H)[\{H\} + K_w/\{H\}](\text{pH Error})$$

TABLE 3. The minimum error in calculated organic anion charge due to neglecting the activity coefficient of H^+.

Ionic Strength	Error in $\Sigma z_i[L_i]$ in μeq/l* pH 2	pH 3	pH 5	pH 7	pH 9	pH 11	pH 12
0.001	*	−37.6	−0.4	0.0	0.4	37.6	*
0.010	−1216.4	−121.6	−1.2	0.0	1.2	121.6	1216.4
0.100	−3803.5	−380.4	−3.8	0.0	3.8	380.4	3803.5
1.000	−5973.0	−597.3	−6.0	0.0	6.0	597.3	5973.0

$$(\text{Error in } \Sigma z_i[L_i]) = -((1-\gamma_H)/\gamma_{H^2})[\{H\} - K_w/\{H\}]$$

* Actual ionic strength exceeds the tabulated value.

[COOH] = 100 μeq/l), it is essentially impossible to obtain reliable data at pH values that are less than 4.0 or greater than 10.0. Without reliable data in these pH regions, the acid-base properties of relatively strong ($K_i > 1.0 \times 10^{-4}$) and relatively weak ($K_i < 1.0 \times 10^{-10}$) organic acids cannot be accurately determined and, of course, cannot be properly modeled. Only a few aquatic ecosystems contain high DOC concentrations, so direct titrations of water samples are not generally useful for determining the acid-base properties of organic acids in aquatic ecosystems. Rather, reliable titration data are best obtained on extracted (and probably fractionated) samples that are titrated at high DOC concentrations. Mathematical models based on such titration data implicitly assume that (*a*) the properties of the extracted DOC are representative of the DOC in the aquatic ecosystem being studied and (*b*) the acid-base properties of the DOC are independent of DOC concentration.

A final source of error is contamination of the sample with titratable weak acids and bases. Aquatic ecosystems contain NH_4^+, Fe(III), Al(III),

$Si(OH)_4$, H_2CO_3, and all the conjugate bases that can be derived from these acids. Unless sample and titrant preparation have completely removed these inorganic substances, the $\Sigma z_i[L_i]$ calculated from Eq. 2 will not equal organic anion charge concentration. The worst problems occur with H_2CO_3, which is very difficult to exclude from samples and titrants at high pH values. All solutions, buffers, and titrants must be prepared with CO_2-free water, and titrations should be conducted under an inert atmosphere.

Model-independent Properties

Assuming that experimental data are accurately obtained and that calculations of $\Sigma z_i[L_i]$ consider ionic strength effects on γ_H, several properties of the mixture of proton binding sites can be estimated or at least constrained by a simple inspection of a plot of $\Sigma z_i[L_i]$/DOC versus pH. Consider the typical plots in Fig. 1, which are proton binding data sets for 100 and 1000 mg C/l solutions of organic acids that were isolated from the Suwannee River in southeast Georgia. Even though $\Sigma z_i[L_i]$/DOC increases continuously as pH increases, the rate of increase is slowest in the pH 7–9 range. Because COOH groups generally have titration endpoints in this pH range, the COOH content of the sample should be less than or equal to the

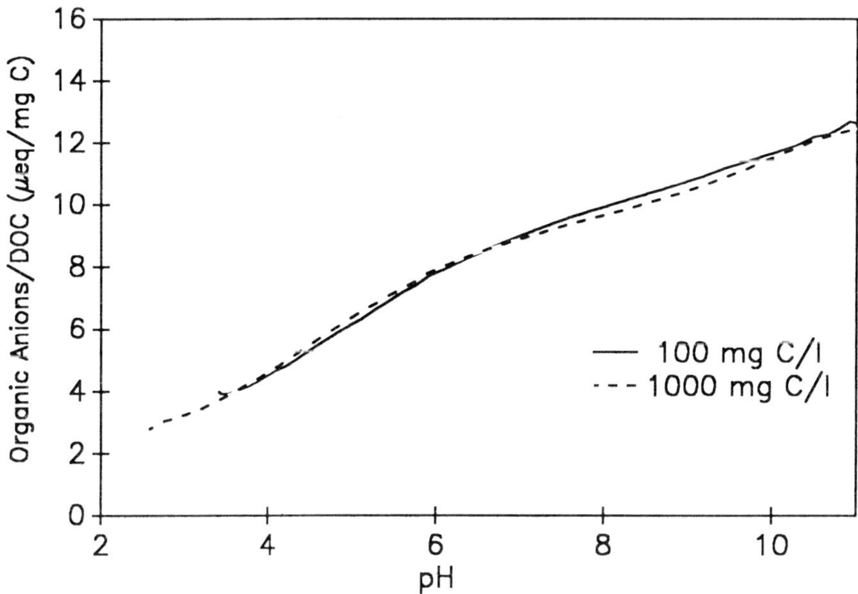

Fig. 1—Titrations of 100 and 1000 mg C/l solutions of Suwannee River dissolved organic matter in deionized water with NaOH at 25.0°C.

$\Sigma z_i[L_i]$/DOC values in the pH 7–9 range, being somewhat less if relatively acidic noncarboxyl groups are also being titrated in this pH range.

Suppose that pH 8.0 is operationally used as the COOH titration endpoint and that the COOH content of the sample is defined as the value of $\Sigma z_i[L_i]$/DOC at pH 8.0. Regardless of the mathematical model that is used to describe the acid-base chemistry of the sample, one half of the COOH groups must have pK_i^c values that lie at or below the pH where one half of the COOH groups are ionized. Consequently, the mean pK_i^c of the COOH groups in this sample must be about 4.2, which is comparable to the mean pK_i^c for most simple carboxylic acids (Perdue 1985).

It is also noteworthy that the two plots coincide very closely over the entire pH range, even though the DOC concentrations differ by an order of magnitude. This common observation is important for two reasons: (*a*) there are no unusual concentration-dependent properties such as solute-solute interactions or conformational equilibria to be explained in acid-base models, and (*b*) modeling results based on titrations at high DOC concentrations can probably be used to predict the acid-base chemistry of aquatic organic acids at natural DOC levels, all other factors (e.g., ionic strength) being equal.

MATHEMATICAL MODELS

A rigorous description of the acid-base chemistry of organic acids in aquatic ecosystems is not attainable at the present time. Such a description would have to include all the parameters given in Table 1 for every proton binding site in the mixture. Because of the large number of organic acids in aquatic ecosystems, the exact description would overwhelm even the most sophisticated computer programs for solving multicomponent chemical equilibria, because most programs are limited to dealing with fewer than 100 components. Even if a rigorous description of the system were available, its complexity could (and almost certainly would) be avoided by the use of simplifying mathematical models that are more amenable to manual or computational manipulation.

A variety of convenient mathematical models have been proposed to describe the acid-base and metal ion complexation chemistry of complex mixtures of organic acids, especially humic and fulvic acids. Such models are used to reproduce titration data that were obtained under a given set of conditions (DOC concentration, ionic strength, etc.) and to generate comparable data for other experimental conditions. The former descriptive task might be carried out to smooth data, interpolate between experimental points, plot data, etc., and the ability of a given model to accomplish such a task depends primarily on the number of adjustable fitting parameters in the model. On the other hand, for a given number of adjustable fitting parameters, the latter predictive task is likely to be accomplished best by mathematical models that explicitly consider the chemical properties of the mixture of organic acids being modeled. All mathematical models must contend with the observation that the average concentration quotients (see $\overline{K}$ in Table 1) of these mixtures are strongly pH-dependent functions, being greatest at low pH.

The most common models can be classified as (*a*) polymer models, (*b*) multi-site models, and (*c*) continuous distribution models. Major reviews of several common models of proton and/or metal ion binding by complex mixtures of organic acids have been given by Perdue (1985) and Sposito (1986). Other less comprehensive reviews, often combined with application of one or more models to specific data sets, have recently been published by Cabaniss et al. (1984), Dzombak et al. (1986), Fish et al. (1986), and Turner et al. (1986). Only the broadest descriptions of the common models will be attempted in this review.

Polymer Models

From Table 1, the quantitative equation for proton binding by a single binding site is

$$[L_i] = C_i \left[\frac{K_i^c}{K_i^c + [H]} \right]. \tag{3}$$

If applied directly to complex mixtures of organic acids such as humic substances, Eq. 3 underestimates $[L_i]$ when $[H] > K_i^c$ and overestimates $[L_i]$ when $[H] < K_i^c$. Similar results have been observed when Eq. 3 is used to describe the acid-base chemistry of synthetic and natural polymers, where the acid-weakening effect of accumulating negative charge on sequential dissociation steps in a polymeric acid is fairly well understood.

Polymer chemists developed modified forms of Eq. 3 to accommodate the unique electrostatic and conformational properties of polymers. Perhaps because organic acids in aquatic ecosystems originate largely from biopolymers, their acid-base properties are frequently described using the polymer-based equations, even though the presence of true polymers in these mixtures has not been demonstrated. The two most common equations are:

1. the Tanford electrostatic equation,

$$\Sigma[L_i] = C_L \left[\frac{K_c^\circ \exp(Z\alpha)}{K_c^\circ \exp(Z\alpha) + [H]} \right], \tag{4}$$

where $\alpha = \Sigma[L_i]/C_L$ and Z is a composite constant containing an electrostatic interaction factor and the average charge of $\Sigma[L_i]$ (so $Z < 0$). This equation attempts to simulate the acid-weakening effect of accumulating negative charge on sequential dissociation steps in a polymeric acid whose binding sites would otherwise have the same intrinsic K_c° value. Because $\Sigma[L_i]$ appears on both sides of Eq. 4, its calibration and use are not straightforward. A rearranged equation in which [H] is the dependent variable would be easier to use. In any case, the fitting parameters C_L, K_c°, and Z are determined by nonlinear regression. If C_L is already known, linear regression can be used with a logarithmic form of Eq. 4 (Wilson and Kinney 1977; Cabaniss et al. 1984).

2. the modified Henderson-Hasselbalch equation,

$$\Sigma[L_i] = C_L \left[\frac{K_h^{\circ 1/n}}{K_h^{\circ 1/n} + [H]^{1/n}} \right], \quad (5)$$

where n is considered either as an empirical fitting parameter (Huizenga and Kester 1979) or as a constant that is "fundamentally related to the number and size of 'monomer' units and the degree of molecular stretch in a hypothetical polymeric molecule" (Varney et al. 1983), and K_h° is an average concentration quotient. The three fitting parameters K_h°, C_L, and n can be determined by nonlinear regression. If C_L is known, a linear logarithmic form of Eq. 5 is commonly used to determine K_h° and n.

In the study of adsorption processes (i.e., gas–solid and solution–solid equilibria), the Langmuir equation, which is formally analogous to Eq. 3, is widely used. In that field of research, frequent deviations from the Langmuir equation are reported. In most instances, an empirical equation known as the Freündlich equation can be successfully fitted to nonconforming experimental data. If the Freundlich equation is expressed in terms of the defined parameters in Table 1, it becomes:

$$\Sigma[L_i] = C_L \left[\frac{K_f^\circ}{K_f^\circ + [H]^{1/n}} \right], \quad (6)$$

where n is an empirical fitting parameter and K_f° is an average concentration quotient. Even though the Freundlich equation was not developed to describe the acid-base properties of polymers, it is evident from Eqs. 5 and 6 that the Freundlich equation and the modified Henderson-Hasselbalch equation are equivalent models. The three fitting parameters K_f°, C_L, and n can be determined by nonlinear regression. If C_L is known, a linear logarithmic form of Eq. 6 is commonly used to determine K_f° and n.

In recent years, the basic electrostatic "polymer" model has been embellished with a variety of more complex features that attempt to incorporate the effects of conformational changes and Donnan potential effects on proton binding equilibria (Varney et al. 1983; Marinsky and Ephraim 1986; Ephraim et al. 1986). Such models are clearly based on the hypothesis that the acid-base chemistry of the complex mixture of organic acids in aquatic ecosystems is similar to that of natural and synthetic polymers.

Multi-site Models

Multi-site models are conceptually the simplest and the most widely used models of the acid-base chemistry of organic acids in aquatic ecosystems (see, e.g., Cabaniss and Shuman 1988; Dzombak et al. 1986; Eberle and Feuerstein 1979; Eshleman and Hemond 1985; Fish et al. 1986; Kramer and Davies 1988; Leuenberger and Schindler 1986; Paxeus and Wedborg 1985). These models simply assume the presence of a small number of binding sites, each of which is described as in Table 1 and Eq. 3. The model takes the simple form:

$$\Sigma[L_i] = C_L \sum_{i=1}^{n} \frac{C_i}{C_T} \cdot \left[\frac{K_i^c}{K_i^c + [H]}\right]. \tag{7}$$

Most authors realize that the C_i and K_i^c values that are obtained when this model is fit to proton binding data are not chemically significant parameters. Some representative results are given in Fig. 2, which shows the cumulative mole fraction of binding sites versus pK_i^c ($pK_i^c = -\log(K_i^c)$) for several recent studies. As previously indicated, the C_i and pK_i^c values that are found are closely related to the pH range covered in a titration. Binding sites whose pK_i^c values are outside the pH range of the titration are simply not found. This forces the cumulative mole fraction of proton binding sites to range from zero to one within the pH range that was examined in a given study.

In all these studies, regardless of the experimental conditions, pH range of titration data, number of binding sites in the model, etc., at least 50 percent of the binding sites have pK_i^c values of less than 4.0. In fact, in all but one study, the mean pK_i^c is in the 3.3–4.0 range. This general agreement

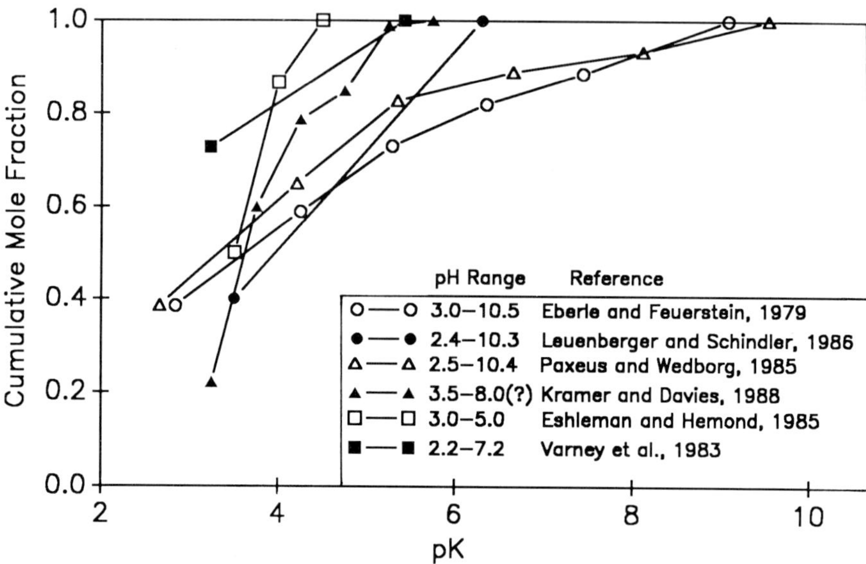

Fig. 2—Cumulative discrete binding site distributions of organic acids from selected aquatic ecosystems.

under such varied conditions strongly indicates that the complex mixtures of organic acids being titrated are quite similar in all these studies and that they contain a large proportion of highly acidic proton binding sites.

Continuous Distribution Models

Continuous distribution models are a logical extension of multi-site models. Rather than attempting to identify a small number of discrete binding sites and determine their C_i and K_i^c values, continuous models assume that the mixture of organic acids in aquatic ecosystems is so complex that it can be modeled as a continuum. Gamble (1970) first described why the $p\overline{K}$ ($p\overline{K} = -\log(\overline{K})$) of a complex mixture of organic acids would vary with pH, and he proposed numerical techniques for estimating the "instantaneous" K_i^c of the binding sites reacting at a given pH. His treatment, which is reviewed by Perdue (1985), Sposito (1986), and Dzombak et al. (1986), does not yield fitting parameters and should be considered more as a mathematical description than as a model.

Oliver et al. (1983) proposed that the kind of variation of $p\overline{K}$ with pH that was described by Gamble (1970) could be more readily incorporated into predictive models if the pH dependence of $p\overline{K}$ were described by a simple equation. From titrations of a number of samples of humic substances from colored natural waters, they obtained the following equation:

$$p\overline{K} = f(pH) = 0.96 + 0.90\,pH - 0.039(pH)^2\,. \tag{8}$$

From the definitions in Table 1 and Eq. 8, it follows that

$$\Sigma[L_i] = C_L\left[\frac{10^{-f(pH)}}{10^{-f(pH)} + H}\right]. \tag{9}$$

This equation, because of its simplicity, is widely used to estimate $\Sigma[L_i]$ in natural waters. Although it is sometimes mistakenly referred to as a one-site model, it is an empirical function that is used to describe a continuum of binding sites. It should be noted that the $p\overline{K}$ values used to calibrate this model were based on a definition of C_L as the concentration of acidic functional groups that are titrated at pH 7.0. The titration curves of organic acids from natural water samples (see Fig. 1) do not have an actual inflection point at pH 7.0, so Eqs. 8 and 9 are not "calibrated" beyond that pH.

Shuman et al. (1983) described a numerical method that can extract an "affinity spectrum" of metal ion binding sites from smoothed titration data. The authors assume a continuous distribution of binding sites, in which the mole fraction of binding sites in the interval (pK, pK + dpK) is described by a continuous function F(pK). When this model is adapted to proton binding data, the organic anion concentration is given by:

$$\Sigma[L_i] = C_L \int_{-\infty}^{+\infty} \frac{K}{(K + [H])} F(pK)\, d(pK)\,. \tag{10}$$

The affinity spectrum, F(pK), can be numerically estimated at specified pK values, allowing Eq. 10 to be evaluated. Like the approach of Gamble (1970), the affinity spectrum model does not yield fitting parameters that can be used predictively. Furthermore, when data for proton binding by a single pure ligand are subjected to the affinity spectrum analysis, the results indicate the existence of a rather broad normal distribution of binding sites. The method has two major disadvantages: (*a*) it requires smoothed data sets and is sensitive to the data-smoothing algorithm used, and (*b*) it is numerically unstable with respect to the log(a) parameter that is used to estimate F(pK) (see Leuenberger and Schindler 1986; Dzombak et al. 1986).

Perdue and co-workers (Perdue and Lytle 1983; Perdue et al. 1984) have proposed a model similar to the affinity spectrum model but with F(pK) defined as the Gaussian distribution function. One such function should be used for each of the major classes of acidic functional groups in organic acids (carboxyl and phenolic groups). A detailed mathematical treatment of continuous distribution models, including the Gaussian model, was recently published by Parrish and Perdue (1989). Substitution of

$$F(pK) = (\sigma\sqrt{2\pi})^{-1}\exp(-0.5((\mu - pK)/\sigma)^2) \quad (11)$$

into Eq. 10 yields the Gaussian distribution model for a class of proton binding sites.

$$\Sigma[L_i] = C_L(\sigma\sqrt{2\pi})^{-1}\int_{-\infty}^{+\infty}\frac{K}{(K + [H])}\exp(-0.5((\mu - pK)/\sigma)^2)\,d(pK) \quad (12)$$

For a bimodal Gaussian distribution model, two expressions are used on the right side of Eq. 12, each with its own C_L, μ, and σ values. The model works best when C_L values are independently determined for one or both classes of proton binding sites, but it can be used to determine C_L, μ, and σ for each class of binding sites using nonlinear regression methods to fit Eq. 12 to experimental data. The model has been used extensively for modeling both proton and metal ion binding, with varying degrees of success (Cabaniss et al. 1984; Cabaniss and Shuman 1988; Fish et al. 1986; Perdue and Lytle 1983; Perdue et al. 1984; Rhea and Young 1987; Turner et al. 1986). The mean pK_i^c for COOH groups reported by Perdue et al. (1984) is 3.7, which corresponds to the mean pK_i^c values of discrete site models (3.3–4.0).

Of the continuous distribution models, only the Oliver model and the Gaussian distribution model can generate model-fitting parameters when fit to experimental data. The former model, however, was calibrated by its authors and is usually used without recalibration. The Gaussian distribution model, like the polymer-based models and the multi-site models, is designed to be calibrated each time it is used. Perdue and Parrish (1987) recently published the computer source code for fitting the Gaussian distribution model to proton or metal ion binding data.

RECOMMENDATIONS

The use of mathematical models to describe the acid-base chemistry of organic acids in aquatic ecosystems will be a necessity for the foreseeable future. The descriptive capabilities of current models are quite adequate, but their predictive capabilities need to be improved. Most current models correctly predict that the degree of dissociation of the mixture of organic acids is a function of pH but not of the DOC concentration. None of these models, however, is especially good at predicting the effects of ionic strength, temperature, and metal ion binding on acid-base chemistry. Model development in these areas is needed.

Much more effort should be made to collect very accurate acid-base titration data, with special attention given to eliminating factors that cause small determinate errors in pH measurements. Titrations should be conducted on isolated samples at *high* DOC concentrations, not on water samples at natural DOC concentrations. The conversion of pH into $[H^+]$ must consider the activity coefficient of H^+, as determined by the ionic strength at each point in a titration. Some of the current confusion over the effects of ionic strength on acid-base equilibria may thus be eliminated.

New techniques for isolation of organic acids from natural waters are needed to avoid the fractionation of DOC and losses of up to 50 percent of DOC that occur when resin adsorption methods are used. If mathematical models can be calibrated with relatively unfractionated DOC samples, then DOC concentrations can be used to estimate more accurately the total concentrations of carboxyl and other acidic functional groups in aquatic ecosystems.

Acknowledgements. Although the research described in this article has been sponsored by the United States Environmental Protection Agency through Cooperative Agreement No. CR813471-01 to Georgia Institute of Technology, it has not been subjected to Agency review and therefore does not necessarily reflect the views of the Agency and no official endorsement should be inferred.

APPENDIX

Ionic Strength and Activity Coefficient Corrections

If the ionic strength (I) is known, the Davies equation can be used to calculate γ_H in dilute aqueous solutions:

$$\ln \gamma_H = -A[(\sqrt{I}/(1+\sqrt{I})) - B \cdot I], \qquad (A1)$$

where A = 1.17 and B = 0.2 in water at 298K. If the background ionic strength in a titration is I_o, then the ionic strength at any point in the titration described by Eq. 2 is:

$$I = I_o + \frac{1}{2}\left[C_B + C_A + [\{H\} + K_w/\{H\}]/\gamma_H + \Sigma z_i^2[L_i]\right]. \quad \text{(A2)}$$

The term representing the contribution of organic anions to the ionic strength cannot be measured, so a rigorous treatment of ionic strength effects on the acid-base equilibria cannot be achieved. The minimum effect of ionic strength on titration results can be calculated, however, by assuming that z_i equals one for all organic anions. Combining Eq. 2 and Eq. A2, the approximate ionic strength at any point in a titration is:

$$I = I_o + C_B + \{H\}/\gamma_H. \quad \text{(A3)}$$

If (C_B + $\{H\}/\gamma_H$) is much smaller than I_o over an entire titration, then I will be constant. Otherwise, I will change continuously during a titration. Equations A1 and A3 are interdependent and must be solved iteratively until consistent values are obtained for both I and γ_H at each point in a titration. Once γ_H is known, Eq. 2 can be solved for $\Sigma z_i[L_i]$.

REFERENCES

Aiken, G.R., D.M. McKnight, R.L. Wershaw, and P. MacCarthy, eds. 1985. Humic Substances in Soil, Sediment, and Water: Geochemistry, Isolation, and Characterization. New York: Wiley.

Cabaniss, S.E., and M.S. Shuman. 1988. Copper binding by dissolved organic matter. I. Suwannee River fulvic acid equilibria. *Geochim. Cosmo. Acta* 52: 185–193.

Cabaniss, S.E., M.S. Shuman, and B.J. Collins. 1984. Metal-organic binding—a comparison of models. In: Complexation of Trace Metals in Natural Waters, ed. C.J.M. Kramer and J.C. Duinker, pp. 165–179. The Hague: Nijhoff.

Dzombak, D.A., W. Fish, and F.M.M. Morel. 1986. Metal-humate interactions. I. Discrete ligand and continuous distribution models. *Env. Sci. Tech.* 20:669–675.

Eberle, S.H., and W. Feuerstein. 1979. On the pK spectrum of humic acid from natural waters. *Naturwissen.* 66:572–573.

Ephraim, J., S. Alegret, A. Mathuthu, M. Bicking, R.L. Malcolm, and J.A. Marinsky. 1986. A unified physicochemical description of the protonation and metal ion complexation equilibria of natural organic acids (humic and fulvic acids). II. Influence of polyelectrolyte properties and functional group heterogeneity on the protonation equilibria of fulvic acid. *Env. Sci. Tech.* 20:354–366.

Eshleman, K.N., and H.F. Hemond. 1985. The role of organic acids in the acid-base status of surface waters at Bickford Watershed, Massachusetts. *Wat. Res. Res.* 21:1503–1510.

Fish, W., D.A. Dzombak, and F.M.M. Morel. 1986. Metal-humate interactions. II. Application and comparison of models. *Env. Sci. Tech.* 20:676–683.

Frimmel, F.H., and R.F. Christman, eds. 1988. Humic Substances and Their Role in the Environment. Dahlem Konferenzen. Chichester: Wiley.

Gamble, D.S. 1970. Titration curves of fulvic acid: the analytical chemistry of a weak acid polyelectrolyte. *Can. J. Chem.* 48:2662–2669.

Huizenga, D.L., and D.R. Kester. 1979. Protonation equilibria of marine dissolved organic matter. *Limnol. Ocean.* 24:145–150.

Kramer, J.R., and S.S. Davies. 1988. Estimation of non-carbonato protolytes for selected lakes in the Eastern Lakes Survey. *Env. Sci. Tech.* 22:182–185.

Leuenberger, B., and P.W. Schindler. 1986. Application of integral pK spectrometry to the titration curve of fulvic acid. *Analyt. Chem.* 58:1471–1474.

Marinsky, J.A., and J. Ephraim. 1986. A unified physicochemical description of the protonation and metal ion complexation equilibria of natural organic acids (humic and fulvic acids). I. Analysis of the influence of polyelectrolyte properties on protonation equilibria in ionic media: fundamental concepts. *Env. Sci. Tech.* 20:349–354.

Oliver, B.G., E.M. Thurman, and R.L. Malcolm. 1983. The contribution of humic substances to the acidity of colored natural waters. *Geochim. Cosmo. Acta* 47:2031–2035.

Parrish, R.S., and E.M. Perdue. 1989. Computational methods for fitting statistical distribution models of multi-site binding equilibria. *Math. Geol.* 21:199–219.

Paxeus, N., and M. Wedborg. 1985. Acid-base properties of aquatic fulvic acid. *Anal. Chim. Acta* 169:87–98.

Perdue, E.M. 1985. Acidic functional groups in humic substances. In: Humic Substances in Soil, Sediment, and Water: Geochemistry, Isolation, and Characterization, ed. G.R. Aiken, D.M. McKnight, R.L. Wershaw, and P. MacCarthy, pp. 493–526. New York: Wiley.

Perdue, E.M., and C.R. Lytle. 1983. Distribution model for binding of protons and metal ions by humic substances. *Env. Sci. Tech.* 17:654–660.

Perdue, E.M., and R.S. Parrish. 1987. Fitting multisite binding equilibria to statistical distribution models: Turbo Pascal program for Gaussian models. *Comp. Geosci.* 13:587–601.

Perdue, E.M., J.H. Reuter, and R.S. Parrish. 1984. A statistical model of proton binding by humus. *Geochim. Cosmo. Acta* 48:1257–1263.

Rhea, J.R., and T.C. Young. 1987. Application of a continuous distribution model for proton binding by humic acids extracted from acidic lake sediments. *Env. Geol. Wat. Sci.* 10:169–173.

Shuman, M.S., B.J. Collins, P.J. Fitzgerald, and D.L. Olson. 1983. Distribution of stability constants and dissociation rate constants among binding sites on estuarine copper-organic complexes: rotated disk electrode studies and an affinity spectrum analysis of ion-selective electrode and photometric data. In: Aquatic and Terrestrial Humic Materials, ed. R.F. Christman and E.T. Gjessing, pp. 349–370. Ann Arbor: Ann Arbor Science.

Sposito, G. 1986. Sorption of trace metals by humic materials in soils and natural waters. *CRC Crit. Rev. Env. Cont.* 16(2):193–229.

Thurman, E.M. 1985. Organic Geochemistry of Natural Waters. Dordrecht: Nijhoff.

Turner, D.R., M.S. Varney, M. Whitfield, R.F.C. Mantoura, and J.P. Riley. 1986. Electrochemical studies of copper and lead complexation by fulvic acid. I. Potentiometric measurements and a critical comparison of metal binding models. *Geochim. Cosmo. Acta* 50:289–297.

Varney, M.S., R.F.C. Mantoura, M. Whitfield, D.R. Turner, and J.P. Riley. 1983. Potentiometric and conformational studies of the acid-base properties of fulvic acid from natural waters. In: Trace Metals in Seawater, ed. C.S. Wong, E. Boyle, W.K. Bruland, J.D. Burton, and E.D. Goldberg, pp. 751–772. New York: Plenum.

Wilson, D.E., and P. Kinney. 1977. Effects of polymeric charge variations on the proton-metal ion equilibria of humic materials. *Limnol. Ocean.* 22:281–289.

Organic Acids in Aquatic Ecosystems
eds. E.M. Perdue and E.T. Gjessing, pp. 127–139
John Wiley & Sons Ltd

Variability of Organic Acids in Watersheds

J.R. Kramer, P. Brassard, P. Collins, T.A. Clair, and P. Takats

McMaster University
Department of Geology
Hamilton, Ontario L8S 4M1, Canada

Gelbstoffe sind nicht Geldstoffe,
aber Gelbstoffe werden sich in Geld verwandeln
wenn wir die Huminstoffe verstehen.

Abstract. It is proposed that formation of low pK_a organic acids at high elevations is replaced by the predominance of higher pK_a acids at lower elevations. Residence time is important also; for example, laboratory incubation of high DOC waters in the dark shows a shift towards basic sites after 28 days. Analysis of bog DOM shows that the hydrophilic and hydrophobic fractions have similar proton binding distributions. Acid binding site analysis results for the eastern U.S. are variable, but the hypothesis of change from low pK_a to high pK_a sites with decreasing elevation is confirmed for the Adirondack region of New York.

INTRODUCTION

Humic substances are ubiquitous and are a major component of low ionic strength surface waters that are found in small watersheds. For example, dissolved organic carbon (DOC) is typically more abundant than dissolved inorganic carbon (DIC) in the U.S. E.P.A. Eastern Lakes Survey (ELS) (Overton et al. 1986).

Dissolved organic matter (DOM) is defined as those organic substances that pass through a 0.45 μm filter. The common measurement for DOM is DOC. Thurman (1985) divides DOM into the following groups with average concentrations of DOC for the U.S. as:

Humic substances:	Others:
40% fulvic acid	10% carbohydrates
10% humic acid	7% carboxylic acids
30% hydrophilic acids	3% amino acids
	<1% hydrocarbons

About 80% of DOM is designated as humic substances. Operational definitions are used to classify humic substances, and the other (simple) organic acids may be measured directly. For the aquatic environment, humic substances have been defined as polyelectrolytes that sorb onto XAD or weak-base ion exchange resins (Thurman 1985, pp. 104–107). Hydrophilic acids, which do not adhere to the XAD resin, have been hypothesized to be a mixture of simple organic acids and polyelectrolytic acids.

Humic substances typically contain 5–15 μeq/mg C (microequivalents per milligram carbon) of acidic functional groups, of which about one-half are carboxylic acids. The remaining acidity is due to phenols, protonated amines, alcohols, etc. The average pK_a values of carboxylic acids and phenols in humic substances are 3.5–4.0 and 9–10, respectively. Some weak carboxylic acids, strong phenols, or other acidic functional groups with pK_a values in the 4–7 region are also present (Perdue 1985). Some of the functional groups of intermediate acidity could be inorganic iron and aluminum species, which invariably occur in humic substances.

TABLE 1. Summary of some low molecular weight acids found in DOC (Thurman 1985). Some acid pK_a's (1st, 2nd, etc.) are added in parentheses to indicate the buffering regions (Martell and Smith 1979).

Group	Examples	Percent of DOC
Volatile fatty acids	acetic acid (4.56)	~2
Nonvolatile fatty acids	C-14, C-16, and C-18 monocarboxylic acids (pK_a = 4.7–4.8), e.g., palmitic, stearic	~4
Hydroxy acids	glycolic (3.63), lactic (3.66), pyruvic (2.26)	~1
Dicarboxylic acids	saturated C-8, C-9, and adipic (4.42, 5.42), fumaric (3.05, 4.49), maleic (1.91, 6.33), malonic (2.65, 5.28), oxalic (3.82), succinic (4.20, 5.64)	~0.5
Aromatic acids	anisic (4.48), benzoic (4.0), p-coumaric, dihydroxy benzoic, ferulic, 3- and 4-hydroxy benzoic (4.08, 4.58), mandelic (3.40), phenylacetic (4.09), phthalic (2.95, 5.41), protocatechuic (4.34, 8.68, 12.8), salicylic (2.97, 13.7), syringic, trimesic, and vanillic acids; hydroxy phenols	~0.5

TABLE 2. Comparison of low molecular weight (LMW) and high molecular weight (HMW) fractions of organic acids.

LMW	HMW
1. Strong acids with pK_a's ~2–5.	1. Less strong acids.
2. Multiple functional groups; only specific substances complex with metals.	2. Excellent complexation with numerous metals, especially Al, Fe, and Cu.
3. Easily biodegradable; turnover times of hours.	3. Stable; turnover times up to hundreds of years.
4. Formation as biological metabolites; high concentrations when high productivity, little soil retention.	4. Induced by chemical and enzymatic processes; formed by reactions between monomers and oligomers.
5. Biological impact mostly pH-related (e.g., un-ionized vs. ionized).	5. Biological impact mostly physiological.
6. Fluctuation can be rapid and variations large.	6. Quality and quantity subject to long-term effects; predictable seasonal repetition.
	7. Influenced by land management practices.

Direct measurement of individual organic acids has not been carried out routinely. Some of the acids that have been identified are shown in Table 1 along with their pK_a values. These simple organic acids contain 10–80 μeq/mg C of acidic functional groups, which is higher than for humic substances. These acids would be found in the hydrophilic acid fraction. Most of these organic acids are relatively strong acids with pK_a values in the 2–5 range.

Humic substances have been separated by size fractionation, with continuous flow filtration units being used to separate humic substances into nominal molecular weight fractions (e.g., 100,000, 10,000, or 1,000 daltons). More traditional gel filtration techniques are also being modified for separations, but not without problems (e.g., Wershaw and Aiken 1985). A combination of these techniques with other separation methods (e.g., gradient HPLC) and modern spectroscopic techniques, such as NMR and ESR, will give more information on the nature and variability of organic substances in aquatic systems. Table 2 provides a comparison of the low molecular weight (LMW) and high molecular weight (HMW) fractions of DOM. The two fractions have molecular weights of <500 and >500 daltons, respectively (Thurman 1985).

While trends in organic matter or DOC concentrations have been determined, the variation in space and time of organic *acids* in an aquatic environment is not well known. Aquatic organic acids are formed by both autochthonous and allochthonous processes. Autochthonous organic acids originate from modification of algal and macrophytic cells, from algal exudates, and from the microbial and photochemical modification of organic matter. Allochthonous organic acids originate from modification of vegetation, from water passing through organic soil, and from biological decomposition of detritus. The importance of the processes producing organic acids varies greatly, but high DOC waters (>10 mg C/l) routinely have a major allochthonous contribution.

Groundwaters and wetlands vary in humic substance composition due to the different processes of formation, transport, and diagenesis. Groundwaters average about 50% hydrophilic acids and 25% identifiable compounds. These LMWs probably result from the preferential removal of HMWs by sorption on soils. Wetlands, however, contain predominantly fulvic acids, averaging about 60%, whereas streams and lakes are more elevated in hydrophilic acids (Thurman 1985). The DOC in groundwater changes about tenfold going from the surface organic layers to the Al/Fe-O-OH layers. Thus the major pathway of organic acids from land to water appears to be from flooding of wetlands or overland flow.

Organic acids vary in watersheds for many reasons: (*a*) the varying production and degradation of organic acids associated with (primary) production in surface waters, mainly lakes, (*b*) the result of mixing different sources of organic acids in a watershed, (*c*) the varying biological degradation products resulting from different chemical (redox, presence or absence of sediment and bacteria) regimes and different residence times, and (*d*) the changes in colloidal chemistry as affected by mineral oxide substrates, ionic strength, and metal chelation. These and other mechanisms vary in time and space, making the organic chemistry of humic substances variable and complex.

DOC concentration increases in warmer climates, but the effect of increased DOC production with climate is minimal when compared to the contributions of DOC from wetlands. The climatic effect is probably small because decomposition rates also increase, resulting in small changes in net productivity.

Seasonal changes in terrestrial production influence the nature of organic acids. Fatty acids generally decrease in the fall with leaf and foliage dieback. Volatile fatty acids are the result of fermentation reactions and hence are elevated in concentrations in anoxic aquatic environments. Volatile fatty acids, especially acetic acid, can increase in surface waters when a lake turns over.

Humic substances can be formed by the oxidative condensation of polyphenols (Flaig 1978, 1988). In aquatic environments, the condensation reaction will lead to polymers resistant to biological attack by enzymes. Alternating aerobic/anaerobic conditions, such as those associated with periodic flooding, are associated with the formation of stable polymers.

Temporal changes occur on diurnal and seasonal cycles. Biological activity gives rise to a maximum in DOC production in the afternoon. Seasonal maxima in DOC concentration occur in summer and fall. Fulvic acids often remain relatively constant, but humic acids have been shown to increase in concentration from spring to autumn and decrease again in winter. If the DOC of an aquatic system is controlled mostly by primary production/degradation, a different pattern may arise. The variation in Fe(III) concentration along with the adsorption of humic substances to iron oxy-hydroxides has been postulated as the reason for minimum humic substance concentration in spring and maximum concentration in fall for a softwater lake in England (Tipping and Woof 1983). None of the above mechanisms has been characterized in depth, perhaps due to the lack of a definitive analytical protocol.

Trivalent ions such as Fe and Al, the pH, metal oxy-hydroxides, and clays affect the coagulation and ultimate fate of humic substances. Cations may sorb to COOH groups at low pH or form true chelates in aromatic ring structures. Schnitzer (1969) suggests the following complex stability at a pH of 3.5 for divalent cations: Cu>Fe>Ni>Pb>Co>Ca>Zn>Mn>Mg, and at a pH of 5: Cu>Pb>Fe>Ni>Mn>Co>Ca>Zn>Mg. Khan (1969) lists the effectiveness of metal ions for coagulation as: Al>Fe>Cu>Zn>Ni>Mn.

Humic substances at aquatic concentrations tend to coagulate at a pH of ≤3. The presence of electrolytes may decrease the pK_a and the pH of peptization may increase to a pH of 4.5–5 (Posner 1964).

Metal oxy-hydroxides and clay minerals are important substrates for humic substances. The reaction of humic substances occurs in a variety of ways (Yariv and Cross 1979) and over a wide range of chemistries.

Organic acids tend to be a significant component in aquatic chemistry when some of the following conditions are met:

- well-weathered, old soil horizons with minimum weathering cation exchange potential;
- significant area of wetlands in the watershed;
- minimum contact of surface waters with soil minerals;
- well-developed riparian zone;
- storm driven hydrology;
- low ionic strength waters;
- inefficient mineralization.

One hypothesis which encompasses many of the above phenomena and processes would predict LMW, low pK_a organic acids in the headwaters of catchments. These LMWs would degrade rapidly and would not be significant downstream without a new source. Higher pK_a phenolic groups would become significant at lower elevations, especially where wetlands are common. Phenols would be associated with HMWs and soil degradation processes.

This chapter first discusses the acid-base characterization of humic substances using a new technique for reduction of titration data. The method is applied to XAD fractions that were obtained using the Leenheer (1981) technique. Then the method is used to analyze the change in humic substances during a laboratory bacterial degradation study. Finally changes in field data are assessed by applying the method to the U.S. E.P.A. National Surface Water Survey (NSWS) Eastern Lakes Survey (ELS) data to determine the pK_a–concentration distribution of humic substances for different regions and elevations.

DISCRETE AFFINITY SPECTRUM ANALYSIS OF TITRATION DATA

Detailed resolution of titration data for organic acids appears to be limited to 3-4 binding constants (e.g., Dzombak et al. 1986; Fish et al. 1986). On the other hand, the potential clearly exists to obtain many binding sites from the 100 or more data sets that can be obtained for a titration profile. Linear programming minimization techniques (Tobler and Engel 1983) can be used on titration data for humic substances to obtain a detailed characterization of pK_a's. In this analysis, we assume that each functional group can be approximated by a monoprotic acid. Thus for the jth site, the concentration, C_j, and conditional acid dissociation constant, K_j, we write:

$$HA_j = H^+ + A_j^-, \qquad C_j = [A_j^-] + [HA_j], \qquad K_j = \frac{[H][A_j^-]}{[HA_j]}$$

and

$$[A_j] = \frac{C_j K_j}{[H] + K_j}. \tag{1}$$

We further assume that at any titration condition, i, the protons bound (or released) are additive. From proton balance conditions, we can then write, incorporating Eq. 1:

$$[BASE]_i - [ACID]_i + [H]_i - [OH]_i = -C_{anc} + \Sigma\, C_j \frac{K_j}{[H]_i + K_j}, \tag{2}$$

where C_{anc} is the initial proton condition concentration (acid neutralizing capacity), and [BASE] and [ACID] are base and acid titrant concentrations, respectively. Equation 2 can be expanded to a multilinear expression if intervals of log K are specified and bounded. Best fits using linear programming (LP) are made for equal pH interval titration data for the conditions, $C_j \geq 0$ and absolute $(\Sigma\, e_i)$ = minimum, where e_i is the error for each data set as formulated in Eq. 2. The primal and dual LP solutions are then obtained using these constraints. We have assessed this procedure for simulated multifunctional group titration data and humic and fulvic acids. A minimum of two equally spaced data sets is required to define one pK_a. The method is quite robust, with similar results being obtained even when reasonable errors occur in the experimental measurements.

Analysis of Fractionated Sample

A 20 l sample from Luther Lake, Ontario (DOC: 14.5 mg C/l), was passed through a 0.45 μm filter, acidified and filtered to remove the humic acid fraction, and concentrated by the DOM fractionation procedure of Leenheer (1981). The hydrophobic acid (HPOA) and hydrophilic acid (HPIA) fractions, as well as the original filtered sample, were then titrated and analyzed by discrete affinity spectrum analysis. The HPIA shows a major concentration peak of 251 μeq/l at a pK_a of 3 not exhibited by the sample or HPOA fraction. The HPOA fraction is very similar to the initial sample with pK_a's at 4.4–4.6, 6.2, 7.4–7.8, and 9–11. The major peak at $pK_a = 3$ appears to be an artifact of the method, because it does not occur in the results for the unprocessed sample. The fractionation scheme does not produce different pK_a spectra for hydrophobic and hydrophilic fractions, as might be anticipated.

BACTERIAL DEGRADATION OF ORGANIC ACIDS

The diagenesis of organic acids in streams and lakes can lead to unexpected changes in the acidic nature of humic substances. As suggested earlier, the simpler or LMW organic acids of low pK_a might be expected to degrade easily. A laboratory study over 100+ days was carried out to assess the change in the pK_a spectra of high DOC aquatic samples. Aerobic laboratory incubations, in the dark at 22°C, of high DOC stream waters from Nova Scotia, Canada, were carried out on both sterile (0.2 μm filtered water) and nonsterile water, with and without sediment that had been previously autoclaved. Figure 1 shows the change in the pK_a spectrum over time of the nonsterile sample with sediment. Under both sterile and nonsterile conditions (with sediment), the concentrations of low pK_a acidic functional groups decreased; however, change was faster and larger for the nonsterile sample, with low pK_a groups disappearing by 28 days compared to over 100 days for the sterile samples. The degradation of pK_a sites under nonsterile conditions can be explained by bacteria, and measured enzyme activity (esterases, lipases, and phosphatases were positive) correlates with pK_a change. Sterile sample degradation has to be explained by another mechanism, presumably associated with the sediment, such as redox reactions, involving Fe(III) which is prevalent in these samples. The

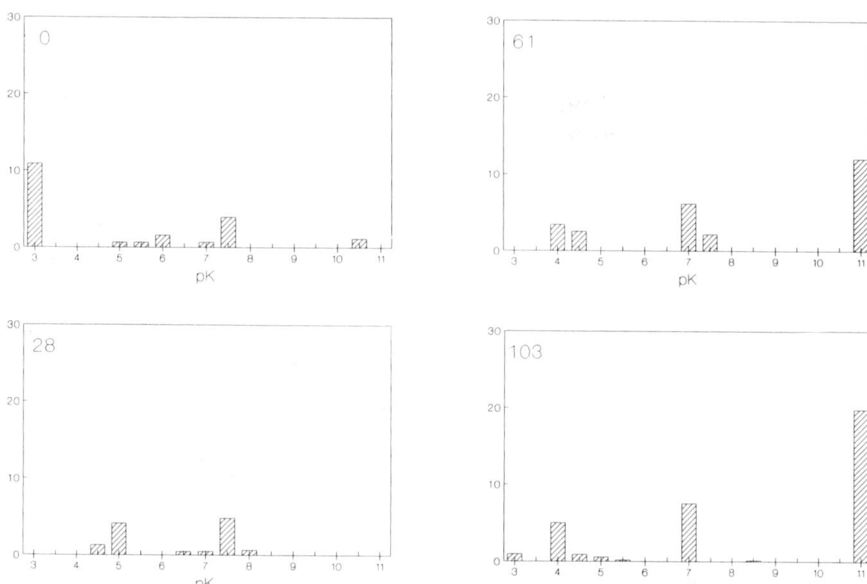

Fig. 1—Change in pK_a spectra of nonsterile organic-rich sample (DOC = 9.6 mg C/l) at different incubation times (in days, shown on each plot) in the dark and in the absence of sediment. Concentrations are normalized to DOC (μeq/mg C). A pK_a interval of 0.5 was used for the discrete affinity spectral analysis of the titration data. Y-axis, left: normalized ligand concentration (μeq/mg C DOC).

nonsterile samples (with sediment) also showed increases in pH (4.8 to 7.0+), ANC (~0 to 200+ μeq/l), and DOC (9 to 15 mg/l) over the time of the experiment. Sterilized water alone showed no change, and sterilized water with sediment showed smaller increases in pH and ANC and negligible increase in DOC (9 to 10 mg/l). Thus bacteria appear to be important in the preferential breakdown of low pK_a groups on the time scale of days. As discussed previously, the loss of low pK_a sites can be explained in part by the preferential degradation of simple LMW organic acids.

An important extrapolation of these results is that residence time of organic-rich waters affects organic acid composition, which is more acidic for waters of short residence time (e.g., headwaters) and more basic for longer residence time waters. Presumably the residence times of different hydrological pathways within a watershed would show similar patterns. The modification of the acid spectra occurs rapidly. The analysis shows a change after 28 days, but in reality modification may commence only after a few days.

VARIATIONS IN EASTERN LAKES SURVEY ORGANIC ACID SPECTRA

The U.S. E.P.A. Eastern Lakes Survey (ELS) (Overton et al. 1986) data base contains acid and base titration, DOC, DIC, and pH measurements. The pH range of these titrations was from about 3.5 to 11. Discrete affinity spectral analysis, as described earlier, was carried out on the titration data.

Although the data base represents one of the better assessments of low ionic strength lakes to date, there are some difficulties in using it to assess the variation in organic acids. One such difficulty is that the sampling was a one-time analysis for individual lakes, containing no watershed studies. Therefore spatial variations can only be estimated by statistical comparison of large regions. Temporal variations cannot be assessed. Most lakes were sampled shortly after fall turnover.

Also, the analytical protocol for titrations was not designed for determination of pK_a distributions. The titrant addition was for an equal volume rather than a constant pH interval; thus the high and low pH regions are heavily weighted. The titrations were not carried out in a closed system; therefore the assumption of constant DIC used (see below) to remove the carbonato portion may not be valid. No corrections were made for contamination of CO_2 in base titrant, a common occurrence. Finally, some of the data in the high and low pH regions are suspect; these data can give false pK's at the tails of the distribution. The titration data were assessed for internal consistency, using criteria inherent in Eq. 2. When the criteria were not met, data were eliminated.

The carbonato portion was subtracted from the titration data base using the DIC value measured just prior to titration. It is assumed that this value is constant over the titration range. If this is so, then

$$[HCO_3^-] + 2[CO_3^{2-}] = \frac{DIC\,([H^+]K_1 + 2K_1K_2)}{12010\,([H^+]^2 + [H^+]K_1 + K_1K_2)}, \qquad (3)$$

and for each titrant data set the value of Eq. 3 is subtracted from the left side of Eq. 2. Assessing an increase in DIC due to contamination of base titrants is impossible except to query large pK_a concentrations in the 10-11 range.

Inspection of Eq. 2 shows that the values should increase or be constant for base titrations, and decrease or be constant for acid titrations. If a significant amount of data are in error, the LP minimization technique tends to generate a sizable artificial concentration at one or both of the limiting pK_a's. Therefore succeeding data were eliminated when they did not meet this criterion. All of this discussion emphasizes the need to have high quality equal pH interval data for determination of the concentration–pK_a distribution.

Figure 2 is a summary of the ELS data for DOC concentrations greater than 5 mg C/l. The data are shown for subregions (1 = New England, 2 = Midwest, 3 = Southeast). Region 3a has been deleted because it contained only three lakes. In total, 271 data (lakes)—about 20%—were assessed. The plots show the average DOC-weighted concentrations for pK_a's at 0.5 intervals from 3 to 11. The results are for data adjusted to meet the criteria inherent in Eq. 2. In addition, all data have had carbonato contribution subtracted.

The buffer capacity, β, is also shown on a logarithmic scale. The contribution of the organic acids to β can be visualized by estimating the difference of the curve for β and the two boundary straight lines (between 2300 μeq/l at pH 3 and 11 and 0.23 μeq/l at pH of 7). The results show a buffering region for all plots between 3.5 or 4 and 7 or 8. Some plots show very large concentrations in the pK_a 9–11 range which may result from CO_2 contamination as well as phenolic groups.

From the plots, we can conclude that the pH will be strongly influenced for lakes with DOCs between 5–10 mg C/l and alkalinities of ~100 μeq/l.

The COOH group concentration density (μeq/mg C) can be estimated by adding the concentrations for pK_a's <7.5. These range between 4.5 and 13.7 (average = 8.2 ± 2.5) which is within most estimates. The lower concentration densities may be due to elimination of some valid data at low pH. For example, the second plot for region 3b (orig.) shows the pK_a spectra obtained when suspect data are not removed. The amount and distribution of low pK_a's are much greater for the unfiltered data than for the corrected data. The actual values may lie between these two limiting estimates, but no further refinement seems justified at this time.

The trend in organic acid distributions was determined for different elevations in region 1. pK_a–concentration spectra for 133 analyses were averaged for three elevations: 15%(>530 m), 35% (363–530 m), and 50% (<363 m). Figure 3 shows the average change in pattern from high elevation to low elevation lakes. The plots show a predominance of low pK_a's at the higher elevations changing to a prevalence of high pK_a's at lower elevations. These results are consistent with the hypothesis of dominance of low pK_a acids at high elevations, the instability of low pK_a LMWs, the degradation of low pK_a organic acid groups, and the production and stability of soil phenolic substances at lower elevations.

Fig. 2—pK_a–concentration relationships for 10 subregions of the U.S. Eastern Lakes Survey. All data have been filtered using the criteria inherent in Eq. 2, except plot 3b (orig.). Bars are one standard deviation. Sample sizes vary from 13 to 37. Y-axis, left: concentration density (μeq/mg C DOC); right: buffer capacity, β (μeq/l).

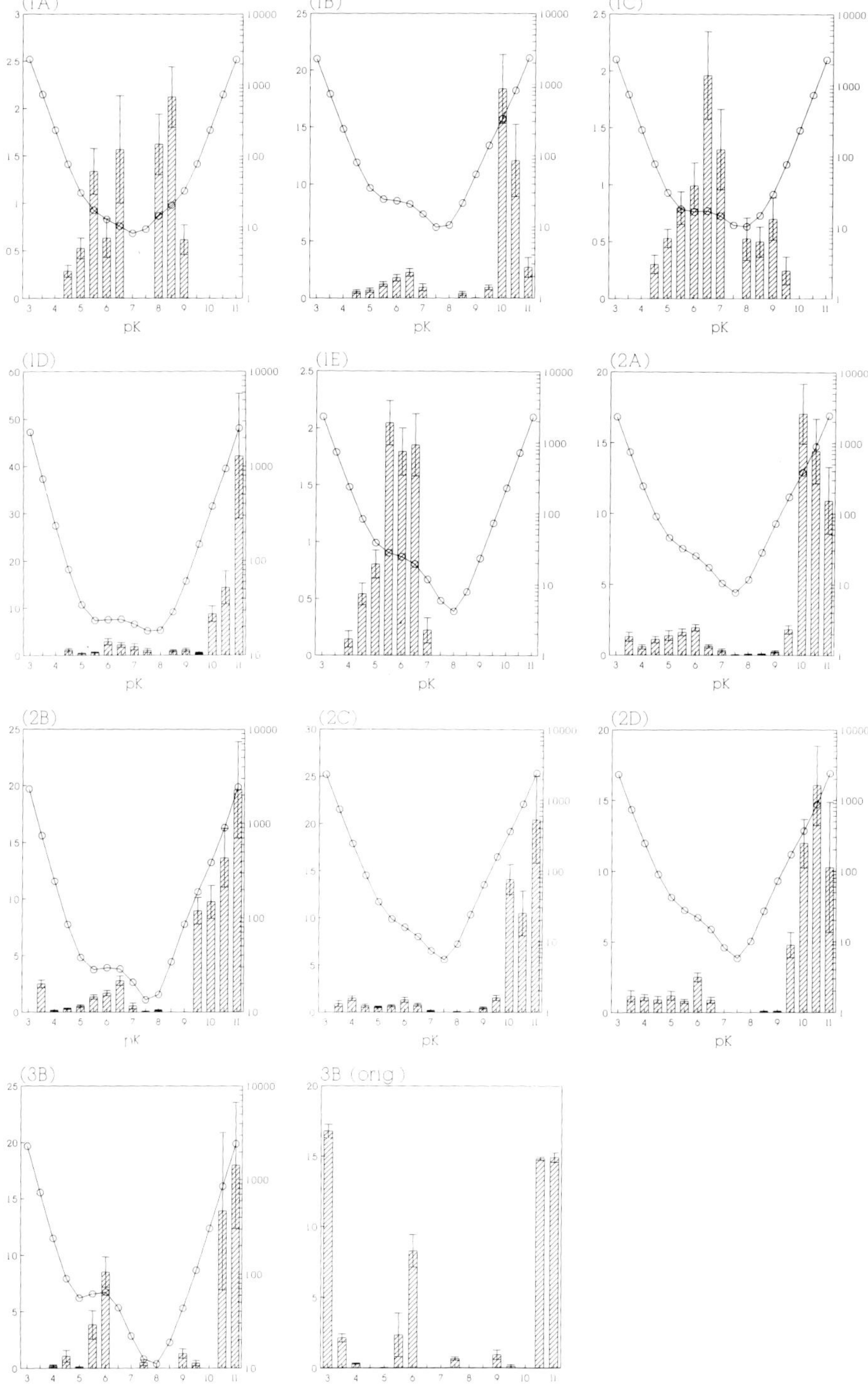

(1A)
(1B)
(1C)
(1D)
(1E)
(2A)
(2B)
(2C)
(2D)
(3B)
3B (orig.)
pK

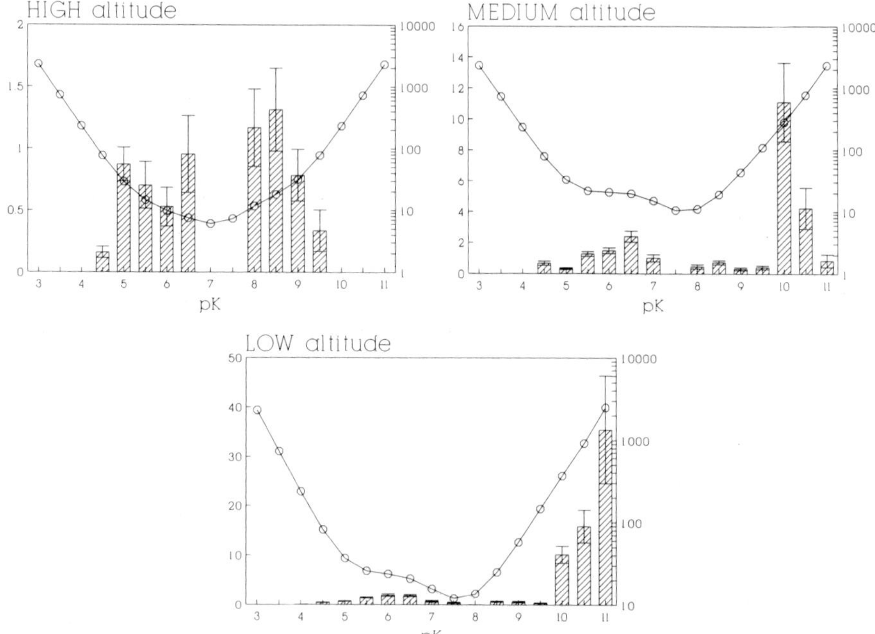

Fig. 3—Change in pK_a concentration spectra for different elevations in region 1, Eastern Lakes Survey. High altitude is > 530 m; medium altitude is 363–530 m, and low altitude is < 363 m. Data have been filtered using the criteria discussed in the text, and the carbonato contribution has been subtracted. Y-axis, left: concentration density (μeq/mg C DOC); right: buffer capacity, β (μeq/l).

CONCLUSION

From literature summaries, an experiment on change in organic acid pK_a spectra over time, and an analysis of changes in pK_a spectra with elevation in the northeastern U.S., we confirm the working hypothesis that low pK_a groups (3–5) predominate in higher elevation surface waters and are probably short-lived. Organic acids with higher pK_a's (9–11) predominate in lower elevation lakes. Undoubtedly seasonal variations and hydrological influences, such as wetland flooding, alter this association.

REFERENCES

Dzombak, D.A., W. Fish, and F.M.M. Morel. 1986. Metal-humate interactions. I. Discrete ligand and continuous distribution models. *Env. Sci. Tech.* 20:669–675.

Fish, W., D.A. Dzombak, and F.M.M. Morel. 1986. Metal-humate interactions. II. Application and comparison of models. *Env. Sci. Tech.* 20:676–683.

Flaig, W. 1978. Contributions of soil organic matter in the system soil-plant. In: The Terrestrial Environment, ed. W.E. Krumbein. Vol. 2 of Environmental

Biogeochemistry and Geomicrobiology, pp. 419–439. Ann Arbor: Ann Arbor Science.
Flaig, W.J.A. 1988. Generation of model chemical precursors. In: Humic Substances and Their Role in the Environment, ed. F.H. Frimmel and R.F. Christman, pp. 75–92. Dahlem Konferenzen. Chichester: Wiley.
Khan, S.U. 1969. Interaction between the humic acid fraction of soils and certain metallic cations. *Soil Sci. Soc. Am. Proc.* 33:851–854.
Leenheer, J.A. 1981. Comprehensive approach to preparation, isolation, and fractionation of dissolved organic carbon from natural waters and wastewaters. *Env. Sci. Tech.* 15:578–587.
Martell, A.E., and R.M. Smith. 1979. Critical Stability Constants, vol. 3. New York: Plenum.
Overton, W.S., et al. 1986. Characteristics of Lakes in the Eastern United States, vol. 2. Lakes Sampled and Descriptive Statistics for Physical and Chemical Variables. EPA/600/4-86/007b. Washington, D.C.: Environmental Protection Agency.
Perdue, E.M. 1985. Acidic functional groups of humic substances. In: Humic Substances in Soil, Sediment and Water, ed. G.R. Aiken, D.M. McKnight, R.L. Wershaw, and P. MacCarthy, pp. 493–526. New York: Wiley.
Posner, A.M. 1964. Titration curves of humic acid. *8th Int. Cong. Soil Sci., Bucharest Trans.* 2:161–174.
Schnitzer, M. 1969. Reactions between fulvic acid, a soil humic compound and inorganic soil constituents. *Soil Sci. Soc. Am. Proc.* 33:76–81.
Thurman, E.M. 1985. Organic Geochemistry of Natural Waters. Boston: Nijhoff/Junk.
Tipping, E., and C. Woof. 1983. Seasonal variations in the concentration of humic substances in a soft-water lake. *Limnol. Ocean.* 28:168–172.
Tobler, H.J., and G. Engel. 1983. Affinity spectra: a novel way for the evaluation of equilibrium binding constants. *Naun. Schmied. Arch. Pharm.* 322:183–192.
Wershaw, R.L., and G.R. Aiken. 1985. Molecular size and weight measurements of humic substances. In: Humic Substances in Soil, Sediment and Water, ed. G.R. Aiken, D.M. McKnight, R.L. Wershaw, and P. MacCarthy, pp. 479–482. New York: Wiley.
Yariv, S., and H. Cross. 1979. Geochemistry of Colloid Systems, pp. 365–368. New York: Springer.

Standing, left to right:
Mark Shuman, Henk De Haan, Herbert Fredrickson, Jürgen Niemeyer, François Morel, Lars-Olaf Öhman, Mike Perdue

Seated, left to right:
Jim Kramer, Wolfgang Calmano, Michael Weis, Arne Henriksen, Jaakko Mannio

Organic Acids in Aquatic Ecosystems
eds. E.M. Perdue and E.T. Gjessing, pp. 141–149
John Wiley & Sons Ltd

Group Report
How Are Acid-base Properties of "DOC" Measured and Modeled and How Do They Affect Aquatic Ecosystems?

M.S. Shuman, Rapporteur
W. Calmano
H. De Haan
H.L. Fredrickson
A. Henriksen
J.R. Kramer
J.Y. Mannio
F.M.M. Morel
J. Niemeyer
L.-O. Öhman
E.M. Perdue
M. Weis

INTRODUCTION

Dissolved organic carbon (DOC) undergoes an extensive number and variety of interactions in aquatic ecosystems. It exchanges protons, metals, and nutrients, it inhibits or stimulates microorganism growth, it attaches to colloid surfaces, etc. These interactions may appear at equilibrium within observation time scales or may appear reactive. They are studied to understand the basic character and function of DOC and to make predictions about its behavior under a wide range of environmental conditions such as concentration, temperature, pH, alkalinity, and ionic strength.

Fundamental to these studies is the ability to make analytical measurements. Two properties limit the straightforward application and interpretation of most analytical techniques: DOC concentration and composition. High concentrations are often needed for precise analytical measurements. Steps taken to concentrate bring about changes in DOC character because the concentrated sample differs from the original in pH, metal and particulate concentration, and ionic strength. Furthermore, DOC is a mixture of molecules with wide molecular size distribution. Any molecule in this mixture may differ greatly in character and reactivity from any other. Only a very small percentage of this mixture, perhaps about 10%, can be separated and identified as individual molecules; most of it cannot be fractionated to this degree. Thus, almost all analytical measurements of DOC properties are measurements of average properties. These measurements, made on altered DOC and reflecting average properties, are those relied upon for understanding DOC and predicting its behavior. It is in the light of these constraints that all topics in this report must be considered.

Analytical measurements provide data which enable hypotheses about the character of DOC to be tested. These hypotheses may take the form of fairly simple descriptive models, like acid-base equilibria, or more complex models, such as one describing DOC behavior in surface waters which combines photoredox, biological, and mass transport processes. Whenever measurements are made in the laboratory, questions arise about their utility to field studies and to the aquatic ecosystem as a whole. Laboratory measurements are made under controlled and simplified circumstances in which conditions such as temperature, pH, and metal content are carefully controlled. A concentration step which will alter the material may be required as mentioned above. The time frame of laboratory observations is normally no shorter than a few seconds and no more than several days. In an ecosystem, conditions may vary considerably within this time frame and DOC residence times may be months or years.

Laboratory experiments are a necessity; they are designed to describe the system in simplest terms, to understand it at the most fundamental level, and to investigate processes in isolation uncoupled as much as possible from a vast continuum of varying conditions over lengthy time scales. One purpose of this reduction to simplest terms is to understand underlying physical/chemical principles. Another equally important purpose is to integrate this knowledge with the environmental ecosystem, i.e., to extrapolate from the simple to the highly complex and variable system. Accomplishment of this latter purpose, extrapolation to the field, is not trivial but is also perhaps not insurmountable. The strategy contains several elements. Whenever possible, properties should be measured on the environmental sample itself, without alteration, and in the field over time. If this is not possible, and it often is not because of some analytical need or purpose that requires concentration, isolation, or fractionation procedures, several procedures should be compared to reveal how the measured properties change and how the sample is altered during sample preparation. Of course, the only assurance that an altered sample has the same properties as the original sample is to compare altered with unaltered samples. If the original sample and its properties can be reconstructed with altered material, the altered material can be used as a surrogate for the original.

This report comments on recent research and on present research needs in the area of proton and metal binding of the organic acid fraction of DOC and discusses application of such research to two problems: the ion balance in colored, low ionic strength lakes and the kinetics of organic acid transformations in aquatic ecosystems. Included is a discussion of concentration methodologies as the first step in a laboratory investigation. Recommendations follow each section. These suggest areas of future research and urge greater cooperation and working arrangements between laboratory-based scientists and scientific investigators in the field.

CONCENTRATION/ISOLATION/FRACTIONATION METHODOLOGIES

The term concentration refers to the necessity to concentrate DOC to a level at which analytical measurements such as IR, NMR, bioassays, etc., are possible. Isolation refers to an additional need, not only to concentrate the DOC but also to separate it from matrix constituents such as metals, silica, or Cl^-. Isolation may be purely an analytical requirement, for example the need to separate DOC from Cl^- to use a Cu ion-selective electrode, or it may be a research design requirement to allow pH or ionic strength to be manipulated. Fractionation refers to a specific research need, as is the need to separate low from high molecular weight acids in order to study their properties separately. The decision to separate DOC from its solvent or its matrix, or to divide it in some way before analysis, depends on the specific purpose in mind. When the purpose is clear, so is the information value of the data. For example, fractionation is required to study low molecular weight organic acids and the data are useful to that fraction only. However, if the purpose is to study the properties of DOC or organic acids as a whole, then the concentration step must not fractionate the DOC unless it can be shown that the fraction has the same properties as the whole.

XAD resins are widely used to concentrate and isolate DOC from aquatic samples. They were quickly adopted to concentrate and desalt gram quantities of dissolved organic matter (DOM) from aquatic samples because they allowed greater convenience and speed than any other method available at the time of their introduction. XAD resin does fractionate DOC and the question remains whether all properties of the XAD extract match the properties of the original DOM. For example, David (unpublished) has found that the carboxyl acidity of the "hydrophilic" fraction (the portion not extracted by the XAD resin) is about 20% higher than the XAD fraction obtained from the same sample. On the other hand, XAD concentrates are found to have the same color and Cl_2 reaction products (Christman, unpublished) as the whole water sample from which they came.

It is unclear exactly how XAD resins fractionate DOC, but the mechanisms appear to be related to the molecular size of the solute and to the exclusion properties of the resin. The reason why "hydrophilic" acids are not extracted may be related to molecular size (sizes greater than about 100,000 daltons may be excluded) or affinities (carboxyl- and hydroxyl-rich molecules with mole ratios of C/COOH less than about 3/1 have low affinities (Thurman et al. 1978; Arhen et al. 1979). Perhaps XAD uniformly extracts only intermediate molecular sizes and does not extract extremely large or extremely small size carboxylic acids. This is speculation, because the molecular size distribution of extracted organic acids is not known precisely. Extraction efficiencies of XAD resins appear to be related to the type of sample which is extracted and this may in turn be related to size distribution (the size distribution of DOC in seawater or groundwater is quite different from the distribution in freshwater).

Recommendations

1. Whenever feasible, analytical measurements on organic acids should be made on unaltered samples.
2. If concentration, isolation, or fractionation is necessary, several techniques should be used and compared for internal consistency of measured properties.
3. Systematic investigations should be undertaken to discover how measured properties change as the sample is modified by pretreatment. Fractions could be added to reconstruct the original material to determine if original properties are obtained.

PROTON AND METAL-DOC BINDING

Data collected from metal-spiked samples indicate that organic complexation of trace metals is prevalent in many aquatic environments, although the specific compounds or binding sites responsible for this complexation are unknown and the actual extent of complexation at natural (unspiked) conditions cannot be directly measured. Empirical thermodynamic models which represent metal chelation in natural waters are available based on titrimetric data, but these cannot be extrapolated confidently to conditions outside the data upon which they are based (Cabaniss and Shuman 1988). A great need exists for data at low natural metal concentrations and within a wide range of pH and ionic strength.

Strong chelation may come either from specific biogenic chelators, such as siderophores or cystein-rich polypeptides, or from humic compounds. While it seems reasonable that the biota should, under some conditions, release specific chelators in their surrounding medium, actual field data on such compounds are lacking. Progress in this field may come from the use of sensitive analytical techniques developed with chelators produced in laboratory cultures of aquatic organisms and applied to field samples.

Strong effective metal chelation by isolated DOM has been demonstrated many times. Once corrected for variations in experimental conditions, the data of many researchers, using various techniques and different samples, are in good agreement. Mathematical representation of these data and hence the chemical interpretation of binding vary widely. Chelation is described equally well as a mixture of discrete ligands, as a continuous distribution of metal affinities, or as a polyelectrolyte (metal affinity of individual ligands enhanced or diminished by long-range coulombic forces from neighboring functional groups). In all cases to date the purpose has been to describe complexation data and their variation with solution conditions (pH, ionic strength, and metal and ligand concentrations) with a minimum set of parameters. Description with the purpose of greatest economy of terms is not very useful in itself, and future progress in thermodynamic models of DOM complexation must be based on a sound understanding of the physical/chemical reality of metal binding. For example, donor atoms involved in reacting ligands should be identified and binding mechanisms understood to distinguish between intrinsic metal affinities (e.g., strong binding attributed to thiol groups) and polyelectrolyte effects. The variation of molecular size distribution with pH, ionic strength, and metal concentrations, as well as the role of polydispersity on metal binding, should also be understood. The need for such fundamental knowledge will probably necessitate new experimental approaches to probe metal binding at the molecular level.

Development of any thermodynamic model of metal-DOM binding is by itself insufficient. The kinetics of coordination reactions in natural waters need not always be rapid. Competing reactions with other ligands and metal exchange reactions may lead to overall kinetics that are slow compared to hydrodynamic residence times (years). For example, metal exchange rates may be very slow if Al and Fe occupy binding sites. In addition, in systems with high alkaline-earth concentrations, such as estuarine and marine water, Ca or Mg in occupied sites may also be slow to exchange (Hering and Morel 1988). Reaction rates will be a function of many parameters including pH (H^+ or OH^- catalyzed reactions), molecular size and charge, etc.

Kinetic methods may be particularly convenient for obtaining thermodynamic data and gaining insight into the nature of the binding process. For example, one may probe the binding properties of humates by following, over a range of concentrations, the kinetics of metal exchange with a strong chelator such as calcein blue or a fluorescent metal such as Europium whose fluorescence is enhanced or quenched over a range of concentrations.

Recommendations

1. New or novel techniques should be applied to probe the molecular size, charge density, and nature of the binding mechanisms of organic acids.

2. New thermodynamic metal-binding models which include metal competition, proton competition, and ionic strength effects are needed that can predict behavior over a wide range of environmental conditions. These models need to be based on the identity of binding groups, molecular configuration, and charge.
3. Actual data sets, not just derived binding parameters, should be reported for future model evaluation and comparison.
4. Laboratory predictions should be tested with field data.

ORGANIC ACID CONTRIBUTION TO CHARGE BALANCE

The ion balance in colored freshwaters that are sensitive to acidification by anthropogenic strong acid inputs can be significantly influenced by the presence of organic acids. Empirical models are used to understand and predict changes from these anthropogenic acidic inputs. The models are based on the ionic composition of the water and require estimating the free organic acid anion concentration (A_f). Organic anions in acidic, organic-rich waters can be a significant or major fraction of the total anions (Hemond 1980; Eshleman and Hemond 1985). Often the models either ignore organic anions or consider them constant during changes in acidic inputs. Major unresolved questions are how organic acid concentrations and their acid-base characters change during anthropogenic acidification, how large these changes are, and whether they are reversible.

A_f is commonly estimated from the difference between the sum of all inorganic cation concentrations (e.g., Ca, Mg, K, Na, NH_4, Al, Mn, and H^+) and the sum of all inorganic anion concentrations (e.g., SO_4, Cl, HCO_3, NO_3, and F). Large errors can be propagated simply from this summing up of 10–15 analytically derived values, each with its own considerable error (typically 2–10% relative error for each [Cronan et al. 1978]). Another way of estimating A_f comes from an empirical expression that attempts to relate A_f to total carboxyl acidity (A_t) and pH (Oliver et al. 1983). However, this approach has problems too. The total concentration of carboxyl acidity (A_t) is the sum of A_f and the concentrations of metal-complexed carboxylic acids (A_c) and protonated carboxylic acids (A_H). Thus $A_t = A_c + A_H + A_f$. The Oliver model does not consider A_c, which leaves this term undetermined unless additional information is available such as independent metal complexation data or a thermodynamic binding model. An alternative method of estimating $A_c + A_f$ based on increase in pH (production of alkalinity) upon photooxidation has also been used (Hemond 1980; Eshleman and Hemond 1985; Gorham et al. 1985). However, interferences need to be considered; for example, sulfuric acid from nonionic sulfur in some bog samples is found to interfere (Gorham et al. 1985).

Recommendations

1. Look for direct, specific analytical methods to determine organic acid anions.
2. Report complete acid-base titrations to better define the protolytic chemistry of whole samples. Report DIC-pH to obtain carbonate species.
3. Investigate the role of metals (e.g., Fe, Al) and their complexes on the proton balance in these aquatic ecosystems.
4. Develop and test empirical relationships to predict organic anion concentrations as a function of pH (i.e., a refined Oliver model).

KINETICS

It is necessary to consider kinetics in all aspects of organic acid chemistry. For example, the effectiveness of ion exchange resins in lowering the ash content of DOC isolates and the protonation of acidic anions by strong acid ion exchange chromatography may be limited by the rate of exchange of proton for metal on metal-occupied carboxylate binding sites. Titration curves of organic acids exhibit irreversible behavior—the so-called "hysteresis effect," in which the base titration is not reproduced when a reverse titration with acid is carried out. The slow exchange or hydrolysis of Al has been hypothesized to explain hysteresis as has bacterial alteration of DOC. Metal exchange reactions may be fast or slow depending on the metals involved, their relative concentrations and binding properties, and the nature of the ligand and its acid-base properties. There may also be specific acid or base catalysis reactions. Water exchange rates may limit the rate of reaction of some metals with donor atoms as they do with $Al(H_2O)_6^{3+}$ at low pH (Eigen 1963). The size, charge density, or molecular configuration of organic acids may affect the rate at which an incoming ion or molecule can approach a reactive site. Kinetics can limit the rate of redox, sorption–desorption, and precipitation reactions. Samples collected in the field may be altered via a number of kinetic processes during transport and storage. The transformation rates of nutrients and other bioactive substances can be affected by pH, light, temperature, and molecular size and shape. In this regard, Jones et al. (1988) demonstrated that a DOC-Fe-PO_4 species, in brownwater lakes that were studied, reduced phosphate availability to aquatic biota by sequestering phosphate and then releasing part of the bound phosphate during illumination.

This discussion of the effects of DOC on the biological availability of nutrients leads to the general question of the relevance of data derived in the laboratory to processes in the natural environment. Environmental samples are manipulated for two major purposes: to facilitate analytical measurements and to eliminate interactive variables. The former has been mentioned above with respect to concentrating DOC for water samples and removing substances which interfere with analytical measurements. The latter manipulation is, in part, necessitated by conceptual considerations. Performing a "controlled" experiment is usually interpreted as the process of holding all known potential variables constant, varying one in a known way, and recording the response of the system. This experimental design has several advantages. The results are easy to interpret and they can be used as the first argument in a logic used to show cause and effect relationships. Results from these types of experiments are valuable. Experiments of this design should and will continue to be performed.

However, if we are to understand the function of natural ecosystems then we must conduct experiments designed to determine which factors dominate in systems governed by interacting variables. One approach, the bottom up approach, is to synthesize models of simple systems. With proper forethought, models derived from "controlled" experiments on environmental processes can be integrated to predict processes in systems of ever-increasing complexity, ultimately ecosystems. The accuracy of ecosystem models can be judged only by comparing model predictions with data derived from long-term ecosystems studies; this is the top down approach.

Recommendations

1. The effects of kinetics on laboratory analytical procedures such as the hysteresis effect on acidity titrations, particularly the possible involvement of Al and Fe, should be studied.
2. The kinetics of reactions involving organic acids themselves (acid-base, complexation, redox, sorption–desorption) should be investigated.
3. Laboratory studies need to be designed so that results can be integrated into ecosystem models.

REFERENCES

Aiken, G.R., E.M. Thurman, R.L. Malcolm, and H.F. Walton. 1979. Comparison of XAD macroporous resins for the concentration of fulvic acid from aqueous solution. *Analyt. Chem.* 51:1799–1803.

Cabaniss, S.E., and M.S. Shuman. 1988. Copper binding by dissolved organic matter. I. Suwannee River fulvic acid equilibria. II. Variation in type and source of organic matter. *Geochim. Cosmo. Acta* 52:185–200.

Cronan, C.S., W.A. Reiners, R.C. Reynolds, and G.E. Lang. 1978. Forest floor leading: contributions from mineral, organic, and carbonic acids in New Hampshire subalpine forests. *Science* 200:309–311.

Eigen, M. 1963. Fast elementary steps in chemical reaction mechanisms. *Pure Appl. Chem.* 6:105–117.

Eshleman, K.N., and H.F. Hemond. 1985. The role of organic acids in the acid-base status of surface waters at Bickford Watershed, MA. *Wat. Res. Res.* 21:1503–1510.

Gorham, E., S.J. Eisenreich, J. Ford, and M.V. Santelmann. 1985. The chemistry of bog waters. In: Chemical Processes in Lakes, ed. W. Stumm, pp. 339–363. New York: Wiley.

Hemond, H.F. 1980. Biogeochemistry of Thoreau's Bog, Concord, MA. *Ecol. Monogr.* 50:507–526.

Hering, J.G., and F.M.M. Morel. 1988. Kinetics of trace metal complexation: role of alkaline-earth metals. *Env. Sci. Tech.* 22:1469–1478.

Jones, R.I., K. Salonen, and H. De Haan. 1988. Phosphorus transformations in the epilimnion of humic lakes: abiotic interactions between dissolved humic materials and phosphate. *Freshwat. Biol.* 19:357–369.

Oliver, B.G., E.M. Thurman, and R.L. Malcolm. 1983. The contribution of humic substances to the acidity of colored natural waters. *Geochim. Cosmo. Acta* 47:2031–2035.

Thurman, E.M., R.L. Malcolm, and G.R. Aiken. 1978. Prediction of capacity factors for aqueous organic solutes absorbed on a porous acrylic resin. *Analyt. Chem.* 50:775–779.

Organic Acids in Aquatic Ecosystems
eds. E.M. Perdue and E.T. Gjessing, pp. 151–166
John Wiley & Sons Ltd

Effects of Ecosystem Changes (e.g., Acid Status) on Formation and Biotransformation of Organic Acids

R.C. Petersen, Jr.

Stream and Benthic Ecology Group
Limnology Institute/Department of Ecology
Box 65, University of Lund
221 00 Lund, Sweden

Abstract. Changes in ecosystems can be caused by chemical and physical disturbances and as such can be shown to affect the transformation and biotransformation of organic acids. The main chemical changes in ecosystems are acidification and eutrophication. The main physical changes are due to mining, forestry, and agriculture. Acidification which lowers pH below 5 causes a measurable decrease in the proportion of organic acids that are dissociated. This results in an increase in octanol solubility and therefore increased lipid solubility and bioavailability of organic acids. This deduction is supported by evidence that organic acids are directly toxic at pH 5 while rather benign at pH 6 and 7. An indirect effect of acidification on organic acids is the increase in aluminum, which is known to complex with and remove organic acids from surface waters. Eutrophication by the addition of nitrogen and phosphorus will increase the mineralization of organic acids although this may be limited to oligotrophic areas. Physical changes due to mining, agriculture, and forestry consist of increasing soil mineralization and increasing the loss of more humified organic acids from deeper in the soil profile. In some surface waters, such as those in Sweden, the increase in organic acids due to land use seems to outweigh the reduction in organic acids due to acidification.

INTRODUCTION

The purpose of this chapter is to review how man-made disturbances in terrestrial and aquatic ecosystems affect the genesis, degradation, and mobility of organic acids. This chapter will address only ecosystem changes that can be linked to man's activity and will not address naturally occurring changes on an evolutionary time scale or over a seasonal cycle.

We will be concerned primarily with organic acids defined as dissolved organic substances that are able to donate hydrogen ions as a conjugated acid-base pair, have a molecular weight from 1000 to 3000 daltons (Malcolm et al. 1989), and as such are one of the major identifiable fractions of humic substances. However, since there is not a clear separation in the available literature between the recalcitrant, relatively stable complex mixtures of humic organic acids and the labile, simpler, and lower molecular weight protohumic organic acids, it will not be possible to clearly separate the biological mechanisms and ecological behavior of humic organic acids from the low molecular weight organic acids fraction.

Concerning the concentration of organic acids that occurs in aquatic systems, there does not seem to be any simple relationship between the total dissolved organic carbon (DOC) and the organic acid fraction. Using the procedure of Leenheer and co-workers (Leenheer 1981; Aiken 1985), the hydrophilic and hydrophobic organic acid fraction ranges from 36 to 92% of the total DOC pool and represents from 2.4 to 35.1 mg/l (Table 1). While there may be a tendency for the organic acid fraction to increase in waters that are draining humic areas (i.e., Suwannee River, Table 1) and to be low in waters influenced by sewage treatment effluents (i.e., South Platte River), there does not seem to be enough information at present to link surface water ecosystem characteristics with the concentration of organic acids.

Organic acids are a mixture of organic compounds. For this reason it is difficult or even impossible to differentiate between changes in chemical structure of an organic acid from changes in composition of organic acids. However, we can divide changes in ecosystems into two broad categories,

TABLE 1. Amount (mg/l) and percent (%) of DOC as hydrophilic and hydrophobic organic acids as determined by fractionation procedures. The percent of combined hydrophilic and hydrophobic bases and neutrals are also reported for the five different water bodies.

Sample	DOC (mg/l)	Organic acids (mg/l)	% of DOC: Acids (%)	Bases (%)	Neutrals (%)	Source
South Platte River	7.5	2.7	36	27	37	Leenheer (1981)
Black Lake	8.3	4.1	49	32	19	Aiken (1985)
Ohio River	3.7	2.5	68	7	25	"
Missouri River	3.4	2.4	70	30	0	"
Suwannee River	38.2	35.1	92	4	3	"

namely those due to chemical disturbance of the ecosystem and those due to physical disturbance. It will then be assumed that the change in the ecosystem will result in a mechanistically similar change in the organic acids. Physical disturbances in the ecosystem such as changes in land use due to forestry practices, agriculture, and peat mining should result in changes in composition of the organic acid pool. Chemical disturbances such as acidification and eutrophication should result in changes in the chemical structure or biological behavior of organic acids.

CHEMICAL CHANGES IN ECOSYSTEMS

There are two major chemical changes occurring in ecosystems that can have an effect on organic acids. One is due to acid rain, which occurs primarily in acid-sensitive areas of North America and northern Europe and can be expected to affect the dissociation and molecular structure, and therefore the aqueous solubility and complexing ability, of organic acids. The second is eutrophication, which is due to increases in nitrogen and phosphorus from both point-source and diffuse effluents and can be expected to affect the microbial degradation of organic material.

Acidification

There is now a long list of direct biological effects of acid rain on the inland waters of Scandinavia and North America (Schindler 1988). While there is considerable information on the relative contribution of organic acids to overall system acidity, there is only scattered information on the direct effect of acidification on the organic acids themselves. For the sake of clarity these two topics will be divided into separate sections, one dealing with the role of organic acids in acidification and the other with the effects of acidification on organic acids.

Role of organic acids in acidification. The relative contribution of anthropogenic acid deposition compared to natural processes of acidification such as regrowth of forests, changing land use patterns, and the natural acidity of forest soils is currently a topic of considerable environmental debate. The key to this discussion is discovering what part of the acidification now being observed is attributable to anthropogenic inputs of mineral acids and what part is attributable to other causes. Krug and Frink (1983), building on earlier work by soil chemists, were correct in their statement that natural processes and changes in land use can result in the low pH of surface waters. However, they were incorrect in their suggestion that the observed widespread acidification is primarily a natural process, since acidification has occurred in remote areas on the North American Precambrian shield where land use patterns have not changed (Schindler 1988) and have been shown to be dominated by mineral acids, not organic acids, in both the northeastern United States (Johnson et al. 1984) and in Norway (Brakke et al. 1987).

The relative contribution of natural compared to anthropogenic sources has been further clarified by Gorham et al. (1986), who examined the causes of acidification in 33 surface waters of Nova Scotia. Using a multiple correlation technique, they found that DOC was the best predictor of H^+, better than SO_4^{2-}, Ca^{2+}, and (Na^+ + Mg^{2+} + K^+), but that acidification in this region, which is dominated by peat bogs and heavily influenced by the sea, is due to both natural and anthropogenic processes. Those waters in peaty areas were naturally acidic due to organic acids while those in forested areas were acidic primarily due to anthropogenic sources. Gorham et al. (1986) suggested that naturally acidic surface waters were more susceptible to further acidification than similar waters not heavily influenced by organic acids. This conclusion has been accepted by Schindler (1988) and supported by Brakke et al. (1987) working in Norway.

Effects of acidification on organic acids. There are two possible direct effects of acidification on organic acids. With a reduction in pH it can be expected that aqueous solubility and complexing ability will decrease, resulting in a decrease in the concentration of organic acids in surface waters. Second, as the aqueous solubility decreases, the lipid solubility will increase, resulting in greater bioavailability of organic acids. Both mechanisms depend on a measurable decrease in the dissociation of organic acids under the prevailing field conditions of acidification.

For naturally occurring organic acids the average pK_a, the pH at which 50% of the acids are dissociated, has been reported to fall within the range of 3.7 to 4.4 (Cronan and Aiken 1985; Wright 1988). The pK_a of individual organic acids has a much wider range, but most of the naturally occurring organic acids have their pK_a's around pH 4. If we assume an average pK_a, then the percent of undissociated organic acids can be calculated at a particular pH from the Henderson-Hasselbalch equation:

$$HA/C = 1/1 + 10^{pH - pK_a} \quad (1)$$

where HA/C is the percent of undissociated material (HA) out of the total material (C), which is dependent on the pH of the sample and the pK_a of the organic acids.

Using Eq. 1, the percent of undissociated organic acid has been calculated for three organic acids having pK_a's of 3.4, 4.0, and 4.4 over a range of pH values from 6 to 3, as plotted in Fig. 1. As can be seen there is a rather steep change in the percent of undissociated organic acids as the pH of the

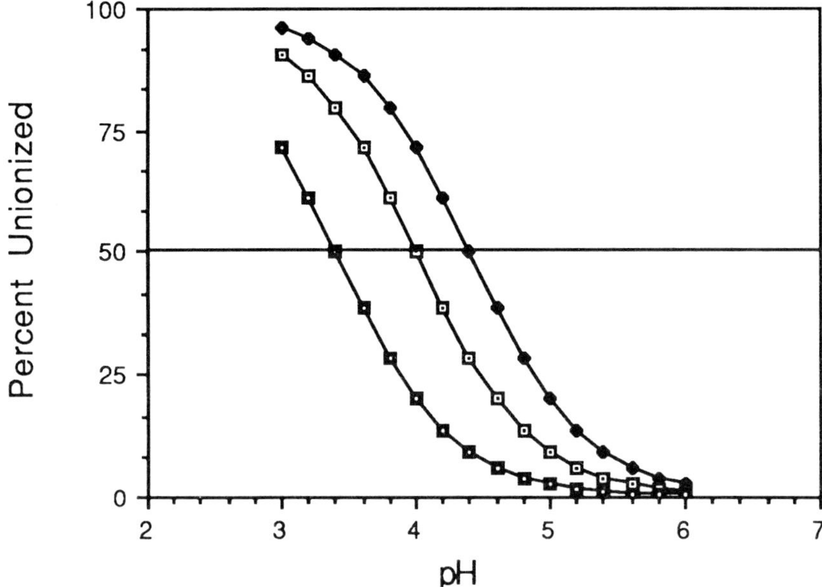

Fig. 1—The undissociated percent of organic acids with pK_a of 3.4, 4.0, and 4.6 (left to right curves) as calculated from the Henderson-Hasselbalch equation using a pH range that occurs in freshwaters undergoing acidification.

sample approaches the pK_a of the organic acid. For all three pK_a values there is very little change in the percent of undissociated organic acid between pH 5 and 6. Below pH 5 a rather large change occurs.

The implication of Fig. 1 is that if acidification of surface waters concerns pH changes above pH 5 then we can predict that only a very small change will occur in the aqueous solubility of organic acids. This implies that the overall effect of acidification above pH 5 on organic acids will probably not be important. However, if acidification causes the pH to go below 5 then there will be a marked increase in the percent of organic acids that are not dissociated. Undissociated organic acids are more hydrophobic and more lipid soluble, and their ability to penetrate or attach to biological membranes will be enhanced compared to the dissociated form. In general, we can expect a much greater direct biological effect of organic acids below pH 5. This may be one reason why it has been suggested (Gorham et al. 1986) that acidification of surface waters may be more harmful in humic areas where the natural pH levels are near the pK_a of the organic acids.

At present there is no reliable estimate of an overall average pH decrease due to acid rain. There are several local estimates of pH decrease ranging from 0.6 pH units in a series of 50 lakes in southern Norway to a decrease of 0.35 units for a series of lakes in Sweden (cited in Petersen et al. 1987).

In general, the major change in surface water chemistry due to acid rain is more a reduction in alkalinity than pH (Schindler 1988). However, short-term pulses of low pH in excess of one pH unit and down to pH 3.6 are commonly observed following snowmelt even in systems which are otherwise well buffered against the effects of acidification (Jacks et al. 1986). It can be expected that these short-term drops in pH will alter dissociation and therefore aqueous solubility of organic acids.

There is some direct evidence that a reduction in dissociation of organic acids actually occurs. Wright et al. (1988) covered an 860 m^2 catchment area with a canopy and excluded acid rain while applying seawater salts at the ambient level with a sprinkler system. Over the four years of acid rain exclusion there was an increase in the concentration of organic anions as measured by charge density (μeq/l) due to an increased dissociation of the organic acids. Over the same period there was no increase in either the pH of the runoff or total organic carbon (TOC).

In order to test whether the observations by Wright and co-workers occur under natural conditions, Petersen and co-workers (unpublished) have measured the charge density, alkalinity, and DOC in a series of 19 southern Swedish streams all receiving acid rain. Charge density was calculated as the summation of cations minus inorganic anions divided by organic carbon in units of μeq/mg C and was a measure of the degree of dissociation of the organic anions (see Shuman et al., this volume). The streams were selected to represent a broad range of pH and organic acid concentrations with pH ranging from 4.2 to 7.5 and DOC from 5.5 to 22.5 mg/l. For statistical reasons they were selected to have equal replication within one of four groups termed clear-acid, clear-neutral, humic-acid, and humic-neutral.

There was no evidence of a dependence of DOC concentration on either pH or alkalinity (Fig. 2A). This should not be interpreted as contradicting the often observed negative correlation between pH and DOC. Gjessing (1976) and others have reported this negative correlation and explained it by the contribution of the organic acids to the sample pH. In the 19 streams presented in Fig. 2, pH is determined both by organic acids and mineral acids in varying degrees. This results in low pH in streams with and without high (>10 mg/l) concentrations of DOC. In fact if the five streams with low DOC, the clear-neutral and clear-acid groups, are left out of the correlation analysis, a significant negative correlation is observed.

There was, however, a significant ($P<0.001$) and nontrivial ($r^2 = 0.52$) positive relationship between charge density (μeq/mg C) and pH or alkalinity (Fig. 2B). These results support the experimental results of Wright et al. (1988) in that the concentration of DOC being exported from a catchment does not seem to be influenced by pH but does affect the degree of dissociation of the organic acids. It is entirely possible that with a lowering of pH the aqueous solubility of organic acids would decrease, resulting in

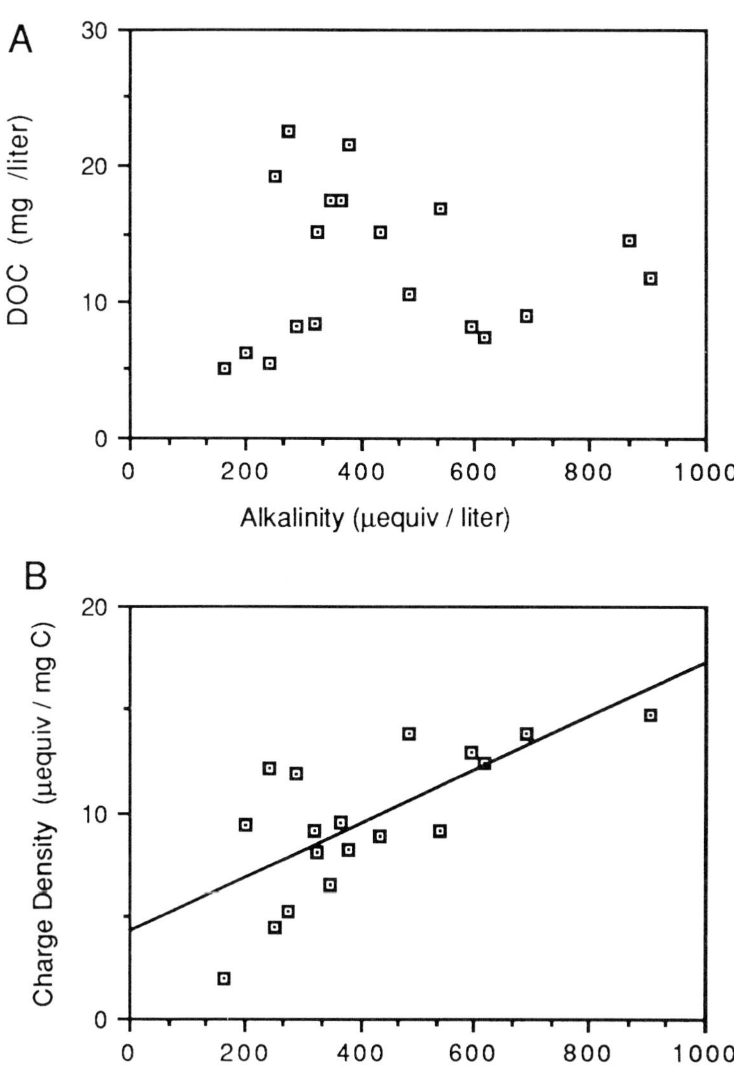

Fig. 2—A: The lack of a significant correlation between DOC (mg C/l) and alkalinity (µeq/l) in 19 streams in southern Sweden. B: The significant ($P < 0.01$) and nontrivial ($r^2 = 0.52$) correlation between charge density (µeq/mg C) and alkalinity (µeq/l) in 19 streams in southern Sweden.

a decrease in DOC. For the range of catchments presented in Fig. 2 this mechanism does not seem to be of sufficient strength to result in a measurable decrease in DOC transport. However, the effect of pH or alkalinity on the dissociation of the organic acids is a strong mechanism.

In addition to a decrease in dissociation, acidification can affect the molecular structure and electrostatic charge of the organic acids. Of 8 of the 19 streams reported in Fig. 2 there is a significant ($P<0.01$) and nontrivial ($r^2 = 0.52$) negative correlation between the extinction coefficient (absorptivity) and alkalinity (Fig. 3). The extinction coefficient is the absorbance at 430 nm corrected to pH 7 and divided by the DOC in mg/l of the sample. The pH correction in this example is not important because the effect of pH on absorbance in the range of pH 4 to 7 represents only a small correction of about 12% and acts to decrease absorption, not increase it. The extinction coefficient tends to be an indication of more humified and complex material, tending to be higher in waters draining peat bogs than in streams or lakes (Thurman 1985). The increase in extinction coefficient with a lower alkalinity and higher hydrogen ion concentration of the waters reported in Fig. 3 suggests that acidification causes the molecules to coil up and contract, as has been suggested by Schnitzer (1980).

Changes in loss from catchments. There has been considerable controversy surrounding the hypothesis that acidification has decreased the DOC of surface waters. Based solely on soil and water chemistry there will be a reduction in the solubility of organic acids which will be measurable as a reduction in DOC leaving acidified catchments. Catchment dynamics is not only dependent on soil and water chemistry but on a combination of many other factors including (*a*) atmospheric deposition of nitrogen, which could

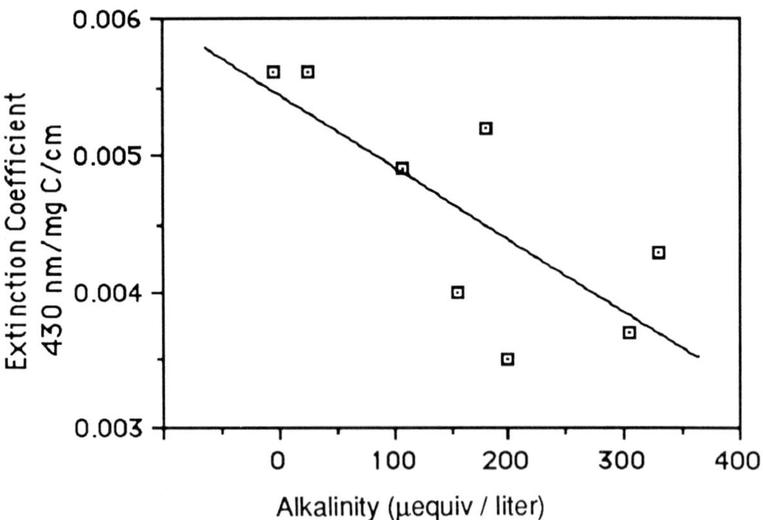

Fig. 3—Correlation between extinction coefficient and alkalinity for eight of the streams presented in Fig. 2.

result in an increase in primary production within the catchment, and (*b*) increased plowing, mining, or disturbance of forest soils, which could increase leakage of organic acids enough to compensate for the decreased solubility.

Regardless of the many factors that could obscure a net decrease in DOC loss from watersheds, there are several lines of evidence which support the hypothesis that acidification has resulted in such a decrease in some catchments. The most compelling of these is the repeated observation that acidification is well correlated with an increase in aluminum concentration, and that there is a measurable decrease in DOC from the inlet to the outlet of lakes which have both a high DOC and aluminum concentration (Driscoll et al. 1986). Another possible mechanism, but for which no reliable field data exist, is the increased mineralization of organic acids by ultraviolet radiation under acidic conditions (Choudhry 1981).

Changes in biological mobility. There is considerable pharmacological and toxicological evidence showing that the biological activity of ionizable organic compounds is pH dependent (Albert 1985). To date this evidence comes primarily from the study of xenobiotics and not from naturally occurring dissolved organic material. It then becomes interesting to speculate that if acid rain is lowering the pH of lakes and streams and results in a decrease in the dissociation of organic acids, then the bioavailability and lipid solubility of these organic acids should increase in accordance with the pH partition hypothesis (Albert 1985).

The standard measure of the bioavailability of an organic material has been the octanol/water partition coefficient (K_{ow}). It has been used since the turn of the century for the study of specific organic compounds. Bioavailability is not, however, related to lipid solubility by a simple linear relationship, because penetration of lipid membranes can be offset by attachment to surfaces. Hansch et al. (1968) have shown that for a series of sedatives the biological response was related to the K_{ow} not by a simple linear relationship but by a second-order polynomial, where a maximum in biological activity is reached at some intermediate level of lipid solubility above which the sedative becomes quickly bound to inert surfaces. A similar mechanism was discussed by Horth et al. (1988), where the K_{ow} values for a series of xenobiotics were proportional to their partitioning between water and soil.

Petersen and Kullberg (1985) defined the octanol/water partition coefficient of humic material (K_{how}) as the ratio of the absorbance at 430 nm of organic material in the octanol phase to the absorbance in the water phase. They examined the increase in K_{how} with the increase in hydrogen activity for three types of humic material, a natural humic water collected from a peaty bog (MDK), a leaf leachate made from silver maple leaves (*Acer platanoides*), and a commercial preparation of humic acid (Fluka). The K_{how} increased with increasing hydrogen activity for all three humic waters. The K_{how} for both Fluka and the maple leaf leachate approximately tripled for each drop per unit pH. The potential bioavailability or lipid solubility of both these humic materials increased by 5000% with a reduction in pH from 6.5 to 3.5.

The increased octanol solubility of humic material is correlated with an increase in direct toxic effect which can be separated from the effects of pH and complexed toxic metals (Petersen et al. 1986; Petersen and Persson 1987). In a series of experiments using the survival of the microcrustacean *Daphnia magna*, a common lake zooplankton routinely used in toxicological studies, Petersen and Persson (1987) found significant differences in the toxicological properties of two humic materials at pH 5. There was no toxic effect at pH 6 or 7. In one series of experiments they used a podzolic (spodosol) soil extract. Survival of *D. magna* was best represented by a second-order polynomial ($r^2 = 0.99$, $P < 0.05$; Fig. 4A) whose second derivative gave maximum survival at 10.7 mg C/l. In a second experiment using humic material extracted from a bog using the International Humic Substances Society (IHSS) procedures (Leenheer 1981; Thurman 1985; and others), survival was also best represented by a second-order polynomial ($r^2 = 0.82$, see Fig. 4B) but whose second derivative gave maximum survival at 27 mg C/l.

The second-order effect observed in these studies (Fig. 4) suggests that at low pH and low humic material, a direct pH effect determines the toxic response most likely due to alteration in membrane permeability and activity of the trace metals present in the culture medium. As the humic material is increased in concentration, the test organism is initially benefited by such mechanisms as the binding or chelation of the trace metals or reduction of membrane permeability. Essentially the humic material tends to protect the organism from direct negative pH effects and from reactive metals.

This protective mechanism occurs until a maximum is reached. The maximum is probably determined by the amount of metals present, pH, and the complex-forming ability of the humic material. It is also determined by the amount of hydrophobic and lipid soluble organic material present. From the relative size of the regression coefficients of the second-order polynomial, the first coefficient is about 10 times the value of the second, implying that only a small amount of the total carbon present is causing the sharp decrease in survivorship.

In a related study, Petersen and Kullberg (unpublished) have determined the amount of carbon that is octanol soluble in a series of water samples from seven locations over four months in southern Sweden. They found that the amount of carbon ranged from 23 to 58 μg/l and represented 0.12 to 0.37% of the total carbon pool depending on the source. The absolute amount of carbon is small but well within the range of toxicities for

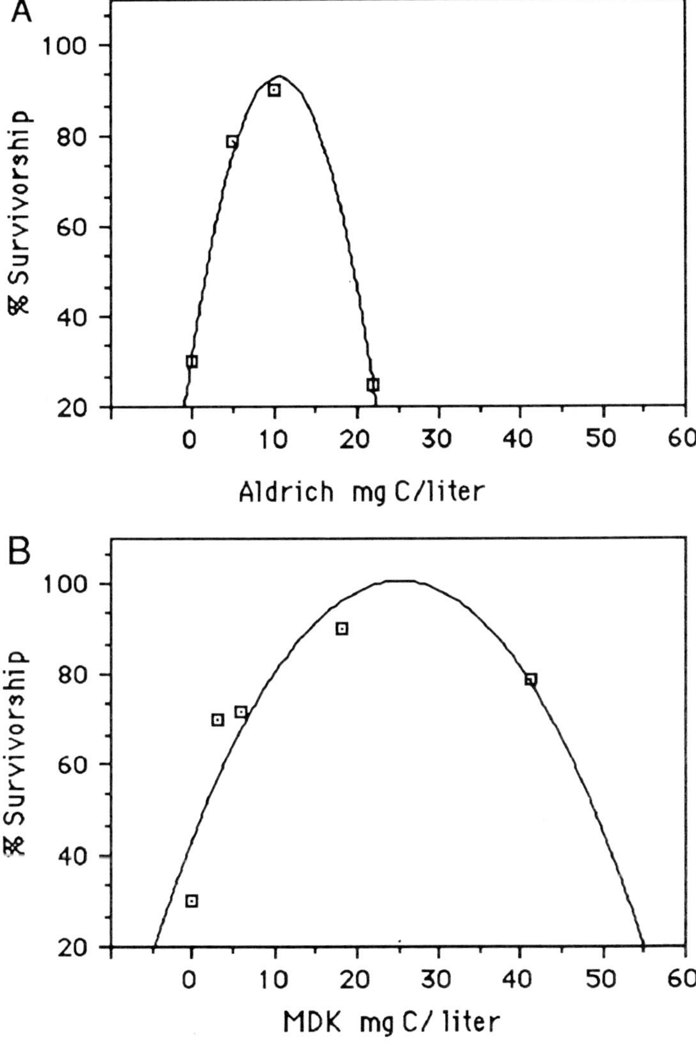

Fig. 4—A: Percent survivorship of *Daphnia magna* at pH 5 with increasing concentrations of Alderich humic material. Plotted curve is the second-order polynomial least squares fit to the data. B: Percent survivorship of *D.* magna at pH 5 with increasing concentrations of a natural bog humus (MDK). Both curves are the second-order polynomial fit to the data. (Figures redrawn using data reported in Petersen and Persson 1987).

insecticides (Mayer and Ellersieck 1986). In other words, the amount of humic material that is participating in the toxic reaction is small, representing only a fraction of the total carbon and organic acid pool, but it is still an amount that is very likely to have a significant biological effect at low pH.

Eutrophication

Most research to date on eutrophication and organic acids has dealt with the role of humics in lowering the availability of phosphorus to primary producers. This comes from several reports that phytoplankton production in dystrophic lakes seems to be far below that expected from information on phosphorus loading (Jackson and Hecky 1980) and from a recent study which demonstrated abiotic uptake of ^{33}P from added PO_4 in waters of high humic content (Jones et al. 1988).

However, there is some information that suggests that eutrophication can lead to an increase in mineralization of organic acids. De Haan (1983) has studied benzoate-metabolizing bacteria in oligotrophic and eutrophic lakes to determine factors affecting their ability to (co)metabolize fulvic acids. He found no real difference in the numbers of benzoate-metabolizing bacteria but did find an increase in benzoate-metabolizing bacteria due to sewage pollution in a section of the oligotrophic lake. He speculated that the "rate of fulvic acid degradation may be related to the trophic level of the aquatic system." This may imply that in a humic oligotrophic system there are microbial systems available to metabolize humic carbon, but these are essentially nitrogen- and phosphorus-limited. The addition of effluents rich in these macronutrients will then lead to increased degradation of humic carbon compounds.

PHYSICAL CHANGES IN ECOSYSTEMS

Changes in land use patterns were suggested by Krug and Frink (1983) as a major cause of surface water acidification. They were incorrect in this suggestion but were quite correct in pointing out the rather large changes that are occurring in drainage pattern and the effect they have on the organic acids of the receiving surface waters (Forsberg and Petersen 1989). In a study of 11 lakes in southern Sweden, L. Colvin (pers. comm.) has compared average water color over the 5-year period of 1977–1981 to the 5-year period of 1983–1987 (Fig. 5A). All 11 lakes increased in color with an average of 20 mg Pt/l. While acidification was occurring in the watersheds of these lakes there was no significant decrease in pH. In addition, a decrease in pH should reduce water color, not increase it. In other words, it is difficult to suggest that this increase in water color is due to acid rain.

What has been changing in the watersheds of this region of Sweden is the ditching of forests and peat bogs. Forests are ditched to reduce ground frost and to increase drainage. Peat bogs are ditched to allow mining. Sweden is currently exploiting its peat reserves for electrical power. Ditching in both areas stimulates the loss of organic material from deeper in the soil

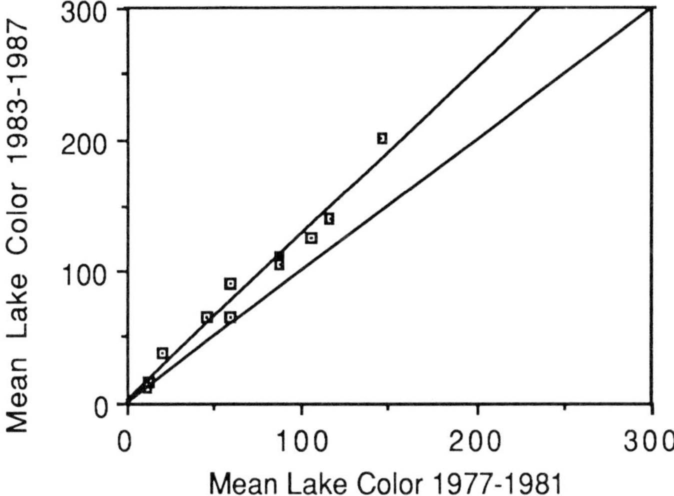

Fig. 5—The mean lake color (mg Pt/l) of 11 southern Swedish lakes during two five-year periods (L. Colvin, unpublished).

layer which, being more humified, will have more color. K. Heikkinen (pers. comm.) has investigated DOC transport from the Kiiminkijoki River in northern Finland and measured the color/carbon ratio, which is an analogue to the extinction coefficient, over a one-year period. She reports high values for waters draining peat bog areas during the summer and low values during the winter and spring. During the summer low-flow period, waters drain from the deeper peat strata well within the bog. There is a rather large contribution of well-humified material which results in a high extinction coefficient (Thurman 1985). During winter and spring, stream discharge is higher and is dominated by precipitation and surficial flows. The extinction coefficient will then be low as surficial flows pick up more recently contributed organic material which is not well humified.

In addition to changing the drainage pattern of soils, agricultural practices will increase the mineralization, degradation, and oxidation of organic acids. Andriulo et al. (1987) have compared conventional tillage, which consists of burning wheat stubble, plowing, harrowing, and then seeding, to a nonconventional tillage practice consisting of direct seeding on wheat stubble without plowing. They then followed the change in soil humus for 11 years and found that the conventional tillage resulted in an increase in oxygen percentage, fulvic acid fraction, pH, and mineral content, while the carbon percentage, aromaticity, and humic acid fraction decreased.

CONCLUSIONS

1. Both chemical and physical changes in ecosystems affect the transformation of organic acids.
2. Chemical changes occurring through acidification are able to affect the dissociation and molecular structure of organic acids directly. An indirect effect of acidification is increased complexation with aluminum, resulting in a decrease in DOC in affected surface waters.
3. Eutrophication and physical changes due to changes in land use patterns such as agriculture, forestry, or peat mining affect organic acids through increased mineralization or by altering the molecular composition through changes in drainage pattern.
4. Ertel et al. (1988) pointed out that the transport between different types of environments is an obvious but largely overlooked aspect. This also applies to organic acids which are transported via hydrological forces from soils through groundwater to rivers through estuaries to the oceans. Most researchers tend to concentrate on only one of these enviroments. Because important changes in both the physical and chemical environment occur at the interfaces between these environments (e.g., air–water, soil–water, freshwater–marine), it is entirely likely that many of the crucial linkages and transformation processes are not known.

Acknowledgements. This paper was written while the author was supported by grants from the Swedish Environmental Protection Board and the Royal Physiographical Society. The paper would not have been possible without the continued support, cooperation, and professional contributions by A. Hargeby, A. Kullberg, and L.B.-M. Petersen, members of the Stream and Benthic Ecology Group. This paper has also benefited from contributions by and discussions with L. Colvin, K. Heikkinen, Prof. O. Sangfors, and C. Wesen.

REFERENCES

Aiken, G.R. 1985. Isolation and concentration techniques for aquatic humic substances. In: Humic Substances in Soil, Sediment, and Water, ed. G.R. Aiken, D.M. McKnight, R.L. Wershaw, and P. MacCarthy, pp. 363–386. New York: Wiley.

Albert, A. 1985. Selectivity Toxicity. New York: Chapman and Hall.

Andriulo, A.E., R.A. Rosell, and M.B. Crespo. 1987. Effect of tillage on organic matter properties of a soil of central Argentina. *Sci. Total Env.* 62:453–456.

Brakke, D.F., A. Henriksen, and S.A. Norton. 1987. The relative importance of acidity sources for humic lakes in Norway. *Nature* 329:432–434.

Choudhry, G.G. 1981. Humic substances. II. Photophysical, photochemical and free radical characteristics. *Tox. Env. Chem.* 4:261–295.

Cronan, C.S., and G.R. Aiken. 1985. Chemistry and transport of soluble humic substances in forested watersheds of the Adirondack Park, New York. *Geochim. Cosmo. Acta* 49:1697–1705.

De Haan, H. 1983. Use of ultraviolet spectroscopy, gel filtration, pyrolysis/mass spectrometry and numbers of benzoate-metabolizing bacteria in the study of humification and degradation of aquatic organic matter. In: Aquatic and Terrestrial Humic Materials, ed. R.F. Christman and E.T. Gjessing, pp. 165–182. Ann Arbor: Ann Arbor Science.

Driscoll, C.T., C.P. Yatsko, and F.J. Unangst. 1986. Longitudinal and temporal trends in the water chemistry of the north branch of the Moose River. *Biogeochem.* 2:347–371.

Ertel, J.R., et al. 1988. Genesis. In: Humic Substances and Their Role in the Environment, ed. F.H. Frimmel and R.F. Christman, pp. 105–112. Dahlem Konferenzen. Chichester: Wiley.

Forsberg, C., and R.C. Petersen. 1989. A darkening of Swedish surface waters due to increased humus inputs during the last 15 years. *Verh. Int. Verein. Limnol.* 24, in press.

Gjessing, E. 1976. Physical and Chemical Characteristics of Aquatic Humus. Ann Arbor: Ann Arbor Science.

Gorham, E., J.K. Underwood, F.B. Martinand, and J.G. Ogden, III. 1986. Natural and anthropogenic causes of lake acidification in Nova Scotia. *Nature* 324:451–453.

Hansch, C., A.R. Steward, S.M. Anderson, and D. Bentley. 1968. The parabolic dependence of drug action upon lipophilic character as revealed by a study of hyponotics. *J. Med. Pharm. Chem.* 11:1–11.

Horth, H., et al. 1988. Environmental reactions and functions. In: Humic Substances and Their Role in the Environment, ed. F.H. Frimmel and R.F. Christman, pp. 245–256. Dahlem Konferenzen. Chichester: Wiley.

Jacks, G., E. Olofsson, and G. Werme. 1986. An acid surge in a well-buffered stream. *Ambio* 15:282–285.

Jackson, T.A., and R.E. Hecky. 1980. Depression of primary productivity by humic matter in lake and reservoir waters of the boreal forest zone. *Can. J. Fish. Aq. Sci.* 37:2300–2317.

Johnson, N.M., G.E. Likens, M.C. Feller, and C.H. Driscoll. 1984. Letters: acid rain and soil chemistry. *Science* 225:1424–1425.

Jones, R.I., K. Salonen, and H. De Haan, 1988. Phosphorous transformation in the epilimnion of humic lakes: abiotic interactions between dissolved humic materials and phosphate. *Freshwat. Biol.* 19:357–369.

Krug, E.C., and C.R. Frink. 1983. Acid rain on acid soil: a new perspective. *Science* 221:520–525.

Leenheer, J.A. 1981. Comprehensive approach to preparative isolation and fractionation of dissolved organic carbon from natural waters and wastewaters. *Env. Sci. Tech.* 15:578–587.

Malcolm, R.L., E.T. Gjessing, and C. Bowles. 1989. Comparison of Nordic and Suwannee reference aquatic fulvic and humic acids. *Sci. Total Env.*, in press.

Mayer, F.L., and M.R. Ellersieck. 1986. Manual of acute toxicity: interpretation and data base for 410 chemicals and 66 species of freshwater animals. Resource Publication 160. Washington, D.C.: U.S. Dept. of the Interior, Fish and Wildlife Service.

Petersen, R.C., Jr., and A. Kullberg. 1985. The octanol/water partition coefficient of humic material and its dependence on hydrogen ion activity. *Vatten* 41(4):236–239.

Petersen, R.C., Jr., B.L. Madsen, M.A. Wilzback, C.H.D. Magadza, A. Paarlberg, A. Kullberg, and K.W. Cummins. 1987. Stream management. Emerging global similarities. *Ambio* 6:166–179.

Petersen, R.C., Jr., and U. Persson. 1987. Comparison of the biological effects of humic materials under acidified conditions. *Sci. Total Env.* 62:387–398.

Petersen, R.C., Jr., L.B.M. Petersen, U. Persson, A. Kullberg, A. Hargeby, and A. Paarlberg. 1986. Health aspects of humic compounds in acid environments. *Wat. Qual. Bull.* 11:44–52.

Schindler, D.W. 1988. Effects of acid rain on freshwater ecosystems. *Science* 239:149–157.

Schnitzer, M. 1980. Effect of low pH on the chemical structure and reactions of humic substances. In: Effects of Acid Precipitation on Terrestrial Ecosystems, ed. T.C. Hutchinson and M. Havas. New York: Plenum.

Thurman, E.M. 1985. Organic Geochemistry of Natural Waters. Dordrecht: Nijhoff/Junk.

Wright, R.F. 1988. Rain Project: role of organic acids in moderating pH change following reduction in acid deposition. Symposium on the Acidification of Waters in Kejimkujik National Park, Nova Scotia, Canada.

Wright, R., F.E. Lotse, and A. Semb. 1988. Reversibility of acidification shown by whole-catchment experiments. *Nature* 334:670–675.

Organic Acids in Aquatic Ecosystems
eds. E.M. Perdue and E.T. Gjessing, pp. 167–177
John Wiley & Sons Ltd

Alteration of Bioavailability and Toxicity by Phototransformation of Organic Acids

D.A. Francko

Department of Botany and Microbiology
Oklahoma State University
Stillwater, OK 74078, U.S.A.

Abstract. Dissolved humic material (DHM) in natural waters is capable of absorbing many important nutrients and toxicants, thus affecting the bioavailability of these substances and the productivity of lacustrine systems. Direct and indirect photodependent reactions involving DHM may alter the ability of this class of dissolved organic matter to sorb dissolved substances or may effect the release of bound constituents back to lake water. Phototransformations of DHM may also decompose high molecular weight DHM into lower molecule weight forms more easily assimilated by microflora. This chapter presents an overview of recent work on the importance of photodependent transformation mechanisms in lacustrine nutrient dynamics and toxicological phenomena.

INTRODUCTION

Dissolved humic material (DHM) comprises the bulk of the dissolved organic carbon (DOC) pool in lake water. Allochthonous DHM formed by the decomposition of terrestrial plant material and autochthonous DHM originating from algal and aquatic plant decomposition both contribute to the DHM pool in lake water. DHM represents a heteropolymeric class of compounds, with a predominance of aromatic ring structures in material of allochthonous origin and aliphatic subunits in autochthonous material (reviewed by Christman and Gjessing 1983). Unless otherwise noted, the designation "DHM" in this chapter will be used generically to include humic and fulvic acids. As a major constituent of the DOC pool, DHM transformations play a central role in the detrital food webs of lacustrine systems (e.g., the "microbial loop" concept of Azam et al. 1983).

However, the limnological significance of DHM transcends the role of these compounds as carbon and energy sources and sinks. Under certain conditions, DHM molecules sorb cations, anions, trace metals, nutrients, and organic contaminants, thus altering the bioavailability and biogeochemical cycling of these substances (reviewed by Steinberg and Münster 1985). The sorption of dissolved constituents by DHM can affect biota in one of three ways. If the sorption process removes an essential nutrient from solution and limits the accessibility of that substance to biotic assimilation, DHM may act as a growth-limiting substance. Alternatively, DHM sorption can increase the bioavailability of some substances (e.g., some trace metals) that are otherwise relatively insoluble in lake water, thereby stimulating production. Finally, DHM sorption can sequester toxicants (e.g., heavy metals or organic residues), rendering them less harmful to biota than unbound forms.

The rate functions that govern the sorption or complexation of dissolved constituents to DHM exist in dynamic equilibrium with mechanisms that act to release bound substances from the DHM moiety. The dynamics of binding and release of biologically important substances directly affects the productivity rate of aquatic primary producers, thereby influencing all higher trophic levels. Progress toward greater scientific understanding of the above dynamic equilibrium represented one of the major goals of this Dahlem Workshop.

This chapter will focus on photodependent DHM transformation mechanisms and their potential role in altering the bioavailability and toxicity of substances bound or complexed to DHM in lake water.

DHM PHOTOCHEMISTRY

DHM is removed from the water column by two physicochemical mechanisms: (*a*) adsorption to and coprecipitation with suspended particles and colloidal aggregates of calcium carbonate or hydrous metal oxides, and (*b*) UV-light-induced photodegradation. Microbial degradation is also responsible for DHM turnover, especially enzymatic conversions of high molecular weight DHM to low molecular weight compounds that can subsequently be used as carbon sources (reviewed by Chróst et al. 1989). The second physicochemical mechanism above, UV-induced photodegradation, not only cleaves large DHM "molecules" into smaller subunits, thereby facilitating their utilization by microorganisms, but also affects the ability of DHM to sequester and release bound constituents.

Sunlight that penetrates the surface of a water body is variously absorbed by water itself and by dissolved and particulate constituents. DHM, in contrast to pure water, strongly absorbs light in the UV and near-UV portion of the spectrum (ca. 230–375 nm) so that in darkly stained lakes short wavelengths of incoming sunlight are completely absorbed within the top few centimeters of the water surface (Wetzel 1983). Upon absorption of light, DHM moieties exhibit fluorescence emissions in wavelengths characteristic of their molecular structure (reviewed in Christman and Gjessing 1983).

UV and near-UV light absorbed by DHM can initiate a series of direct and indirect photochemical reactions. Direct reactions involve changes in DHM or DHM-associated constituents upon exposure to light. For example, exposure to light can "bleach" colored DHM, thus changing its structure. DHM can also be photooxidized into smaller degradation subunits, a process that may greatly accelerate the utilization of DHM by microbes (reviewed by Steinberg and Münster 1985). The relative importance of photodependent degradation versus microbial enzymatic breakdown of high molecular weight DHM to smaller subunits in natural aquatic systems remains an open question, but light clearly plays a role in organic carbon transfer from DHM into heterotrophic microbes.

Under the heading of indirect photochemical events, irradiated humic materials produce a host of reactive oxygen species including singlet oxygen, hydrogen peroxide, peroxide radicals, and other incompletely characterized energy donor and electron acceptor transients (reviewed by Zika and Cooper 1987). Although the subject will not be discussed in depth in this review, DHM has been implicated in the photodependent reactions that form trihalomethanes from chlorine used to disinfect water supplies (reviewed by Christman and Gjessing 1983).

The chemically reactive intermediates produced by interactions between DHM and sunlight may have profound effects on other dissolved substances affecting biota. For example, organic contaminants with redox potentials lower than those of DHM-produced reactive transients above may be photooxidized and inactivated. Fischer et al. (1987) suggested that DHM in natural waters may serve as a redox "buffering" system analogous to the bicarbonate pH buffering system.

In this review, discussion and speculation will focus on two classes of DHM interaction that have received the most attention from researchers in recent years: (*a*) the dynamic interactions between DHM and key micronutrients, especially phosphorus, and (*b*) interactions between DHM and metal ions that alter metal availability and toxicity.

PHOSPHORUS CYCLING

The cycling of phosphorus in epilimnetic waters, characterized by the biotic assimilation of ionic orthophosphate (ortho-P; PO_4-P), the transfer of P between various biologically important dissolved forms in lake water, and the release of P-containing compounds back to lake water by cellular release or decomposition, remains an incompletely understood phenomenon of considerable limnological significance. Ambient ortho-P concentrations are often insufficient to support planktonic demand (reviewed by Wetzel 1983; Cembella et al. 1984), so attention has focused on processes that might regenerate ortho-P from various complexed P pools present in lake water, especially colloidal P and P esters (reviewed by Cembella et al. 1984; Francko 1986). Recent work indicates that DHM may interact with the epilimnetic P cycle at several control points.

The concentration of DHM may directly affect the rate at which ionic ortho-P can be taken up by planktonic cells. In work on Canadian Shield lakes, Brassard and Auclair (1984) found that planktonic ortho-P uptake rates were directly mediated by DHM in the 1,000–10,000 dalton molecular weight range. In experiments on four tropically distinct lakes and reservoirs in Oklahoma, Francko (1986) found that radiolabeled $^{32}P\text{-}PO_4$ uptake in planktonic particles could be dramatically repressed or stimulated by the addition of small amounts of *Typha* DHM (14–70 mg/l), alone and in concert with ferric iron addition (0.1 or 1.0 mg/l). The degree of stimulation or repression was apparently particle size class and lake specific.

DHM also affects the ability of plankton to access ortho-P through the alkaline phosphatase (AP)-mediated hydrolysis of phosphomonoester (PME) substrates. Stewart and Wetzel (1982) reported that exogenously applied DHM markedly enhanced AP activity in natural bacterial/algal assemblages from Lawrence Lake, Michigan, especially under low light regimes. Their work indicated that low molecular weight DHM was more stimulatory than high molecular weight moieties. DHM stimulation of AP has also been documented by other workers (reviewed by Jones et al. 1988). Francko (1986) reported that exogenously applied DHM either stimulated or repressed planktonic-associated and -dissolved AP, depending on the size class of particles considered and the lake from which samples were collected. Stewart and Wetzel hypothesized that DHM may affect the AP system either through the sequestering of PME compounds or through stimulation of bacterial or algal growth, both of which would increase the competition for P and stimulate AP synthesis. Francko (1986) described evidence that contradicts the former thesis in that DHM addition to whole water samples from four Oklahoma lakes resulted in increased PME concentrations in lake water filtrates after 1 h of incubation.

Perhaps the most well-studied interaction between DHM and phosphorus is the complexation and sequestering of ortho-P by DHM, especially in the presence of iron. Numerous studies have shown that under conditions of low pH and low redox potential DHM, ferric iron, and ortho-P form colloidal aggregates and that P bound to these aggregates may be unavailable to biota through direct assimilation (Francko and Heath 1979, 1982, 1983; reviewed by Steinberg and Münster 1985 and Jones et al. 1988). Magnesium may also interact with DHM and P to form a colloidal aggregate (Steinberg and Baltes 1984), although DHM alone is apparently incapable of significant

P binding. In darkly stained, iron-rich natural waters, the majority of ortho-P may be sorbed to high molecular weight aggregates into a form that is largely unreactive to acid molybdate reagents.

The physicochemical nature of metal–P–DHM binding is not understood, nor has the relationship between DHM–P complexes and non-DHM colloidal P (e.g., calcium carbonate colloids or algal microfibrils [reviewed by Cembella et al. 1984]) been examined. A linear relationship between DHM color ($A_{400\ mm}$) and P complexations has been reported (Francko and Heath 1982) for one bog system; however these results have not been generalized by other workers. Stewart and Wetzel (1981) attempted to estimate the affinity of naturally occurring DHM for ^{32}P-PO_4 and were unable to demonstrate movement of P to DHM aggregates. However, as pointed out by the authors, high levels of calcium carbonate and low levels of soluble iron in their hardwater, alkaline study lake (Lawrence Lake, Michigan) may have interfered with the ortho-P binding process because calcium carbonate colloids also bind P (Wetzel 1983).

The most direct evidence for DHM–iron–P complexation in stained waters has been presented by De Haan et al. (1990) in work on several lakes in southern Finland. Using a double isotype labeling technique these workers showed that, in the presence of DHM, radiolabeled Fe and PO_4 added to epilimnetic samples simultaneously became incorporated into a fraction with a nominal molecular weight of 10,000–20,000 avograms. In the absence of DHM, complexation did not occur. By disturbing the equilibrium between the isotopic concentrations and DHM, labeled PO_4 could be released from its high molecular weight complexed form via a presumed displacement reaction. Whether these complexes were photosensitive was not investigated.

The importance of DHM–P complexes as both a source and sink for ortho-P has generated considerable interest. A generally accepted literature model suggests that ortho-P could be released from high molecular weight bound moieties via a displacement reaction involving a low molecular weight P ester (reviewed in Wetzel 1983). Alternatively, as suggested by De Haan et al. (1990), ortho-P may be released from DHM complexes by another class of displacement reactions. Other researchers, however, have reported that P release may also be mediated by abiotic photodependent reactions.

Francko and Heath (1979) reported that P co-chromatographing with high molecular weight DHM (ion exchange and gel permeation) could be converted to ionic ortho-P upon short-term, low-dose exposure to UV light. This phosphate fraction, termed UV-sensitive P, comprised roughly one-half to two-thirds of the total soluble phosphorus pool in the moderately stained bog study site (Crazy Eddie Bog, Ohio) and was not reactive to acid molybdate reagents or alkaline phosphatase hydrolysis. Irradiation of filtered water samples with an artificial UV light source delivering radiation intensities (ca. 1.4×10^1 J $m^{-2}h^{-1}$) an order of magnitude less than that reaching the surface of a lake on a cloudless summer day induced a linear decrease in DHM absorbance (87% loss) and quantitative conversion (>98%) of UV-sensitive P to ortho-P within 3 h.

In another study on this bog lake, Francko and Heath (1982, 1983) reported evidence consistent with the view that ferric iron was also associated with UV-sensitive P, and that UV light may release P from its complexed form by photoreducing iron to the ferrous state, thereby altering the ability of the DHM–Fe complex to bind P. In this study, the photoreduction process and release of P was reversible: both UV-sensitive P and ferric iron were regenerated when irradiated samples were incubated in the dark. The turnover time for UV-sensitive P release and iron photoreduction in water samples irradiated in the laboratory with either an artificial UV source or by natural sunlight irradiation was about 1 h, even though sunlight delivered far more energy (ca. 2.1×10^2 J $m^{-2}h^{-1}$) to the surface of samples than the artificial UV lamp. This reversible ferric iron photoreductive process has been observed in other waters (Brezonik and Miles 1981). The plausibility of this mechanism in nature was supported by a diel study conducted on water from the study lake in which the concentration of DHM and UV-sensitive P was markedly higher during the nighttime hours than during daylight. In addition to the postulated iron photoreduction release reaction above, it is possible that UV light disrupts the phenolic groups of DHM in the colloid, thereby reducing the affinity of iron for P (Steinberg and Münster 1985).

Much remains to be learned about the role of photodependent P regenerative mechanisms in lake water. Most importantly, very few studies have examined natural waters for the presence of UV-sensitive P, so that the overall importance of this proposed source of planktonic P cannot yet be evaluated. Jones et al. (1988) found that $^{32}P\text{-}PO_4$ added to water from small Finnish lakes became adsorbed to two higher molecular weight fractions (>100,000 and 10,000–20,000 daltons), especially in waters high in DHM and iron. Like the above studies, these bound fractions were largely unreactive to acid molybdate reagents. Irradiation with sunlight resulted in only a modest release of P from bound forms. However, the authors indicated that their photorelease values may have been reduced by the presence of plate glass windows between incoming sunlight and their samples and the presence of azide added to inhibit nannobacterial metabolism. Francko and Heath (1983) examined the fate of $^{32}P\text{-}PO_4$ added to water samples from Triangle Bog, Ohio, and found that while added P sorbed to a high molecular weight fraction and could be released by UV treatment, the released P was chromatographically distinct from ortho-P. Francko (1986) examined waters from four trophically distinct, unstained, alkaline lakes and reservoirs in Oklahoma and found that while UV-sensitive P could be detected in these systems, the amounts were relatively small (e.g., a few

$\mu g\ l^{-1}$). However, even these small amounts of P released could be ecologically relevant in that increases in soluble ortho-P after irradiation ranged from 10–33%. There was no definable relationship between P release, DHM content, and ferric iron photoreduction, in contrast to linear relationships between these variables described earlier. This study supported the view that UV-sensitive P could occur in unstained waters low in iron, as well as iron-rich bog lakes.

In short, work on modeling the many potential influences of DHM on epilimnetic phosphorus cycling remains in its infancy. The most frequently cited models presented in the literature have proposed several ortho-P regenerative mechanisms that are viewed as being mutually exclusive (reviewed by Wetzel 1983). For example, data sets reported by some authors proposed that colloidal displacement reactions represent the only major source of ortho-P regeneration from complexed sources. Other workers, investigating different systems, have proposed that enzymatic hydrolysis of relatively low molecular weight P esters may be of overriding importance. As described here, photodependent transformations involving DHM-Fe-sequestered P have also been reported.

In proposing the "continuum" model of epilimnetic P recycling (Fig. 1), Francko (1986) attempted to reconcile these seemingly divergent views by suggesting that there is no single "epilimnetic phosphorus cycle." In this model, the three major sources of complexed P (PME, non-DHM colloidal P, and DHM-Fe-colloidal P) and major ortho-P regenerative mechanisms (AP activity, colloidal displacement, and UV light) are seen to co-occur in natural waters. The "position" of a given system within the continuum, i.e., those P sources and regenerative mechanisms that are most important in overall P dynamics, would be dynamic rather than fixed in time and influenced by changes in biotic (e.g., community structure) and physicochemical parameters. In this construct, DHM and iron concentrations represent two of the most important physicochemical parameters that could influence the position of the P cycle in the continuum. Francko (1986) provided support for this model with preliminary evidence that changes in the DHM and Fe content of water samples altered biotic ortho-P uptake rates, the relative importance of P regenerative mechanisms, and the pool sizes of various P sources in four lake and reservoir systems in the southcentral United States.

The continuum model may provide a framework useful for organizing future studies on the influence of DHM P recycling. Many fundamental questions remain. As noted earlier, direct photolysis of DHM yields smaller subunits, and we know nothing about the role of such subunits in P recycling. We do not know how UV light absorption changes the structure of DHM. Though P uptake by biota appears to be influenced by DHM, the underlying mechanism for this putative interaction remains unknown as well as

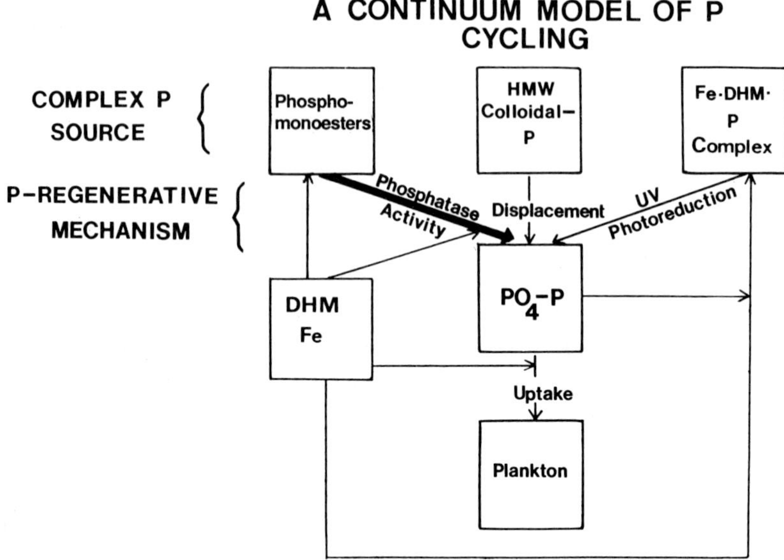

Fig. 1—A conceptual diagram of the continuum model of epilimnetic P cycling. In this model, three groups of complexed phosphorus sources and P-regenerative mechanisms may co-occur in a given lake system. The relative importance of each regenerative mechanism is not fixed in time, but rather dependent on dynamic changes in physicochemical and biotic parameters. The example given depicts a lake in which AP hydrolysis of PME substrates represents the major P recycling mechanism. Central to this chapter is the notion that DHM not only sequesters P into a complexed form but also interacts through unknown mechanisms with AP activity and ortho-P uptake, as well as with the pool size of PME. In this manner and perhaps other interactions not shown in the model, DHM controls multiple aspects of the P cycle, determining the position of a lake in the hypothetical continuum.

the potential effect of light irradiation. The mechanistic and ecological relationships, if any, between colloidal, non-DHM P compounds and DHM–Fe–P remain unknown. UV light cannot penetrate very far into a water column before being absorbed by organic matter. Are photodependent phenomena thus confined to the top few centimeters of the water column, or do circulation patterns within the epilimnion allow such processes to impact the whole epilimnion? Finally, we are still largely ignorant of the basic photochemical reactions that govern both the sorption and release of P from DHM complexes, and whether ortho-P is the only form of P sorbed to these complexes. Clearly, aquatic chemists can contribute much to the evaluation of the above binding and release processes.

METAL AND ORGANIC CONTAMINANT COMPLEXATION

Although the previous section dealt with the interaction between DHM and iron, other species of trace metals are also complexed by DHM. DHM–metal complexation leads to two general phenomena. Complexation may either reduce or enhance the bioavailability of metals required as planktonic nutrients. Alternatively, DHM can complex with toxic metals, rendering them more or less toxic to biota.

Iron uptake in algae is enhanced by the presence of DHM, which apparently prevents the precipitation of insoluble iron that occurs without DHM (Anderson and Morel 1982). In this sense, binding enhances the bioavailability of iron to epilimnetic biota. However, Jackson and Hecky (1980) provided contradictory evidence that DHM may depress phytoplankton productivity by making iron less available for assimilation.

Several studies have indicated that copper toxicity in a wide variety of aquatic plants and animals is reduced by complexation with naturally occurring DHM (reviewed by Nor 1987). Hongve et al. (1980) found that zinc, lead, and mercury were detoxified in bog waters. Cadmium toxicity in the alga *Selenastrum* was markedly reduced by the addition of DHM to growth media (Sedlacek et al. 1983). However, in at least one long-term chronic bioassay, DHM was shown to increase the toxicity of cadmium to *Daphnia*, especially in hard water (Winner 1986).

There are no reports in the literature on the potential effects of photodependent processes on either the binding, release, or toxicity of the above metals. If, as indicated earlier, direct or indirect photochemical events either lyse DHM into smaller subunits or produce photochemical energy donor or electron acceptor transients, sunlight energy could have a significant effect on DHM–metal interactions.

Comparatively little is known about the photodependent, DHM-mediated decomposition of toxic organic compounds in natural waters. Photochemical intermediates formed upon the exposure of DHM to sunlight have been implicated in the photolytic breakdown of polyaromatic hydrocarbons from oil spills (Fischer et al. 1987). Liu et al. (1983) reported that DHM could photocatalyze the degradation of a wide variety of synthetic organic compounds. Conversely, Shimp and Pfaender (1985) conducted chemostat experiments on the effects of long-term exposure of natural microbial communities to additions of DHM and found that when organisms adapted to a high-DHM environment they were less able to degrade a variety of monosubstituted phenol derivatives. Work on the mechanisms involved in the above interactions and their overall ecological significance remains in the early stages.

CONCLUSIONS AND FUTURE DIRECTIONS

Work on DHM and its potential importance to lacustrine dynamics began over 50 years ago (reviewed in Steinberg and Münster 1985). However, it has only been in recent years that limnologists have been able to use modern techniques to probe specific interactions involving DHM. It is clear that photochemical processes may play an extremely important role in planktonic productivity. DHM in the presence of iron and perhaps other metals sorbs critical nutrients, and photolytic release of these bound nutrients represents a key mechanism by which these materials can be made available for biotic utilization. DHM may well be involved in other indirect photochemical processes that have not yet been discovered.

It is also clear, however, that DHM photochemistry operates in a complex system. The concentration and types of DHM, the amount and quality of UV irradiation, water pH, alkalinity, temperature, and a host of other physicochemical variables all appear to affect the interaction between DHM, associated metals, and sorbed species. Further work is clearly required on the stability, reactivity, and photosensitivity of DHM and DHM associations in unaltered lake water. Additional work is also required to characterize more fully those DHM moieties within the whole DHM fraction in natural waters that are most responsible for significant ecological processes. Continued application of sensitive analytical and radiotracer techniques should resolve some of the "next questions" suggested within the body of this paper.

REFERENCES

Anderson, M.A., and F.M.M. Morel. 1982. The influence of aqueous iron chemistry on the uptake of iron by the coastal diatom *Thalassiosira weissflogii*. *Limnol. Ocean.* 27:789–813.

Azam, F., T. Fenchel, J.G. Gray, L.A. Meyer-Reil, and F. Thingstad. 1983. The ecological role of water column microbes in the sea. *Mar. Ecol. Prog. Ser.* 10:257–263.

Brassard, P., and J.C. Auclair. 1984. Orthophosphate rate constants are mediated by the 10^3–10^4 molecular weight fraction in Shield lake waters. *Can. J. Fish. Aq. Sci.* 41:166–173.

Brezonik, P.L., and C.J. Miles. 1981. Oxygen consumption in humic-colored waters by a photochemical ferrous–ferric catalytic cycle. *Env. Sci. Tech.* 15:1089–1095.

Cembella, A.D., N.J. Anita, and P.J. Harrison. 1984. The utilization of inorganic and organic phosphorus compounds as nutrients by eukaryotic microalgae: a multidisciplinary perspective. Part 1. *CRC Crit. Rev. Microbiol.* 10(4):317–391.

Christman, R.F., and E.T. Gjessing. 1983. Aquatic and Terrestrial Humic Materials. Ann Arbor: Ann Arbor Science.

Chróst, R.J., U. Münster, H. Rai, D. Albrecht, K. Wetzel, and J. Overbeck. 1989. Photosynthetic production and exoenzymatic degradation of organic matter in the euphotic zone of a eutrophic lake. *J. Plank. Res.* 11:223–242.

De Haan, H., R.I. Jones, and K. Salonen. 1990. Abiotic transformations of iron and phosphate in humic lake water, revealed by double isotopic labelling and gel

filtration. *Limnol. Ocean.* 35, in press.
Fischer, A.M., J.S. Winterle, and T. Mill. 1987. Primary photochemical processes in photolysis mediated by humic photochemistry of environmental aquatic systems. In: Photochemistry of Environmental Aquatic Systems, ed. R.G. Zika and W.J. Cooper, pp. 141–173. Washington, D.C.: American Chemical Society.
Francko, D.A. 1986. Epilimnetic phosphorus cycling: influence of humic materials and iron on coexisting major mechanisms. *Can. J. Fish. Aq. Sci.* 43:302–310.
Francko, D.A., and R.T. Heath. 1979. Functionally distinct classes of complex phosphorus compounds in lake water. *Limnol. Ocean.* 24:463–473.
Francko, D.A., and R.T. Heath. 1982. UV-sensitive complex phosphorus: association with dissolved humic material and iron in a bog lake. *Limnol. Ocean.* 27:564–569.
Francko, D.A., and R.T. Heath. 1983. Abiotic uptake and photodependent release of phosphate from high-molecular-weight humic-iron complexes in a bog lake. In: Aquatic and Terrestrial Humic Materials, ed. R.F. Christman and E. Gjessing, pp. 467–480. Ann Arbor: Ann Arbor Science.
Hongve, D., O.K. Skogheim, A. Hindar, and H. Abrahamsen. 1980. Effects of heavy metals in combination with NTA, humic acid, and suspended sediment on natural phytoplankton photosynthesis. *Bull. Env. Contam. Tox.* 25:594–600.
Jackson, T.A., and R.E. Hecky. 1980. Depression of primary productivity by humic matter in lake and reservoir waters of the boreal forest zone. *Can. J. Fish. Aq. Sci.* 37:2300–2317.
Jones, R.I., K. Salonen, and H. De Haan. 1988. Phosphorus transformations in the epilimnion of humic lakes: abiotic interactions between dissolved humic material and phosphate. *Freshwat. Biol.* 19:357–369.
Liu, D., J. Carey, and K. Thomson. 1983. Fulvic-acid-enhanced biodegradation of aquatic contaminants. *Bull. Env. Contam. Tox.* 31:203–207.
Nor, Y.M. 1987. Ecotoxicity of copper to aquatic biota: a review. *Envir. Res.* 43:274–282.
Sedlacek, J., T. Källqvist, and E. Gjessing. 1983. Effect of aquatic humus on uptake and toxicity of cadmium to *Selenastrum capricornutum* Printz. In: Aquatic Terrestrial Humic Materials, ed. R.F. Christman and E.T. Gjessing, pp. 495–516. Ann Arbor: Ann Arbor Science.
Shimp, R.J., and F.K. Pfaender. 1985. Influence of naturally occurring humic acids on biodegradation of monosubstituted phenols by aquatic bacteria. *Appl. Env. Microbiol.* 49:402–407.
Steinberg, C., and G.F. Baltes. 1984. Influence of metal compounds on fulvic acid/molybdenum blue reactive phosphate associations. *Arch. Hydrobiol.* 100:61–71.
Steinberg, C., and U. Münster. 1985. Geochemistry and ecological role of humic substances in lakewater. In: Humic Substances in Soil, Sediment and Water: Geochemistry, Isolation and Characterization, ed. G.R. Aiken, D.M. McKnight, R.L. Wershaw, and P. MacCarthy, pp. 105–145. New York: Wiley.
Stewart, A.J., and R.G. Wetzel. 1981. Dissolved humic materials: photodegradation, sediment effects and reactivity with phosphate and calcium carbonate precipitation. *Arch. Hydrobiol.* 92:265–286.
Stewart, A.J., and R.G. Wetzel. 1982. Influence of dissolved humic materials on carbon assimilation and alkaline phosphatase activity in natural algal-bacterial assemblages. *Freshwat. Biol.* 12:369–380.
Wetzel, R.G. 1983. Limnology, 2nd ed. Philadelphia: Saunders.
Winner, R.W. 1986. Interactive effects of water hardness and humic acid on the chronic toxicity of cadmium of *Daphnia pulex*. *Aquat. Tox.* 8:281–293.
Zika, R.G., and W.J. Cooper. 1987. Photochemistry of Environmental Aquatic Systems. Washington, D.C.: American Chemical Society.

Organic Acids in Aquatic Ecosystems
eds. E.M. Perdue and E.T. Gjessing, pp. 179–187
John Wiley & Sons Ltd

Mechanisms and Effects of Reactions of Organic Acids with Anions

E.T. Gjessing

Norwegian Institute for Water Research
NIVA
P.O. Box 68
0808 Oslo 8, Norway

Abstract. In the natural environment, organic acids (OA) appear as aggregated OA units with a negatively charged surface. The degree of aggregation depends on several factors, such as concentration, pH, and ionic strength.

The aggregated OA have hydrophobic and hydrophilic parts and an uncharged interior into which neutral solutes may be incorporated. Anions as such cannot associate with OA. In order to couple with OA, negatively charged solutes have to "complex" to become neutral or positively charged.

Under acidic conditions sulfate apparently reacts with OA and results in a "new" OA with various other properties.

INTRODUCTION

The idea behind the title of this chapter was based on some preliminary results that indicated an interaction between organic acids (OA) in water and sulfate in acid rain-affected areas (Gjessing et al. 1988). A general literature search revealed that little has been published about the association between OA and negatively charged solutes. This chapter has therefore been given a somewhat broader frame than originally planned, with an introduction emphasizing size and shape properties under natural and artificial conditions. Due to the lack of information on the subject, some presently unpublished results are reported and discussed.

It is essential, when discussing and clarifying the role of OA in the aquatic ecosystem, to define as well as possible what we are dealing with. In this chapter OA are limited to organic compounds in water which have a terrestrial origin. This excludes metabolic compounds from organisms as well as extracts and decomposition products of aquatic animals and plants. Even though this limitation includes water extracts from terrestrial plant material, these are generally water soluble compounds that are fairly bioavailable, thereby having a relatively short detention time in the aquatic phase. In addition, in most water they represent only a small part of the dissolved organic carbon.

This leaves us with the most stable form of *terrestrially originated OA*. The size and shape of these OA, as they appear in the aquatic environment, are important issues still disputed in the literature.

In relation to the key questions for this workshop, which are concerned with variability in properties, a size/shape model would be of great help.

It is important to emphasize that the physical and chemical characterization of constituents in water generally have a biological purpose. This means that the membrane penetration potentials of OA are important to elucidate, because the effects and consequences to the biota often depend on these properties. The size/shape relationships are therefore important for predicting and explaining biological effects.

It has been suggested that OA in water have a spherical form, made by coiled chains (R. Swift, pers. comm.). R.L. Malcolm (pers. comm.) states that the molecular weights/size of OA in water range between 1000 (or less) and 3000 daltons (dalton = an atomic mass unit equivalent to 1/12 of the mass of a carbon atom). Prior to these relatively recent reports on size, which are mostly based on drastic pretreatment of the sample before applying the various physical and chemical methods, the size was thought to be at least one order of magnitude higher. The latter results are generally based on more natûral, conditional methods.

Regardless of the discrepancy in the reported size of OA in water, Wershaw (1986) and others favor the view that we are dealing with organic substances which are based on a relatively small compact unit. The degree of multiplicity will depend on the *composition of the ambient water* and *the "age" of the OA in this element* (water).

The unit proposed by Wershaw (1986) is a nonlipid, amphiphile OA (having both hydrophobic and hydrophilic parts) with carboxylic groups and un-ionized polar groups. He proposed that there is a dynamic equilibrium between this "monomeric" unit and micellar aggregates in the natural environment. Even though this model is for soil and sediments, it fits well when explaining many of the observed properties and "reactions" of OA in water. The distribution between the unit molecule and aggregate *will be a function of the concentration of the OA and the pH and ionic composition of the ambient water*.

Physical/chemical work on characterizing the OA unit has made much progress during recent years. The OA unit is relatively uniform, composed of C, O, H, N, and S, with a number of other associated elements such as Ca, Fe, Zn, Br, Sr, Ba, Ce, Ag, Cr, As, La, Co, Sb, Th, Sm, Se, Sc, Eu, Hg, and Cs (Riise and Salbu 1989).

Given an OA unit in the range of 1,000 to 3,000 daltons, it is easy to accept and understand the variation in size, as there are numerous possibilities for these organic substances to decompose or degrade in the aquatic environment on their way from the top of the catchment to the sea. Both photooxidation and microorganisms play a role in this connection. The aggregate theory is, however, disputed. E. Tipping (pers. comm.) states that even though some of the reported molecular weight (MW) data (based on vapor pressure osmometry with high concentrations of the material) show MW as low as 1,000, there is little support for the aggregation theory.

It is important to realize, however, that most methods used in describing the chemical and physical nature of OA in water are based on different kinds of pretreatment, such as isolation, acidification, and concentration. This means that the OA units discussed above are "determined" after a relatively drastic change in conditions from the natural environment. With Wershaw's aggregate theory in mind, it is essential to realize that in natural systems the biological properties in particular are different from what we observe in the laboratory. As indicated above, the nature, shape, and size of the OA in water depend on its age in the water phase; the degree of aggregation will vary with its concentration and the general composition of the ambient water. The stability of the aggregates, therefore, will have a spatial and temporal variability.

Considering a different-size OA aggregate in the micelle-like form (with "sites" for built-in hydrophobic agents, cations, and metal complexes), it is easy to understand the discrepancy in the reported metal complex constants, charge density, and acidic properties.

From a biological effect perspective, a consequence of the model considerations of OA in water, as outlined above, may be that *physical/chemical characterization* (where relatively drastic pretreatments of the sample are involved) *is of limited value*!

ASSOCIATION WITH UNCHARGED AND NEGATIVELY CHARGED SOLUTES

Organic (Polar and Nonpolar)

Lipophilic, nonpolar substances, such as DDT, are held in the interior (hydrophobic parts) of the OA aggregate (Wershaw 1986). Carter and Suffet (1985) found that the binding constant between OA and DDT varied substantially (tenfold) with the type of OA. They also found that the binding constant between OA and various lipophilic organic micropollutants (OMP) differed over several orders of magnitude. The more lipophilic, the higher the binding constant. Reports from Carter and Suffet (1985) and MacCarthy and Jimenez (1985) suggest a positive correlation between the octanol/water partition coefficient of a compound (lipophilicity) and the binding constant for OA.

Wershaw et al. (1969) found that the presence of NaCl together with DDT did not alter the solubility of DDT. Carter and Suffet similarly found that neither the H^+ concentration nor Ca was important for the binding constant. Wershaw et al. concluded from their observations that the solubilization of DDT is not due to coulombic attraction of DDT to the highly charged surface of the aggregate/micelles, but to partitioning into the hydrophobic interiors of the OA.

The association between OA and OMP may be so strong that the OMP becomes analytically unavailable (Gjessing and Berglind 1981, 1982). The OA–OMP complexes are, however, reported to be biologically unavailable (MacCarthy and Jimenez 1985; Carlberg et al. 1986).

Organic (Anions)

Under natural pH conditions most of the OA in water have a net negative charge. According to Gjessing and Gjerdahl (1975) more than 80% of the UV-absorbing matter in water have isoelectric points below pH 2 (distributed into 3 points: 1.25, 1.50, and 1.75). The residual 20% apparently have isoelectric points within the pH range of 2–12. As a result, essentially all OA will act as anions under natural pH conditions.

However, there have been reports on an OA–anion association. De Haan and De Boer (1978) have demonstrated the presence of amino acids in the high MW gel filtration fractions (Sephadex G-25) of OA. The distribution of the amino acids, however, is strongly pH dependent. They concluded that more than 70% of the amino acids present are very weakly bound to the OA.

Inorganic

Anions in water, such as Cl, Br, S, N, and P, have all been identified as part of the OA. However, it is not known in which form these elements are present.

Cl, Br, N. Wigilius et al. (1988) have shown, by adding ^{35}Cl in the form of NaCl to water with OA, that adsorbed organic chlorine (AOCl) does not increase to any significant degree. They found, however, that natural organic acids (NOA) do contain a significant amount of AOCl. The source of this chlorine, the mechanisms for its formation, and the form in which it is present are not known, but the source may be halogen-containing organic micropollutants from atmospheric deposition (Strachan 1988) incorporated into the OA aggregate. This would then imply that there are apparently no sites in OA that will react with chloride ions.

Riise and Salbu (1989) have found relatively high concentrations of bromine (≈200 μg Br/g ⊃: ≈4 μg Br/l) in the Nordic reference fulvic acid and standard fulvic acid of the International Humic Substances Society (IHSS). As with Cl there is little knowledge about the source, formation, and form of this bromine.

With regard to nitrogen in relation to OA, R.L. Malcolm (pers. comm.) found about 0.7 mg N/100 mg in the reference and standard fulvic acids of the IHSS. This is equivalent to 150 μg N in the original waters. Most likely some of this is amino acid N, which is incorporated in the interior of the OA. Standard methods for the determination of nitrate in water, performed on aqueous solutions of IHSS Nordic and standard fulvic acids, suggest about 10 μg NO_3/l (E.T. Gjessing, unpubl. data). At present, information on the association between OA and this N, which presumably is NO_3, is lacking.

P/PO₄. The phosphorus content in the IHSS reference and standard fulvic acids is reported to be relatively low: < 0.05% (R.L. Malcolm, pers. comm). Gjessing (1976) found total phosphate in the high molecular weight OA fraction to be 15 μg/l (1 μg PO_4/mg C).

Jones et al. (1988) have shown that phosphate (^{33}P) interacts with the high molecular weight and medium molecular weight OA (using Sephadex). Their results imply that the kinetics differ, indicating at least two different mechanisms of association. The interaction is stimulated by addition of Fe and appears to be higher in water rich in OA. The high molecular weight bound PO_4 is also less analytically available using the molybdate method. The apparently less available PO_4 that is connected to the high molecular weight OA, and the positive relationship between "uptake" and concentration of OA, may suggest that some of the PO_4 is incorporated into the aggregated OA micelle.

The findings of Jones et al. (1988) do not allow us to conclude that the PO_4 anion as such interacts with OA. The positive influence of Fe on the uptake into the aggregated OA favors the view that the negative charge of the PO_4 is "neutralized" by first "complexing" with hydrated iron.

SO₄. More than 30 years ago, Shapiro (1957) reported that yellow OA in lake water contained no sulfur. Today, most reports on element composition of OA demonstrate a sulfur content of the same order of magnitude as nitrogen, ranging between 0.5 and 1%. The IHSS Nordic reference and standard fulvic acids contain 0.64% and 0.55%, respectively. Regarding the Nordic fulvic acid, it represents 24% of the sulfur in the "whole water" (NIVA-concentrate).

Regardless of whether there are spatial or temporal differences or whether the discrepancy between Shapiro's data and the data of the IHSS Nordic reference and standard fulvic acid is due to improvement in analytical techniques, it is tempting to direct our attention towards the increase in atmospheric deposition of sulfuric acid during the last several decades.

Gjessing et al. (1988) have demonstrated that the analytically available SO_4, after equilibrium dialysis of water from an acidified area, is higher in this water compared to water with OA from a less acidified area. Further, they found that the content of "organic sulfur," which is the difference in sulfate determined before and after removal of the organic matter by UV/H_2O_2 oxidation, is significantly higher in water from the acidified area: 0.08 mg SO_4/mg C compared to 0.04 mg SO_4/mg C. This work also demonstrates in a series of experiments that SO_4 may interact with OA when acidified with H_2SO_4. It is important to note the following:

1. Results comparing the SO_4 content in artificially acidified water (HCl, HNO_3, H_2SO_4), before and after the removal of organic matter with UV/H_2O_2 oxidation suggest an analytically unavailable SO_4 fraction in the H_2SO_4-treated water sample.
2. A significant part of the analytically available SO_4, determined by both colorimetry and ion chromatography, is found in the high molecular weight fraction using gel filtration (Sephadex G-10). The experiments all suggest that sulfuric acid reacts with OA to form some organically associated sulfate (sulfur). The degree of association may depend on the reaction conditions (time, temperature).

EFFECTS AND CONSEQUENCES

The effects and environmental consequences of the complexation between OA and metal, and with organic micropollutants, are increasingly being discussed in the literature. With regard to association products between OA and anions, little is known. Knowledge about the chemical and biological properties is generally lacking.

Two of the major pollution problems that exist today, relative to surface water, are *eutrophication* and *acidification*. Along with eutrophication, the occurrence of "new" toxin-producing species of algae is of increasing concern (see Steinberg, this volume).

With regard to the effect of acid precipitation on surface water, the acid that presently receives the most attention is sulfuric acid. In this section the discussion on effect will be limited to the potentials of reactions between OA and PO_4 and H_2SO_4.

Phosphate

The classic models of Vollenweider (1968) for predicting phytoplankton biomass production based on external nutrient load of elements (such as P) is reported to be inappropriate for dystrophic lakes. Such observations have induced speculation that OA may influence an algal–phosphorus interaction, either directly or indirectly (Steinberg and Münster 1985).

Jones et al. (1988) reviewed the present knowledge of phosphorus transformation in humic lakes and discussed the hydrobiological and environmental potential of the interactions between OA and phosphorus. It should be emphasized that the association between OA and PO_4 is very "dynamic." In their article, Jones et al. state that Fe promotes the association between OA and PO_4, and that UV radiation may enhance the release of PO_4 and its bioavailability.

Both in relation to acidification, with an increased content of Fe in surface water, and in relation to the anticipated increase in global UV radiation, these "OA–PO_4 complexes" might be important to the understanding of the bloom of unwanted algal species.

Sulfate

OA-containing dialyzed water samples from areas receiving acid precipitation are apparently more toxic to Atlantic salmon than similarly treated water from less acidified regions. To remove excess mineral acids, OA-containing water samples from an acidified area (southernmost Norway) and from a less acidified area (eastern Norway) were dialyzed against distilled water: the water samples were recirculated through a dialyzing tube dipped into a compartment containing distilled water. The recirculation of the sample was continued until there was only a small change in the conductivity of the sample with time (< 0.01 mS/m/h) (Gjessing et al. 1988).

In similar experiments, other OA-containing natural waters artificially acidified with H_2SO_4 have also proven to be more toxic to Atlantic salmon and *Daphnia magna* than unacidified water dialyzed in the same way (Gjessing, unpubl. data).

The observations cited above indicate that sulfuric acid interacts with OA in water, apparently forming an OA with a higher sulfur content. The association or the mechanism of the reaction is not known. With the aggregate model of Wershaw in mind, it is unlikely that the association between OA and SO_4 is a matter of ion binding.

With regard to phosphate, the stimulating effects of Fe and Mn for the association with OA may be explained by the formation of a positively charged Fe–Mn PO_4 "complex" that binds to OA through metal binding. Such mechanisms are, however, less likely for sulfate. A possible explanation for the apparent increased amount of sulfur in OA from acidified areas and in H_2SO_4-acidified OA water is that an actual reaction between OA and sulfuric acid takes place.

A formation of a new organic compound by sulfuric acid acidification of surface water could have considerable environmental consequences. Even though the toxic action is not clarified, it has been observed that the charge densities of dialyzed water samples from acidified areas are significantly lower than those from less acidified areas. This may imply that "the sulfonated OA" with lower charge density is less rejected at the membrane surface of the cell and thereby more biologically active.

With regard to OA in relation to drinking water and health, it has been demonstrated that the octanol/water partition coefficient of humic material increases with decreasing pH (Gjessing et al. 1989). As this implies increased biological availability at low stomach pH, the indicated change in properties of the "sulfonated OA" requires an evaluation of the health significance.

REFERENCES

Carlberg, G.E., K. Martinsen, A. Kringstad, E. Gjessing, M. Grande, T. Källqvist, and J.U. Skåre. 1986. Influence of aquatic humus on the bioavailability of chlorinated micropollutants on Atlantic salmon. *Arch. Env. Contam. Tox.* 15:543–548.

Carter, C.W., and I.H. Suffet. 1985. Quantitative measurements of pollutant binding to dissolved humic materials compared with bulk properties of humic materials. *Org. Geochem.* 8(1):145–146.

De Haan, H., and T. De Boer. 1978. A study of the possible interactions between fulvic acids, amino acids and carbohydrates from Tjeukemeer based on gel filtration at pH 7.0. *Water Res.* 12:1035–1040.

Gjessing, E.T. 1976. Physical and Chemical Characteristics of Aquatic Humus. Ann Arbor, MI: Ann Arbor Science.

Gjessing, E.T., and L. Berglind. 1981. Adsorption of PAH to aquatic humus. *Arch. Hydrobiol.* 92:24–30.

Gjessing, E.T., and T. Gjerdahl. 1975. Electromobility of aquatic humus. Fractionation by the use of the isoelectric focusing technique. Humic Substances: Their Structure and Function in the Biosphere, ed. D. Poveledo and H.L. Golterman, pp. 43–51. Wageningen: Pudoc.

Gjessing, E.T., M. Grande, and E. Røgeberg. 1988. Natural organic acids: their role in freshwater acidification and aluminium speciation. Acid Rain Research Report 15/1988, Norwegian Institute for Water Research, Blindern Oslo.

Gjessing, E.T., G. Riise, R.C. Petersen, and E. Andruchow. 1989. Bioavailability of aluminium in the presence of humic substances at low and moderate pH. *Sci. Total Env.* 81/82:683–690.

Jones, R.I., K. Salonen, and H. De Haan. 1988. Phosphorus transformation in epilimnion of humic lakes: abiotic interaction between dissolved humic materials and phosphate. *Freshwat. Biol.* 19:357–369.

MacCarthy, J.F., and B.D. Jimenez. 1985. Effect of humic substances on the bioaccumulation of organic contaminants by aquatic organisms. *Org. Geochem.* 8(1):141.

Riise, G., and B. Salbu. 1989. Major and trace elements in standard and reference samples of aquatic humic substances determined by instrumental neutron activation analysis (INAA). *Sci. Total Env.* 81/82:137–142.

Shapiro, J. 1957. Chemical and biological studies on yellow organic acids of lakewater. *Limnol. Ocean.* 11:161–179.

Steinberg, C., and U. Münster. 1985. Geochemistry and ecological role of humic substances in lakewater. In: Humic Substances in Soil, Sediment and Water: Geochemistry, Isolation and Characterization, ed. G.R. Aiken, D.M. McKnight, R.L. Wershaw, and P. MacCarthy, pp. 105–145. New York: Wiley.

Strachan, W.M.J. 1988. Toxic contaminants in rainfall in Canada: 1984. *Env. Tox. Chem.* 7:871–877.

Vollenweider, R.A. 1968. Scientific fundamentals of the eutrophication of lakes and flowing waters with particular reference to nitrogen and phosphorus as factors in eutrophication. Report No. DAS/CSI/68.27, OECD, Paris.

Wershaw, R.L. 1986. A new model for humic materials and their interactions with hydrophobic organic chemicals in soil-water or sediment-water systems. *J. Cont. Hydrol.* 1:29–45.

Wershaw, R.L., P.J. Burcar, and M.C. Goldberg. 1969. Interaction of pesticides with natural organic material. *Env. Sci. Tech.* 3:271–273.

Wigilius, B., H.O. Borén, and A. Grimvall. 1988. Determination of adsorbable organic halogens (AOX) and their molecular weight distribution in surface water samples. In: Isolation, Characterization and Risk Analysis of Organic Micropollutants in Water, ed. B. Wigilius. *Linköp. Stud. Art Sci.* 20. Linköping, Sweden: Linköping Univ. Press.

Organic Acids in Aquatic Ecosystems
eds. E.M. Perdue and E.T. Gjessing, pp. 189–208
John Wiley & Sons Ltd

Alteration of Organic Substances during Eutrophication and Effects of the Modified Organic Substances on Trophic Interactions

C.E.W. Steinberg

Fraunhofer-Institut für Umweltchemie und Okotoxikologie
P. O. Box 1260
5948 Schmallenberg, F.R. Germany

Abstract. As far as investigated, eutrophication of water bodies changes the quality of organic substances with special respect to volatile odorous compounds and toxic substances. Among the volatile odorous compounds, dienals, alkenols, alkenones, alcohols, and ketones, which are produced mainly by chrysophytes, dominate in noneutrophic environments. Some of these substances are known to affect other microorganisms. Under eutrophicated conditions, when cyanobacteria gain dominance, geosmin, 2-methylisoberneol, and certain nor-carotenoids are the odorous compounds. Several strains of cyanobacteria are known to produce alkaloid or peptide toxins that negatively affect invertebrates, fish, and other vertebrates, whereas a direct effect on bacteria remains uncertain. By excreting siderochromes that selectively chelate ferric iron in laboratory cultures, cyanobacteria can outcompete eukaryotic algae under conditions of iron limitation. Field evidence, however, is lacking.

The community effect of the eutrophication-mediated change in organic substances is a decreasing significance of large-bodied grazers which, however, possess the highest filtering efficiency and the widest spectrum of ingested particles. This community shift, in turn, seems to promote the cyanobacterial maintenance.

INTRODUCTION

The notion that aquatic organisms can modify their environment by the production and release of organic compounds, and in this way exert a direct influence on the growth of other species, was put forth by Pütter (1907, referred to by Jüttner 1983) at the beginning of this century. However, even now, detailed knowledge of this phenomenon is restricted to a very few cases. Different assemblages of groups of algae, which are characteristic of different trophic states, will surely have characteristic differences in their patterns of extracellular substances. These organic substances, in turn, are expected to have influence on the organisms, including the producers themselves.

Concerning only bulk concentrations of dissolved organic carbon (DOC), Thurman (1985) states that DOC varies with the trophic state of a lake, from oligotrophic lakes having the lowest concentrations (1–3 mg/l) to eutrophic lakes having larger concentrations (3–34 mg/l). Among the best-studied classes of organic substances, amino acids and sugars, the same seems to apply for amino acids (Steinberg 1977; Thurman 1985), but obviously not for sugars (Thurman 1985).

The identification of patterns of individual organic substances or classes of organic substances (other than amino acids or sugars) in water bodies of different trophic degrees, however, is still full of gaps. Initial investigations were principally concerned with excreted substances of nutrient character such as amino acids, peptides, sugars, and organic acids. Early work also established the production of autoinhibitors by the cyanobacterium[1] *Nostoc* and the green alga *Chlorella*. Most of the present studies, however, were initiated by practical problems, e.g., with drinking water processing or with water purification and fish breeding, when odorous substances limit the use of the water. Consequently, in many studies, only the nuisance organic substances have been subject to analytical efforts.

In the following, knowledge on geochemical as well as ecological aspects of organic substances under noneutrophic and eutrophic conditions will be summarized.

Eutrophication of water bodies involves the enrichment of plant nutrients, especially phosphorus compounds, often followed by significant shifts in the phytoplankton community towards cyanobacteria, e.g., *Microcystis, Anabaena, Aphanizomenon,* or *Oscillatoria.* When studying single water bodies on the long-term scale, an increase or decrease of cyanobacteria may be considered a significant sign of eutrophication or oligotrophication, respectively (Steinberg and Hartmann 1988).[2] Consequently, "noneutrophic aspects" refer to conditions that are not dominated by cyanobacteria, and "eutrophic aspects" refer to conditions in aquatic water bodies where cyanobacteria clearly maintain dominance in the phytoplankton.

---

[1] These microorganisms are called cyanobacteria (formerly blue-green algae or cyanophyta) due to their blue pigment and bacterial prokaryotic form of cell structure.

[2] However, when comparing *different* aquatic systems (even those which have similar nutrient contents and are in the same climatic region), the inverse deduction is not valid, i.e., the presence of cyanobacteria does not necessarily indicate eutrophic conditions. These differences are caused by different physical properties of the water body, especially by different water column stability (for details see Steinberg and Hartmann 1988).

ORGANIC SUBSTANCES IN WATERS OF DIFFERENT TROPHIC STATES

Volatile Organic Substances

From several oligotrophic as well as eutrophic lakes or reservoirs, which provide raw water for drinking water purposes, there are reports on odorous substances produced by noncyanobacteria as well as cyanobacteria in the phytoplankton (Jüttner 1983; Nicholls and Gerrath 1985; among many others).

Noneutrophic aspects. An early example is the Wahnbach drinking water reservoir (F.R. Germany). In this reservoir, which was mesotrophic at the time of investigation (1978), a musty odor phenomenon was observed and identified which was connected with the bloom of the chrysophyte *Synura uvella*. When studying this bloom and its volatile odorous organic substances, Jüttner (1983) determined a wide spectrum of different kinds of compounds, such as dienals, alkenols, alkenones, alcohols, and ketones. He identified penten-3-one, pentanone-3*, octanone-3*, octanol-1, octanol-3, oct-1-en-3-ol*, 2-pentylfuran, *trans*, *cis*-2,4-decadienal, β-cyclocitral, 6-methylhept-5-en-2-one, and β-ionone. The substances marked with asterisks were the most abundant compounds. Similar patterns of organic volatile substances, especially dienals, appear to be characteristic of chrysophyceae in general (Jüttner 1983).

Diatoms, which are phylogenetically closely related to chrysophytes, are also able to produce odorous substances. Every time the diatom *Asterionella formosa* appeared in the waters, although often in rather low numbers, octadiene and octatriene were detected. Both were released in high amounts by this organism and resulted in a pungent geranium odor (Jüttner 1983). *A. formosa* is an unusually potent generator of these substances. It was shown that octatrienes and octadienes are sexual agents for marine brown algae; however, the ecological role of these substances with *A. formosa* is obscure. Another diatom, *Stephanodiscus hantzschii*, of which a bloom in Lake Constance has been studied, exhibited hexanal release, but no octatrienes were detected. Besides these substances, dimethylsulfide (which causes odors like barberry or trimethylamine) was determined in all diatoms studied (Jüttner 1983).

In most cases the compounds identified can be assigned to a group of compounds originating from a common precursor or homologous series of precursors. For example, β-cyclocitral and β-ionone can easily be recognized by their structure as oxidative cleavage products of carotene. The position of double bonds and oxygen-containing substituents of several C_5- and C_8-alkenes, -alcohols, -aldehydes, and -ketones suggests that these compounds are degradation products of unsaturated fatty acids (Jüttner 1983). There is evidence that even axenic algal cultures are able to produce those odorous substances (Persson 1988).

Eutrophic aspects. As stated above, eutrophic (and polytrophic) phytoplankton assemblages are often characterized by dominance of cyanobacteria. In temperate climates, cyanobacteria are the most prominent producers of (extracellular) organic substances, which can negatively affect other aquatic organisms and cause changes in the trophic interactions as well as impair raw water qualities.

Tastes and odors in water are a worldwide problem which probably existed well before reports began appearing in the scientific literature. For instance, Farlow (1883, referred to in Slater and Blok 1983) recognized that cyanobacteria were responsible for tastes and odors in water. Since then many other reports have been presented. The odor of water contaminated with cyanobacteria is generally described as "muddy," "musty," or "earthy." In addition, odors described as "grassy," "geranium," "vile," "moldy," "sewage," and "nasturtium" have been reported for members of the genera *Anabaena, Aphanizomenon, Microcystis, Nostoc, Oscillatoria,* and *Rivularia* (cf. Slater and Blok 1983). The two compounds most frequently identified as causing off-flavors in water are geosmin and 2-methylisoborneol (MIB) (Fig. 1). Geosmin production has been proven for the cyanobacteria *Symploca muscorum, Oscillatoria tenuis, O. prolifica, O. cortiana, O. variabilis, O. agardhii, O. spendida, O. bornetii f. tenuis, Lyngbya aestuarii, Anabaena scheremetievi, A. circinalis,* and *Schizothrix muelleri*, and the benthic *O. brevis* (prominent in cyanobacterial mats of polluted rivers and eutrophic rivers), whereas MIB has been found with *O. tenuis*, *O. curviceps*, *O. brevis*, and *L. cryptovaginata* (Slater and Blok 1983; Berglind et al. 1983).

It seems likely that these odorous substances are products of the secondary metabolism (see Skulberg 1988). For geosmin, Naes et al. (1988) assume that it seems to function as an overflow metabolite in the isoprenoid pathway.

Figures on naturally occurring concentrations of MIB and geosmin, which are produced by some actinomycetes, streptomycetes, and myxobacterales as well, are rather rare. For example, in Lake Perris (southern California), MIB reached maximum concentrations of approximately 100 ng/l during September and then dropped to about 2 ng/l by early November (Izaguirre et al. 1988). In a eutrophic lake with an anaerobic hypolimnion (Schleinsee, F.R. Germany), Henatsch and Jüttner (1986) observed two independent sources of geosmin. Most geosmin (maximum concentrations of 950 ng/l) was produced under anaerobic conditions in the hypolimnion in the autumn. A much smaller source of geosmin production (maximum concentrations of 190 ng/l) was actinomycetes in the epilimnion.

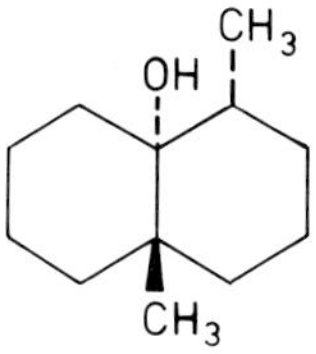

(1)

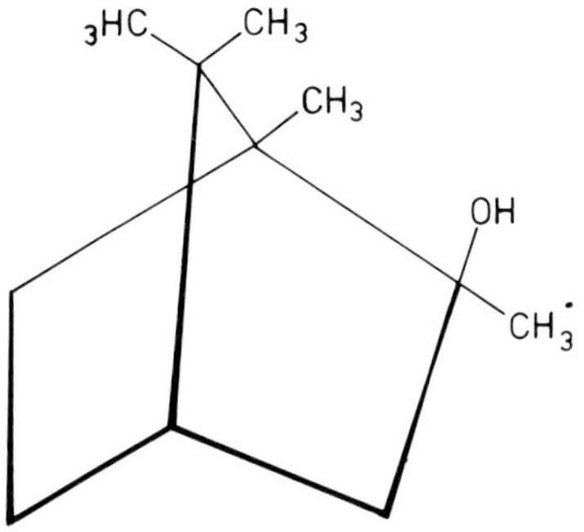

(2)

Fig. 1—Structures of the most frequent odorous cyanobacterial compounds: (1) geosmin, and (2) methylisoborneol.

Geosmin and MIB are resistant to chemical oxidation and are difficult to remove by conventional water treatment methods (cf. Izaguirre et al. 1988). However, biodegradation occurs. For example, MIB was degraded predominantly by *Pseudomonas* spp. within several days (5 to 14) even at microgram-per-liter levels, when organic nutrients (sugars or acetate) were present. It appeared to be rather recalcitrant in the absence of these substances (Izaguirre et al. 1988). With geosmin, the previously reported biodegradation rates of *Bacillus* spp. appear to be overestimated due to artifacts, such as adsorption onto culture vessels (Danglot et al. 1983). In the Schleinsee, with the onset of the autumnal circulation of the water body, rapid aerobic degradation of geosmin to threshold concentrations of 2.7 ng/l was thought to be the reason for the observed decrease (Henatsch and Jüttner 1986).

In addition to taste and odor problems caused by these two agents, several hydrocarbons ranging from C_{14} to C_{20} were isolated from cyanobacteria.[3] The predominance of *n*-heptadecane in the hydrocarbon fraction appears to

[3] *Anabaena oscillarioides*, however, differs quite drastically from other cyanobacteria studied thus far in its pattern of odorous compounds. The formation of high concentration of C_8-compounds with different unsaturations and substituents is remarkable (Möhren and Jüttner 1983).

be a characteristic of the cyanobacteria since this compound is the main hydrocarbon produced by the majority of these organisms which have been specifically investigated for this class of compounds. As the hydrocarbons isolated from cyanobacteria do not have pronounced tastes or odors, their effect on water quality is probably not significant (Slater and Blok 1983).

Further odorous substances of cyanobacteria are nor-carotenoids such as β-ionone, β-cyclocitral, 6-methylhept-5-en-2-one (Henatsch and Jüttner 1983), alkanes, alkylbenzenes, acetophenone, cyclohexanone, and phenylacetonitrile (Slater and Blok 1983).

However, when studying the volatile organic substances in a eutrophic shallow lake (Federsee, F.R. Germany), Jüttner (1984) detected several substances which could not be related to the appearance of cyanobacteria. The major compounds found were: β-cyclocitral, α-cyclocitral, β-ionone, 1,3,3-trimethylcyclohexene, 2,2,6-trimethylcyclohexane, 2,6,6-trimethylcyclohex-2-en-1-one, eucalyptol, geosmin, two argosmin isomers, pent-1-en-3-ol, pent-1-en-3-one, heptadec-1-ene, heptadec-*cis* 5-ene, heptadeca-1, *cis* 8-diene, decanal, and hexanal. The most frequent nor-carotenoids in the water of the Federsee are presented in Fig. 2. The occurrence of geosmin, the argosmin isomers, heptadec-cis 5-ene, and an unknown sesquiterpene could clearly be correlated with the appearance of *Aphanizomenon gracile*, as could the occurrence of β-cyclocitral with the appearance of *Microcystis wesenbergii*. Most of the substances, however, have not been detected before in natural waters.

Other Organic Compounds in Eutrophic and Polytrophic Environments

Toxic cyanobacteria. Toxic genera of cyanobacteria are found worldwide. Carmichael et al. (1985) present an excellent review on this issue. When freshwater cyanobacterial blooms become concentrated inshore by gentle wind and wave action, the toxins can cause illness or death in almost any mammal, bird, or fish which ingests enough of the toxic cells or extracellular toxin. Major losses of animals include cattle, sheep, hogs, birds (domestic and wild), and fish, while minor losses are found among dogs, horses, small wild animals, amphibians, and invertebrates. However, the effects on invertebrates have a significant impact on the trophic structures (see below). Acute lethal oral toxicity to humans has not been documented, but there is increasing evidence that the toxins cause gastroenteritis and contact irritation in uses of certain recreational and municipal water supplies. Lipopolysaccharide endotoxin is also produced by some cyanobacteria and has been implicated in certain water-based outbreaks of gastroenteritis among humans. At present at least two reports exist from Australia which show that increases in liver damage took place during periods with heavy

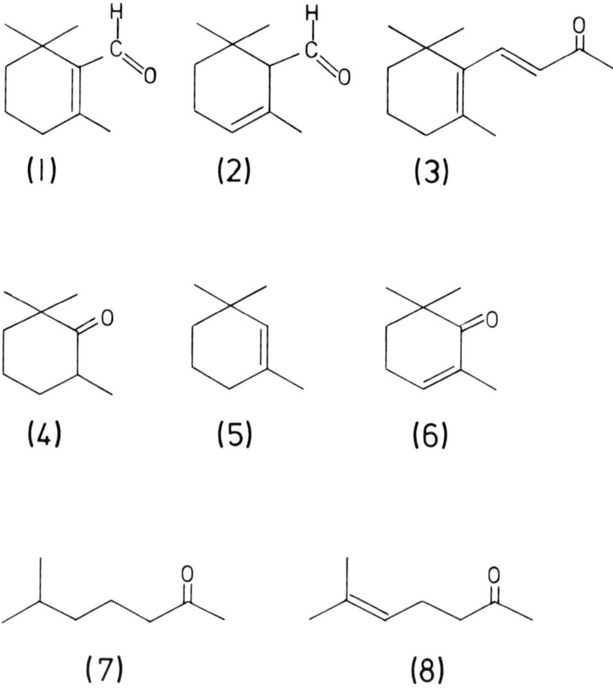

Fig. 2—Nor-carotenoids detected in Federsee lake water (Jüttner 1984): (1) β-cyclocitral, (2) α-cyclocitral, (3) β-ionone, (4) 2,2,6-trimethylcyclohexanone, (5) 1,3,3-trimethylcyclohexene, (6) 2,6,6-trimethylcyclohex-2-en-1-one, (7) 6-methylheptane-2-one, and (8) 6-methylhept-5-en-2-one.

blooms of *Microcystis aeruginosa* (Skulberg et al. 1984; Carmichael et al. 1985; Gorham and Carmichael 1988).

Recent regional surveys have shown that toxic cyanobacterial development in the temperate climates is much more widespread than that which is indicated by animal losses alone (Skulberg et al. 1984). Surveys in Scandinavia (Berg et al. 1986) revealed that about 50% of the occurring cyanobacterial blooms produce toxins (demonstrated by mouse-bioassay). The organisms which predominate in most toxic blooms belong to the genera *Microcystis*, *Anabaena*, *Aphanizomenon*, *Oscillatoria*, and *Nodularia*. Not all members of these genera, however, are toxic, and it is not possible to tell by looking at a bloom or by microscopic examination of the water whether it is toxic or not. At present, it is not known what conditions influence the toxicity of a bloom. Neither the biochemical nor genetic regulation of toxin formation is understood. Biological factors, such as the possibility of selective grazing by zooplankton, may also have an effect on the relative levels of toxic and nontoxic cells in a bloom (Skulberg et al. 1984).

The known features of the chemical nature of cyanobacterial toxins (based on data from Skulberg et al. 1984; Carmichael et al. 1985; and Gorham and Carmichael 1988) are shown in Table 1.

The alkaloid toxins are the most rapidly acting ones. They function as *neurotoxins* paralyzing peripheral skeletal muscles, then respiratory muscles, with death resulting due to respiratory arrest occurring between a few minutes to a few hours. These toxins are produced by various strains and species of *Anabaena* and are generally referred to as "anatoxins." Anatoxin-a (Fig. 3) is the only alkaloid toxin from this group which has been chemically characterized. Poisoning of domestic and wild animals by anatoxins comes from ingestion of the toxin cells and extracellular toxin from a waterbloom.

Known toxins of *Aphanizomenon flos-aquae* are also neurotoxic alkaloids, generally referred to as aphantoxins, with chemical structures similar to the marine saxitoxin (Fig. 3).

The *hepatotoxins* are peptides (Fig. 3) and are primarily found in various strains of *Microcystis aeruginosa*, but also with *Anabaena, Aphanizomenon*, and *Oscillatoria*. These toxins are responsible for most cyanobacterial poisonings, because *M. aeruginosa* is more common in its worldwide distribution than the other toxigenic cyanobacteria (Carmichael et al. 1985).

TABLE 1. Cyanobacteria toxins and their known features.

Cyanobacteria species	Toxin	Structural group
Microcystis aeruginosa:		
Strains of the Northern Hemisphere	microcystins	cyclic heptapeptides
Strains of the Southern Hemisphere	cyanoginosins	cyclic heptapeptides
Anabaena flos-aquae	anatoxin-a	alkaloid
	other anatoxins	unknown
Strain S-23-g-1	microcystin	cyclic heptapeptide
Aphanizomenon flos-aquae	aphantoxins	alkaloid (saxitoxin and derivatives)
	toxin-like anatoxin-c	unknown
Synechocystis sp.	toxin-like anatoxin-c	unknown
Oscillatoria agardhii	oscillatoxins	unknown
Oscillatoria rubescens	oscillatoria toxin	unknown
Oscillatoria agardhii var. *isothrix*	oscillatoria toxin	unknown
Oscillatoria brevis UTEX B1567	LPS-endotoxin	lipopolysaccharide
Schizothrix calcicola	LPS-endotoxin	lipopolysaccharide
Pseudanabaena catenata	unknown toxin	unknown

(1)

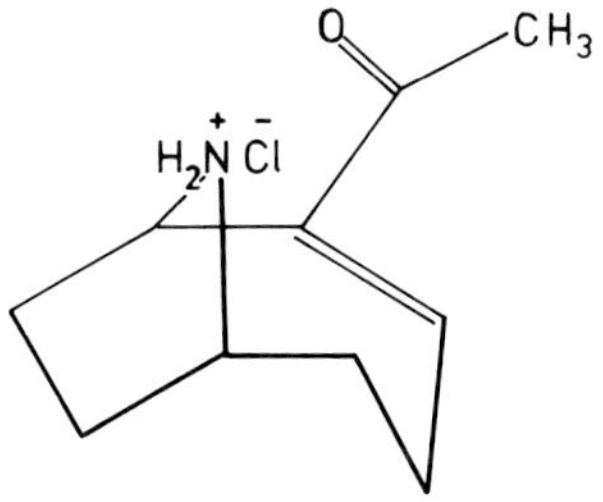

(2)

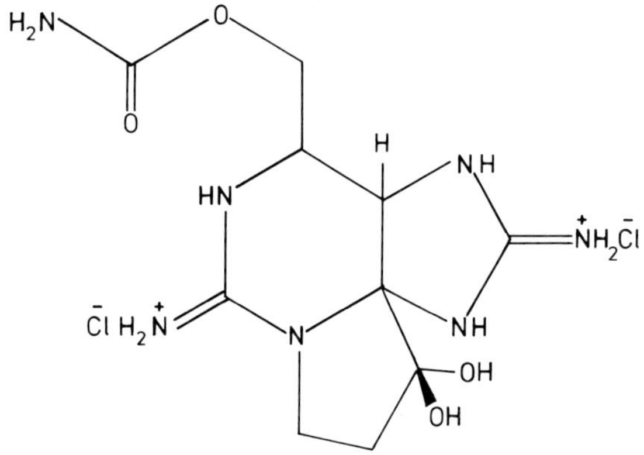

(3)

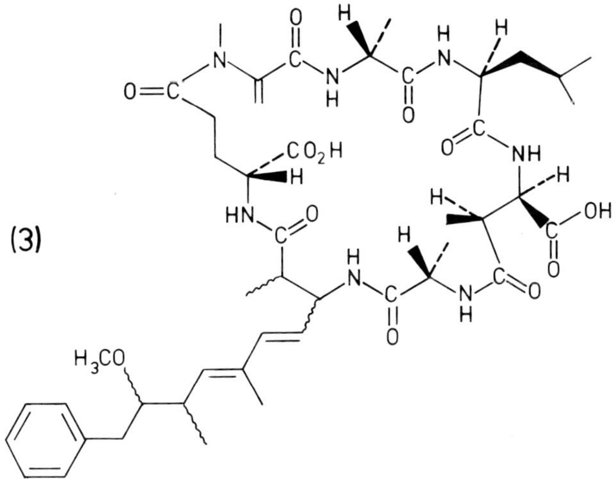

Fig. 3—Structures of cyanobacterial toxins: (1) anatoxin-a hydrochloride, (2) saxitoxin dihydrochloride, and (3) cyanoginosin-LA as an example of cyclic heptapeptide toxins.

Skulberg et al. (1984) give an impression of the toxicities of cyanobacterial toxins. Aphantoxins, for instance, are as toxic as saxitoxins from marine dinoflagellates and more toxic than cobra snake venom. Microcystin and oscillatoria toxins possess only one-fifth the toxicity of saxitoxins. Anatoxins are still less toxic.

Toxic freshwater dinoflagellates. Blooms of dinoflagellates (pyrrhophyta) can be considered signs of eutrophication as well. Dinoflagellates contain the most species documented for causing outbreaks of algal disease ("red tides," causing a disease termed "paralytic shellfish poisoning"). They exist almost exclusively in eutrophicated marine environments. Only very few nonmarine dinoflagellates are known to produce toxic substances, e.g., those dinoflagellates which can cause so-called "freshwater red tides." For example, in the autumn of 1962, the occurrence of mass fish kills was observed in Lake Sagami, near Tokyo, and the appearance of blooms of *Peridinium polonicum* was documented prior to these kills (cf. Shilo 1972). The toxic substance was found to be extracellular, soluble in water and alcohol, not absorbed by active charcoal, and not destroyed in alkaline medium at room temperature. The toxin was called "glenodine," which is highly toxic to fish only in alkaline media. This substance was found to be an alkaloid resembling ibogaalkaloid in its mass spectrum and it contained a sulfhydryl radical in the molecule (references in Shilo 1972). Again, under alkaline conditions "glenodine" inhibited growth and photosynthesis of eukaryotic algae (*Scenedesmus*- and *Dunaliella*-assays) as well.

Shilo (1972) reports that mortalities of fish populations caused by toxic dinoflagellates have also been described for freshwater lakes in Louisiana and Texas. The first recorded instance of a "blood-red" bloom was in 1953 when many thousands of fish succumbed in Lake Austin. The causative organism was shown to be a *Glenodinium* species.

Since blooms of toxic dinoflagellates are relatively rare, the knowledge of the ecophysiology of these substances is limited to the few studies referred to above.

Toxic nonmarine Haptophyceae.[4] In brackish environments, blooms of Haptophyceae are clear signs of eutrophication. It is noteworthy that a small phytoflagellate toxin was first implicated in the sudden mass mortality of fish in nonmarine habitats, namely in the Workumsee (adjoining the Zuidersee, Netherlands), in the second decade of this century. Later, periodic deaths of fish populations in brackish water bodies in Schleswig-

[4] The genus *Prymnesium* belongs, together with the genus *Chrysochromulina* (*C. polylepis* was the alga which caused the death of fish and other vertebrates as well as invertebrate organisms in the eastern parts of the North Sea during early summer 1988), to the newly constructed class "Haptophyceae" (and no longer to the class "Chrysophyceae").

Holstein and Denmark were described (references in Shilo 1972). The responsible organisms were *Prymnesium parvum* and sometimes *P. saltans* as well. They are small (10 μm), free-living, motile phytoflagellates with two flagella and a typical haptonema.

Although fish intoxication due to *Prymnesium* has been reported from various parts of northern Europe, England, and Bulgaria, the problem did not become acute until soon after the introduction of fish-breeding based on inland brackish water sources in Israel, when sudden severe fish kills occurred in a number of such fishponds. The chemical nature of the *Prymnesium* toxin resembles that of a proteolipid, and the lipid part alone is sufficient for ichthyotoxic activity (Shilo 1972). From recent reports (cf. Carmichael et al. 1985) it seems that there is a family of compounds with similar composition but different toxic effects rather than a single compound with different toxic effects. These effects are generally classified as being cytotoxic, hemolytic, and ichthyotoxic.

EFFECTS OF ORGANIC SUBSTANCES ON TROPHIC INTERACTIONS

Effects of Volatile Substances

Several of the volatile odorous compounds identified in water during algal blooms or in algal cultures have been proved in algal and bacterial bioassays to be biologically highly active. Geranylacetone, for example, found in cultures of the primitive Rhodophyceae *Cyanidium caldarium*, was shown to be a new specific inhibitor of carotene synthesis with an inhibition site between phytofluene and zeta-carotene. The bioassay with the cyanobacterium *Synechococcus* 6911 showed an inhibition threshold of 10–20 ppm for 2-pentyl-furan and ppm for *trans*, *trans*-2,4-decadienal. Dihydroactinidiolide (2,4,5,7,7a-hexahydro-4,4,7a-trimethylbenzofuranone) was found to be a dominant compound in the medium of *Synechococcus* strain (= *Anacystis nidulans*)(Jüttner 1983). The presence of this substance in the macrophyte *Eleocharis* sp. is of biological significance; it is released from the rhizomes immediately adjacent to the aquatic habitats as a growth inhibitor for submerged macrophytes. This substance probably has a similar effect within the (phyto)plankton community; however, evidence is lacking. The norcarotenoid 6-methylhept-5-en-2-one is known as an inhibitor of bacterial growth (Henatsch and Jüttner 1983 and references therein). The community effect of volatile organic substances is unknown, because it was not the subject of the cited studies. We are still in the learning stage regarding this issue.

Cyanobacterial Interactions with Prokaryotic Organisms

There are contradictory reports on the *direct effects* of toxins, especially microcystin, on (aquatic) microorganisms. Russian authors in particular reported a wide spectrum of antimicrobial activity against *Escherichia coli*, *Shigella flexneri*, *Salmonella typhimurium*, *Staphylococcus aureus*, *Enterococcus*, and *Candida*. Further, it was found that *Microcystis* toxin extracts inhibited the development of saprophytic microflora in artificial ponds, whereas other authors did not find any effects of cyanobacterial toxins (references in Carmichael et al. 1985).

Cyanobacterial Interactions with Eukaryotic Phytoplankton

The way by which eukaryotic algae can be nearly completely suppressed during cyanobacterial blooms appears to be an *indirect* one. Many authors present evidence that this suppression is due to competition for available iron. When sufficient free ferric iron is available, cyanobacteria can acquire ferric iron through a nonspecific transport system. However, despite the natural abundance of iron in aqueous oxic environments, its low solubility at neutral pH can limit the primary production by phytoplankton. Many organisms must adapt their physiology to accommodate iron-limited environments. Under low ferric iron concentrations a high affinity siderophore-mediated iron transport system appears to be induced especially in cyanobacteria (Kerry et al. 1988). These transport systems are often siderochromes, which are trihydroxamates of low molecular weight that can selectively chelate ferric iron. For example, ferric iron is chelated by acetohydroxamic acid by several orders of magnitude greater than other cations. These complexes are available only for prokaryotic organisms. Thus, they appear to be the agent suppressing eukaryotic algae rather than allelopathic or toxic substances. Murphy et al. (1976) demonstrated that with many eukaryotic algae the iron uptake systems, via solubilization by small peptides, cannot compete with the hydroxamate system that is found in many cyanobacteria (especially nitrogen-fixing cyanobacteria, which possess an extremely high iron requirement) and probably many planktonic bacteria as well.

These statements, however, are based on laboratory studies, and field evidence is still lacking.

Cyanobacterial Interactions with Invertebrates

Do cyanobacteria interact with invertebrates in such a way that negative effects on the trophic relationships in the plankton occur? Are grazing species within the zooplankton selectively favored or handicapped so that the community efficiency is consequently drastically altered? Or, oppositely, do specific grazers exist within the zooplankton community which are capable of controlling cyanobacterial development (under certain circumstances, which can possibly be maintained by directed management strategies)? These are some of the questions being investigated by studies on trophic interactions under eutrophied conditions.

There are many papers on this subject, but our knowledge is not unequivocal. For example, there has been a debate over whether or not nuisance phytoplankton (cyanobacteria) can be grazed and thus controlled in its development by zooplankton, and if so, which (harmful) effects can be observed within the grazers. Lampert (1987), for instance, writes that some features of cyanobacteria can be regarded as antiherbivore defenses. Large filaments of *Aphanizomenon*, *Anabaena*, and large colonies of *Microcystis* cannot be handled by zooplankton but do not interfere seriously with the filtering process. Small colonies and filaments, however, may cause severe inhibition of the feeding process by mechanical interference. This reduces zooplankton growth, reproduction, and survival. Copepods, rotifers, and small-bodied cladocerans such as *Bosmina* are less affected by mechanical disturbance than large-bodied cladocerans (especially the *Daphnia* species). If ingested, some cyanobacteria may be poorly digested or may not provide essential nutrients. Some cyanobacteria, such as *Microcystis* strains, are reported to be toxic to zooplankton. Zooplankton encountering toxic cells ceases to feed. Some evidence of an extracellular herbivore deterrent was found for *Anabaena*. Thus, cyanobacteria may either be individually protected or create an environment that is unfavorable for efficient grazers (e.g., *Daphnia*). These findings are supported for example by Nizan et al. (1986), who found different short-term toxic effects of *Microcystis aeruginosa* on *Daphnia magna*, including markedly reduced food uptake and lethality for juvenile and adult daphnids.

On the other hand, there are several reports presenting clear evidence that even small colonies of *Microcystis* strains are of some nutritional value to several, but not all, grazing zooplankton species. In addition, in tropical Lake Valencia, Venezuela, it was found that cladoceran populations develop well while ingesting *M. aeruginosa* (Infante and Riehl 1984). De Bernardi et al. (1981) demonstrated experimentally that small colonies of *Microcystis* may be suitable food for three species of *Daphnia*. These contradictions are probably due to the fact that the amount of toxin produced differs according to the strains and condition of *Microcystis* (Lampert 1987).

Fulton and Paerl (1988) observed a significant shift in the zooplankton assemblage when the animals fed on *M. aeruginosa*. In laboratory competition experiments, increasing the proportion of *M. aeruginosa* in the algal food supply resulted in a shift from dominance by the relatively *large-bodied* cladoceran, *Daphnia ambigua*, to dominance by the *small-bodied* cladocerans,

Diaphanosoma brachyurum and *Bosmina longirostris*. Their results indicate that there are a variety of mechanisms which allow herbivorous zooplankton to coexist with blooms of *Microcystis aeruginosa*, including resistance to toxic chemicals (*Brachionus calyciflorus*), avoidance of consumption of *M. aeruginosa* by chemosensory means (copepods), or the inability to consume large colonies (some small cladocerans).

Under field conditions as well, a similar shift to small-bodied cladocerans (*Moina* and others) was observed both in the hypereutrophic Hartbeespoort Dam, South Africa (Jarvis et al. 1987), and in Lake Kasumigaura, Japan (Hanazato and Yasuno 1987). Under these conditions, fish predation preferentially on large-bodied cladocerans can also cause the described shift in the zooplankton assemblage. In these two papers, however, the shift is attributed to the indirect decomposition pathway via bacteria and suspended detritus which, as a source of nutrients, become increasingly important to zooplankton during *Microcystis* blooms. This means that decomposed *Microcystis* material, not the intact *Microcystis* cells themselves, was the utilizable food.

This shift in favor of small-bodied grazers was observed not only with *Microcystis* blooms, but also with *Anabaena* blooms (Burns et al. 1987). The authors observed that the plankton in Lake Rotongaio, New Zealand, was often dominated by high densities of the cyanobacterium *Anabaena* sp. yet also contained moderately dense populations of small-bodied herbivorous crustacean grazers. The authors also report on an unexpected finding: dissolved substances in Lake Rotongaio also reduced zooplankton filtering rates. Because the phytoplankton is composed almost completely (ca. 98%) of *Anabaena* sp., the authors suspect the inhibitory substances were produced by this cyanobacterium. However, characterization of these substances is lacking.

Cyanobacterial Interactions with Fish

As stated above, *Microcystis aeruginosa* is the most common cyanobacterium of nutrient-rich freshwaters and has been implicated in a large number of poisoning incidents involving wild and domesticated terrestrial animals. One can expect that there are intoxication effects on fish as well. However, a study by Phillips et al. (1985) revealed that the used pure strain of *M. aeruginosa* was toxic to mice, but was nontoxic to the widely farmed rainbow trout, *Salmo gairdneri*, immersed in a culture of this *Microcystis* species for several days. In contrast, intraperitoneal injections of *Microcystis* caused an acute toxic response in the fish, with 100% mortality within 36 hours. The histopathology showed that within 2 hours, a severe acute necrotizing peritonitis impinging on visceral organs, generalized vasodilation, and early generalized toxic degeneration of hepatic tissue occurred. After 5 hours this response was severe, with foci of massive necrosis within the liver, generalized loss of liver architecture, and generalized pooling of blood within engorged vessels throughout the body.

In view of the widespread occurrence of *M. aeruginosa* in freshwaters and the lack of evidence implicating this organism in rainbow trout poisonings, there seems to be little doubt that the *M. aeroginosa* hepatotoxin(s) is unable to gain entry into the interior of rainbow trout via the intestinal or branchial epithelium in sufficient quantities to cause such rapid death.

Decomposing matter of another species, *Aphanizomenon flos-aquae*, proved toxic as well to a range of tank-held fish species in two classical studies from the 1940s (references in Phillips et al. 1985). A history of fish kills in several freshwaters containing blooms of *M. aeruginosa* and other cyanobacteria nevertheless exists (references in Phillips et al. 1985), and the possibility of chronic effects of *Microcystis* toxin to rainbow trout, beyond Phillips et al.'s (1985) trial of several days, remains to be investigated.

Though most of the studies of cyanobacterial interactions with fish were performed using fish of commercial value, it remains unknown whether, under field conditions, a cyanobacteria-mediated fish kill is selective for species or fish size classes. In a rather speculative manner, one could anticipate kills of planktivorous fish, which may result in decreased predation upon the large-bodied cladoceran grazers. Finally, such a selective fish kill could have a positive effect on the community by indirectly supporting the populations of the most efficient phytoplankton grazers (see following section on COMMUNITY EFFECTS).

Community Effects

Evidence has been presented that the dominance of cyanobacteria as food for grazers causes a significant shift from large-bodied cladocerans to small-bodied cladocerans and rotifers. What is the community effect of this shift? Large herbivorous cladoceran species are characterized by two qualities: they possess (*a*) the highest filtering efficiency and (*b*) the widest spectrum of ingested particles. Their grazing is not selective as with most small-bodied grazers. Consequently, one could expect that a cyanobacteria-caused shift in favor of small-bodied grazers would lead to a somewhat higher instability within trophic relationships. This means that under eutrophic conditions, primary producers appear to be less controlled than under oligo- or mesotrophic ones. Phytoplankton blooms in eutrophic lakes are most commonly composed of algal species that cannot be efficiently controlled by filter-feeding zooplankton. This is because such algae have a form difficult to handle (large-cell or colony-size), or they are directly or indirectly (reduced fecundity and reproduction rates) toxic to animals. Some authors believe that algal populations in eutrophic lakes cannot be controlled by zooplankton, because large filter-feeding cladoceran species, e.g., *Daphnia magna*, *D. pulex*, or *D. pulicaria*, are never abundant, being under pressure exerted by planktivorous fish which select large prey. As a consequence, large cladocerans contribute only marginally to zooplankton community grazing in highly eutrophic lakes. Because of their high filtering rate and wide spectrum of ingested particles, these large cladoceran species should be able to control large and filamentous algae to a greater extent than small cladocerans and rotifers, which are common in eutrophic lakes (Dawidowicz et al. 1988 and references therein). Among several other authors (e.g., Goad 1984; Benndorf 1987; Carpenter and Kitchell 1988), Dawidowicz et al. (1988) tested whether large *Daphnia* can control the density of cyanobacterial filaments. *Daphnia* ingested filaments of all cyanobacterial species present (*Oscillatoria limnetica*, *O. agardhii*, *O. redekei*, *Aphanizomenon flos-aquae*, and *Lyngbya* sp.) and selected the shortest ones (<33 μm). The effect of *Daphnia* on filament density was increased with decreased algal abundance.

Furthermore the authors tried to find out the maximum concentration of filamentous cyanobacteria at which *D. magna* was able to survive and increase its population density. It seems that *D. magna* is able to control population growth of filamentous cyanobacteria in nutrient-rich lake water. However, it can do so only if its population density is high enough to prevent cyanobacteria from reaching the concentration at which filtering rate and food ingestion of *Daphnia* are reduced below the threshold level necessary for the minimum reproduction required for maintaining its dense population. When a natural community of filamentous cyanobacteria approaches a concentration corresponding to about 150 μg/l chlorophyll a, the reproduction of *Daphnia* ceases and its population disappears from the environment within a few days.

By altering the ambient light and nutrient environment, large zooplankton species may suppress cyanobacteria, which mainly prefer low-light conditions. Many cyanobacterial blooms develop in the shade of dense populations of eukaryotic algae (cf. Steinberg and Hartmann 1988). Increasing light conditions provide a negative factor in cyanobacterial competition. Evidence for the importance of grazers in promoting cyanobacterial dominance by removing competing phytoplankton species is equivocal. Zooplankton may suppress nitrogen fixation by cyanobacteria through ammonia excretion and may promote a change in dominance from diatoms to cyanobacteria through recycling phosphorus but not silicon (Burns 1987). However, the other effect seems to be probable as well: high populations of grazing zooplankton with high respiration rates may cause lowering of the pH at least for short periods of time. The experiments of Shapiro et al. (1982) have shown that a lowering of the pH causes a weakness in cyanobacterial competition, which may be mainly due to increased virulence of cyanophages. Most likely both explanations are valid and occur at different times of the annual plankton cycle: (*a*) the suppression of cyanobacterial growth by zooplankton during high grazing activities after the spring plankton bloom and (*b*) the promotion of cyanobacterial growth by phosphorus pulsing during the summer, when zooplankton development is limited by low algal biomass.

CONCLUSIONS

Though there has been some recent progress in identifying organic substances in lakes of different trophic states, the ecophysiological role of organic substances released by both oligotrophic and eutrophic phytoplankton is only fragmentarily understood.

The reasons for the studies of these substances, however, have been rather anthropocentric: substances such as the odorous compounds of chrysophytes or diatoms, or the toxins of dinoflagellates, haptophytes, and cyanobacteria as well, were studied mainly due to their impairment of water quality for use by man. Surely it is not the purpose of *Prymnesium* or *Microcystis* to kill fish or cattle via intra- or extracellular toxic substances. This means that possible ecological benefits for the microorganisms which produce odorous or toxic substances remain almost completely obscure, except as shown in the few studies of Jüttner (1983 and references therein). Even if future studies reveal that most of the questionable substances are nothing but metabolic scum, one should recognize that the environmental conditions under which the "nuisance" microorganisms develop are far from natural. Although eutrophication itself is a natural process of lakes during ontogenesy of lake ecosystems in temperate climates, neither the velocity nor the extent of the recent eutrophication is natural. When environments are anthropogenically undisturbed, for example in deep lakes, cyanobacterially dominated stages within the phytoplankton will rarely occur. These stages are limited to the late periods, when sedimentation has produced shallow lakes in which the internal nutrient loading is sufficient for cyanobacterial development and maintenance. A similar statement is true for the *Prymnesium* dominance in brackish waters: the abuse of small water bodies as fish breeding containers increases the nutrient contents drastically. In conclusion, one can state that anthropogenic activities produce conditions under which "nuisance" microorganisms with narrow ecological niches can increase dramatically and become a nuisance with respect to human purposes. It seems paradoxical that the attention of ecologists is called to these chemical interactions only under anthropogenically changed ecological conditions. The objective of future studies should be to gain knowledge about the basic ecological meaning of these interactions.

There is no doubt that the most effective countermeasure of anthropogenic eutrophication is watershed management, such as phosphorus reduction in sewage effluents or decreasing phosphorus loads from nonpoint sources. However, there is some evidence that eutrophication can be—at least partly—counteracted additionally by manipulative in-lake measures. The paper of Dawidowicz et al. (1988), referred to in detail, is optimistic and should induce water management through the use of ecological principles rather than engineering or fish breeding methods. Their observations may have important consequences for applying biomanipulation methods to hypereutrophic lakes for their recovery. The essence of the biomanipulation approach is to reduce planktivorous fish pressure to allow an increased growth in populations of large planktonic filter-feeding zooplankters. Due to their high filtering rate and the wide spectrum of particles they can ingest, large-bodied cladocerans should keep the density of phytoplankton at low levels, even in nutrient-rich environments (Shapiro et al. 1982). This control of algal abundance, though not efficient in the case of some totally ungrazeable algae such as flake forms of *Aphanizomenon* or big colonies of *Microcystis*, can sometimes be strong enough to support clear water phases. This can be accomplished if large-bodied *Daphnia* populations become abundant and persist in the lakes from early spring on, before the bloom starts building up. However, this is possible only if the fishstock is managed ecologically and does not wipe out the *Daphnia* populations during the summer period (Dawidowicz et al. 1988). Such controls can rarely be balanced to suppress cyanobacteria for long periods. Nutrient regulation of phytoplankton succession is a much more effective control procedure, though not always practical.

REFERENCES

Benndorf, J. 1987. Food web manipulation without nutrient control: a useful strategy in lake restoration? *Schweiz. Z. Hydrol.* 49:237–249.

Berg, K., O.M. Skulberg, R. Skulberg, B. Underdal, and T. Willen. 1986. Observations of toxic blue-green algae (cyanobacteria) in some Scandinavian lakes. *Acta Vet. Scan.* 27:440–452.

Berglind, L., I.J. Johnsen, K. Ormerod, and O.M. Skulberg. 1983. *Oscillatoria brevis* (Kütz.) Gom. and some other especially odouriferous benthic cyanophytes in Norwegian inland waters. *Wat. Sci. Tech.* 15:241–246.

Burns, C.W. 1987. Insights into zooplankton-cyanobacteria interactions derived from enclosure studies. *N.Z. J. Mar. Freshwat. Res.* 21:477–482.

Burns, C.W., D.J. Forsyth, J.F. Haney, M.R. James, W. Lampert, and R.D. Pridmore. 1987. Mechanisms of co-existence and exclusion of zooplankton by cyanobacteria in Lake Rotongaio. *N.Z. J. Mar. Freshwat. Res.* 21:537–538.

Carmichael, W.W., C.L.A. Jones, N.A. Mahmood, and W.C. Theiss. 1985. Algal toxins and water-based diseases. *CRC Crit. Rev. Env. Cont.* 15:275–313.

Carpenter, S.R., and J.F. Kitchell. 1988. Consumer control of lake productivity. *BioScience* 38:764-769.

Danglot, C., G. Amar, and R. Vilagines. 1983. Ability of *Bacillus* to degrade geosmin. *Wat. Sci. Tech.* 15:291–299.

Dawidowicz, P., Z.M. Gliwicz, and R.D. Gulati. 1988. Can *Daphnia* prevent a blue-green algal bloom in hypertrophic lakes? A laboratory test. *Limnologica* 19:21–26.

De Bernardi, R., G. Giussani, and E.L. Pedretti. 1981. The significance of blue-green algae as food for filterfeeding zooplankton: experimental studies on *Daphnia* spp. fed by *Microcystis aeruginosa*. *Verh. Int. Verein. Limnol.* 21:477–483.

Fulton, R.S., III, and H.W. Paerl. 1988. Effects of the blue-green alga *Microcystis aeruginosa* on zooplankton competitive relations. *Oecologia* 76:383–389.

Goad, J.A. 1984. A biomanipulation experiment in Green Lake, Seattle, Washington. *Arch. Hydrobiol.* 102:137–153.

Gorham, P.R., and W.W. Carmichael. 1988. Hazards of freshwater blue-green algae (cyanobacteria). In: Algae and Human Affairs, ed. C.A. Lembi and J.R. Waaland, pp. 403–431. Cambridge: Cambridge Univ. Press.

Hanazato, T., and M. Yasuno. 1987. Evaluation of *Microcystis* as food for zooplankton in eutrophic lake. *Hydrobiol.* 144:251–259.

Henatsch, J.J., and F. Jüttner. 1983. Volatile odorous excretion products of different strains of *Synechococcus* (cyanobacteria). *Wat. Sci. Tech.* 15:259–266.

Henatsch, J.J., and F. Jüttner. 1986. Production and degradation of geosmin in a stratified lake with anaerobic hypolimnion (Schleinsee). *FEMS Microbiol. Lett.* 35:135–139.

Infante, A., and W. Riehl. 1984. The effect of cyanophyta upon zooplankton in a eutrophic tropical lake (Lake Valencia, Venezuela). *Hydrobiol.* 113:293–298.

Izaguirre, G., R.L. Wolfe, and E.G. Means, III. 1988. Degradation of 2-methyliso-borneol by aquatic bacteria. *Appl. Env. Microbiol.* 54:2424–2431.

Jarvis, A.C., R.C. Hart, and S. Combrink. 1987. Zooplankton feeding on size fractionated *Microcystis* colonies and *Chlorella* in a hypertrophic lake (Hartbeespoort Dam, South Africa): implications to resource utilization and zooplankton succession. *J. Plank. Res.* 9:1231–1249.

Jüttner, F. 1983. Volatile odorous excretion products of algae and their occurrence in the natural aquatic environment. *Wat. Sci. Tech.* 15:247–257.

Jüttner, F. 1984. Dynamics of the volatile organic substances associated with cyanobacteria and algae in a eutrophic shallow lake. *Appl. Env. Microbiol.* 47:814–820.

Kerry, A., D.E. Laudenbach, and C.G. Trick. 1988. Influence of iron limitation and nitrogen source on growth and siderophore production by cyanobacteria. *J. Phycology* 24:566–571.

Lampert, W. 1987. Laboratory studies on zooplankton-cyanobacteria interactions. *N.Z. J. Mar. Freshwat. Res.* 21:483–490.

Möhren, S., and F. Jüttner. 1983. Odorous compounds of different strains of *Anabaena* and *Nostoc* (Cyanobacteria). *Wat. Sci. Tech.* 15:221–228.

Murphy, T.P., D.R.S. Lean, and C. Nalewajko. 1976. Blue-green algae: their excretion of iron-selective chelators enables them to dominate other algae. *Science* 192:900–902.

Naes, H., H.C. Utkilen, and A.F. Post. 1988. Factors influencing geosmin production by the cyanobacterium *Oscillatoria brevis*. *Wat. Sci. Tech.* 20:125–131.

Nicholls, K.H., and J.F. Gerrath. 1985. The taxonomy of *Synura* (Chrysophyceae) in Ontario with special reference to taste and odour in water supplies. *Can. J. Botany* 63:1482–1493.

Nizan, S., C. Dimentman, and M. Shilo. 1986. Acute toxic effects of the cyanobacterium *Microcystis aeruginosa* on *Daphnia magna*. *Limnol. Ocean.* 31:497–502.

Persson, P.-E. 1988. Odorous algal cultures in culture collections. *Wat. Sci. Tech.* 20:211–213.

Phillips, M.J., R.J. Roberts, J.A. Stewart, and G.A. Codd. 1985. The toxicity of the cyanobacterium *Microcystis aeruginosa* to rainbow trout, *Salmo gairdneri* Richardson. *J. Fish Dis.* 8:339–344.

Shapiro, J., B. Forsberg, V. Lamarra, G. Lindmark, M. Lynch, E. Smeltzer, and G. Zoto. 1982. Experiments and experiences in biomanipulation. Studies of biological ways to reduce algal abundance and eliminate blue-green. Environmental Protection Agency, Report 600/3-82-096. Washington, D.C.

Shilo, M. 1972. Toxigenic algae. In: Progress in Industrial Microbiology, ed. D.J.D. Hockenhull, vol. 11, pp. 233–265. Edinburgh: Churchill Livingstone.

Skulberg, O.M. 1988. Chemical ecology and off-flavour substances. *Wat. Sci. Tech.* 20:167–178.

Skulberg, O.M., G.A. Codd, and W.W. Carmichael. 1984. Toxic blue-green algal blooms in Europe: a growing problem. *Ambio* 13:244–247.

Slater, G.P., and V.C. Blok. 1983. Volatile compounds of the cyanophyceae—a review. *Wat. Sci. Tech.* 15:181–190.

Steinberg, C. 1977. Vergleich der gelösten organischen Stoffe verschiedener Holsteinischer Seen. *Arch. Hydrobiol.* 80:297–307.

Steinberg, C.E.W., and E.M. Hartmann. 1988. Planktonic bloom-forming cyanobacteria and the eutrophication of lakes and rivers. Opinion. *Freshwat. Biol.* 20:279–287.

Thurman, E.M. 1985. Organic Geochemistry of Natural Waters. Dordrecht: Nijhoff/ Junk.

Organic Acids in Aquatic Ecosystems
eds. E.M. Perdue and E.T. Gjessing, pp. 209–221
John Wiley & Sons Ltd

Interactions of Organic Acids with Inorganic and Organic Surfaces

E. Tipping

Institute of Freshwater Ecology
Windermere Laboratory
Far Sawrey
Ambleside, Cumbria LA22 0LP, U.K.

Abstract. The main processes that may be involved in the adsorption of organic acids (especially humic substances) to natural surfaces are the hydrophobic effect, ligand exchange, electrostatic interactions, van der Waals interactions, hydrogen bonding, coadsorption, competition, and conformational changes in the adsorbed acid. Information is available on the adsorption of organic acids to oxides and the charged siloxane surface and, to a lesser extent, on adsorption to carbonates and sulfates. Ideas about adsorption to natural organic surfaces have to be speculative. Adsorption of organic acids might affect their environmental behavior by immobilization, control of solid-phase and solution concentrations, fractionation, cosedimentation with the adsorbent, protection from microbial attack, and catalytic reaction.

INTRODUCTION

In most natural aquatic environments (soils, waters, sediments) the majority of the organic acids are humic substances. These ubiquitous and plentiful decomposition products of plants and other biological material are composed mainly of C, H, and O, bear acid functional groups (chiefly carboxyl and phenolic OH), and are heterogeneous with respect to structure and to molecular weight, which may range from several hundred to several hundred thousand daltons. The humic substances most likely to be involved in adsorptive interactions are the humic and fulvic acids, the fulvic acids tending to be of lower molecular weight, less hydrophobic, and richer in acidic groups than the humic acids (see, e.g., Aiken et al. 1985). This review focuses on the interactions of humic substances with surfaces, although reference is also made to simpler organic acids such as benzoic and phthalic acids, to acidic polysaccharides, to synthetic polyelectrolytes, and to proteins.

The inorganic surfaces considered here are those of oxides, the siloxane surface of aluminosilicates, and the surfaces of carbonates and sulfates. When it comes to natural organic surfaces there is relatively little direct information on the adsorption of organic acids, and consequently the discussion is more speculative. Mention is also made of the self-association of organic acids and their accumulation at the air–water interface.

ADSORPTION PROCESSES

A number of processes may occur when a solute—in this case an organic acid—is transferred from bulk solution to an interface. Some contribute favorably, others unfavorably, to the net free energy of adsorption. Not all of them involve the acidic functional groups directly. The main processes are discussed below; the reader is referred to the works of Mortland (1970), Theng (1979), Norde (1980), and Birchall et al. (1981) for further information.

The Hydrophobic Effect (Tanford 1980)

This arises when apolar moieties that do not interact favorably with water are removed from contact with the aqueous environment. The relaxation in the imposed water structure around such moieties results in a large positive entropy change. If adsorbent and adsorbate each have apolar regions, the hydrophobic effect may be the major driving force for adsorption. It may also be involved in interactions among adsorbed molecules.

Ligand Exchange

Oxide surfaces become hydroxylated when exposed to aqueous solutions, and the surface hydroxyl groups, represented here by $-SOH$, can be exchanged for acid functional groups such as $-COOH$, by reactions like:

$$-SOH + HOOCR \rightleftharpoons -SOOCR + H_2O \tag{1}$$

For organic acids with more than one functional group, more than one such reaction can take place per molecule, steric effects permitting. The reactions are highly dependent upon pH, since both reactants can bind and dissociate protons (see below).

Electrostatic Interactions

If the surface and the organic acid bear charges, then adsorption will involve, depending on the signs of the charges, either attractive or repulsive electrostatic interactions. Additional (repulsive) interaction may occur between the ionized groups of adsorbed molecules. These electrostatic effects result in the redistribution of co- and counterions, as discussed below.

Van der Waals Interactions and Hydrogen Bonding

Van der Waals interactions will always occur providing the adsorbate and adsorbent come sufficiently close, while surfaces and organic acids invariably bear groups that can form hydrogen bonds. It must be borne in mind, however, that these interactions are exchanges, in which relatively weak bonds to water are replaced by interactions involving the reactants, together with water–water bonding. Adsorption driven purely by these forces may therefore be weak and is more important for large adsorbates which can make many contacts with the adsorbent.

Coadsorption and Competition

When the organic acid is adsorbed, other components of the system will respond. The response may be entirely due to electrostatic effects. The coming-together of unlike charges on the reactants will cause the release of counterions, while the adsorption of organic acid anions by a negative surface will often result in the uptake of additional cations to prevent the build-up of too high an electrical potential at the interface. This might also be achieved by proton- and cation binding at specific sites on the adsorbent and adsorbate. Di- and trivalent cations may form ternary complexes in which they bridge the surface and the organic acid. Competition may occur between the organic acids and other solutes. In the case of the oxide surface, for example, likely competitors for ligand exchange sites are phosphate, silicate, and hydroxyl ions. Because natural organic acids are heterogeneous, the different kinds will compete with each other for adsorption, and this may result in fractionation (Davis and Gloor 1981).

Conformational Changes

Changes in the shape, size, and flexibility of relatively large organic acid molecules when they adsorb may contribute significantly to the net free energy of adsorption. The conformation of the adsorbed molecules may depend upon pH, ionic strength, interactions with other components, and adsorption density.

Given the prevailing conditions, i.e., concentrations of reactants, ionic strength, temperature, etc., the extent of adsorption in a particular system is determined by the interplay among some or all of the above processes. Although some quantitative data on organic acid adsorption are available, they are nearly all for simplified laboratory systems (see below), and mechanistically based quantitative predictions for environmental situations are not generally possible at present. Major requirements for the development of an appropriate model are the definition of natural surfaces, the taking into account of coadsorption and competition reactions (i.e., the inclusion of the complete "surface speciation"), and a proper description of the physical properties of (heterogeneous) organic acids. Ideally, the model should include kinetic aspects of adsorption and desorption, but this would be a formidable task on the basis of current knowledge. A further problem would be the prediction of rewetting of soil adsorption complexes that had dried out. Thus, present understanding is at a qualitative level, and its main use is the rationalization of field observations, rather than quantification and prediction.

ADSORPTION BY OXIDES

As mentioned above, the oxide surface in contact with water bears hydroxyl groups. These can take up and yield protons, according to the reactions

$$-SOH_2^+ \underset{}{\overset{-H^+}{\rightleftharpoons}} -SOH \overset{-H^+}{\rightleftharpoons} -SO^- \quad (2)$$

Each reaction can be characterized by proton dissociation equilibrium constants. The oxide surface can be characterized by the "point of zero charge" (pzc), which is the pH at which the net surface charge is zero. The lower the pzc, the more acid are the surface hydroxyl groups. In the absence of specific adsorption the surface bears a net positive charge below the pzc and a net negative charge above it. Values of the pzc vary from ~1 (SiO_2, δ-MnO_2) to ~12 ($Mg(OH)_2$). Oxides of Al, Fe, and Ti, and some Mn oxides, have pzc's in the neutral range.

Consideration of reaction (1) in terms of the Law of Mass Action shows that adsorption of an organic acid by an oxide surface is maximal when the product of the concentrations of −SOH and RCOOH (or RϕOH in the case of phenols) is maximal. The pH of maximum adsorption depends on the acidity of the oxide surface and on the pK_a's of the organic acid groups; for a fixed organic acid pK_a, it increases with oxide pzc. This partially explains why organic acid adsorption by silica is very weak at usual environmental pH values, while that by oxides of Al and Fe is strong.

In the model of Kummert and Stumm (1980), ligand exchange reactions such as Eq. 1 are combined with the surface protonation reactions of Eq.

2 and with a constant capacitance description of the electrical double layer, which takes purely electrostatic interactions into account. The model gives satisfactory explanations of the uptake of simple mono- and diprotic organic acids by γ-Al_2O_3.

Fig. 1 compares the pH dependences of adsorption by δ-Al_2O_3 of the simple organic acids studied by Kummert and Stumm (1980) with that of USOM (Urnesee Sediment Organic Matter, humic-type material extracted from a lake sediment), studied by Davis (1982). It is evident from the results for benzoic acid, phthalic acid, and salicylic acid that adsorption can be

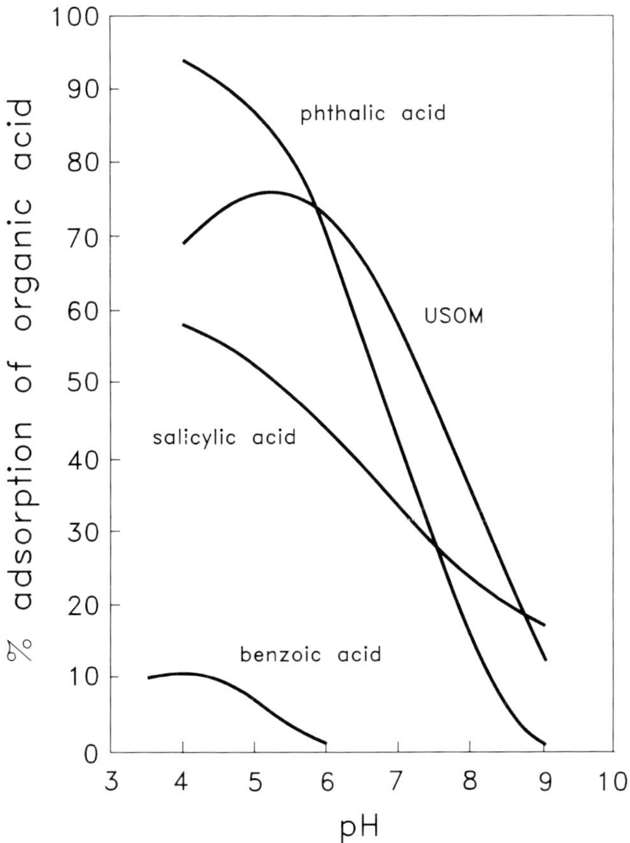

Fig. 1—Adsorption of organic acids by γ-Al_2O_3. Curves for benzoic, salicylic, and phthalic acids are redrawn with permission from Kummert and Stumm (1980) and refer to 2.2 g l^{-1} γ-Al_2O_3 (specific surface area 130 m^2 g^{-1}) with concentrations of organic acids, expressed in mg C l^{-1}, of 16.8 (benzoic, salicylic) and 19.2 (phthalic). The curve for USOM is adapted with permission from Davis (1982) and refers only to adsorbable material; the concentration of γ-Al_2O_3 (120 m^2 g^{-1}) in this case was 1.0 g l^{-1} and that of USOM was 9.4 mg C l^{-1}. The background electrolyte was 0.1 M NaCl and the temperature 22°C in all cases.

enhanced considerably by introducing a second acidic functional group able to interact with the oxide surface (see also Davis and Leckie 1978). This being so, the fact that the plots for phthalic acid and USOM are similar suggests that although the USOM has more acid functional groups per molecule than the simple acids, only a few of them are involved in coordinative links with the oxide surface. A direct comparison simply in terms of numbers of bonds is not fully justified, however, because other factors are probably involved in the adsorption of USOM due to its larger molecular size (see ADSORPTION PROCESSES). Perhaps most important from an energetic viewpoint is the appreciable uptake of protons by acidic functional groups not involved in direct binding to the oxide (Davis 1982). Such protonation is probably necessary to overcome electrostatic repulsion between adjacent adsorbed organic acid molecules, especially at high adsorption densities. However, not all the "excess" acidic functional groups are protonated: adsorbed organic acids confer a net negative charge on their adsorption complexes with iron and aluminum oxides, as revealed by electrophoretic studies (Davis 1982; Tipping 1981).

From the work of Ulrich et al. (1988) with fatty acids of various chain lengths, hydrophobic interactions do not seem to be directly important in the adsorption of organic acids by γ-Al_2O_3, except for long chain lengths (C_{12}) where interaction among adsorbed molecules may be important.

The divalent cations (chiefly Mg^{2+} and Ca^{2+}) present in natural solutions can have marked effects on the adsorption of humic substances by oxides. Adsorption can be considerably enhanced by the coadsorption of these cations, as illustrated by the results for Mn oxides shown in Fig. 2.

Desorption of USOM from γ-Al_2O_3 could be achieved fairly rapidly (within 24 hours) by raising the pH of a suspension initially at pH 5 (Davis 1980). A similarly rapid desorption of lakewater humic substances from a natural iron oxide precipitate has been observed by Tipping et al. (1989). However, repeated washing (at constant pH) of haematite particles to which natural riverine humic substances had been absorbed appeared to bring about little or no desorption, as judged by electrophoretic measurements (A.D. Findlay, E. Tipping, and D.W. Thompson, unpublished). It appears that desorption mediated by a competing adsorbate—OH^- in the above cases—is considerably more efficient than desorption by dilution, perhaps because in the competitive mechanism the adsorptive contacts are replaced one by one, and protons are displaced from noncoordinating sites on the adsorbed organic acids, whereas desorption by dilution requires several links to break simultaneously, a less likely, and therefore slower, process.

ADSORPTION BY SILOXANE SURFACES

Aluminosilicate particles provide large areas of surface in soils, sediments, and waters. Some of the surface is of the hydrous oxide type—mainly SiO_2

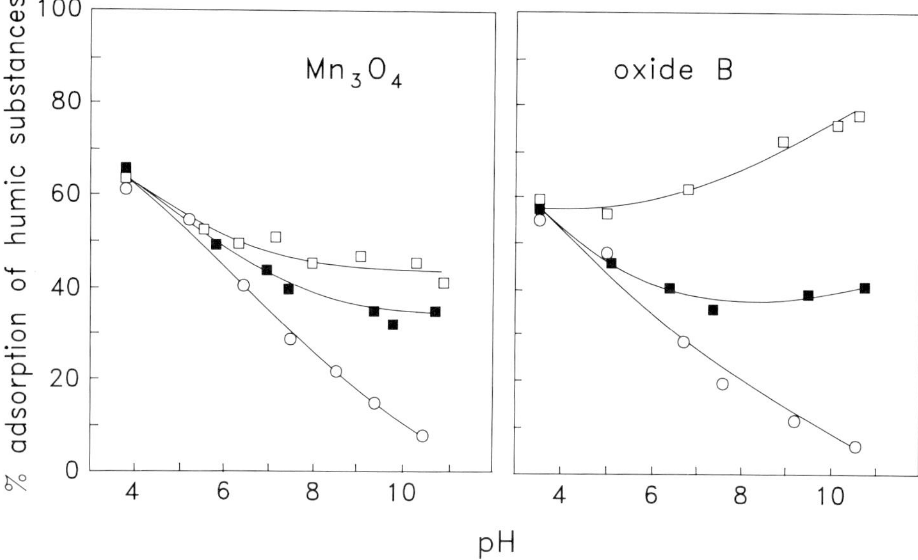

Fig. 2—Adsorption of Penwhirn Reservoir humic substances by Mn oxides: dependence on pH and $CaCl_2$ concentration. The Mn_3O_4 had a specific surface area of 65 m^2 g^{-1} and a pzc of 5.4; the corresponding values for oxide B (stoichiometry $MnO_{1.7}$) were 75 m^2 g^{-1} and 2.8. The concentrations of humic substances and oxides were 8 and 70 mg l^{-1}, respectively. Background electrolyte: 0.01 M NaCl (○), 0.01 M NaCl + 0.0001 M $CaCl_2$ (■) and 0.01 M NaCl + 0.001 M $CaCl_2$ (□).

and $Al(OH)_3$—and is expected to interact with organic acids similarly to pure oxides (see above), but most of it is the siloxane surface, which is composed of a nearly planar layer of oxygen atoms underlain by silicon atoms (see, e.g., Brown et al. 1978; Greenland and Mott 1978). In itself the siloxane surface is hydrophobic, but the isomorphous replacement of Si^{4+} by Al^{3+} usually results in a surface with a fixed negative charge balanced by mobile cations.

Theng and Scharpenseel (1975) studied the adsorption of humic acid by montmorillonite, an aluminosilicate clay the surface of which is almost entirely of the charged siloxane type. At PH 7 adsorption depended on the nature of cations present, being weakest for monovalents, stronger for divalents, and strongest for trivalents, although in the last case hydrolysis of the metal ions (Al^{3+}, Fe^{3+}), accompanied by a fall in pH, complicated interpretation of the results. According to Theng (1979) the role of the cations is to bridge negative sites on the clay and ionized acidic functional groups of the humic acid. It seems possible that the hydrophobic effect also operates.

Adsorption at the charged siloxane surface is pH dependent, decreasing with increasing pH due to increasing electrostatic repulsion as the humic acid groups ionize to a greater degree. The same effect is found for acidic polysaccharides (Theng 1979). At low pH (<4) electrostatic effects are much less, and fulvic acid (low molecular weight, water soluble) is even able to penetrate the interlayer spaces of montmorillonite, the spacing nearly doubling as a result (Schnitzer and Kodama 1966).

Although proper comparisons have not been made, the available evidence suggests that humic adsorption to the charged siloxane surface is considerably weaker than that to metal oxides. For example the plots in Fig. 3 show that adsorption of lakewater humic substances to goethite occurs in the solution

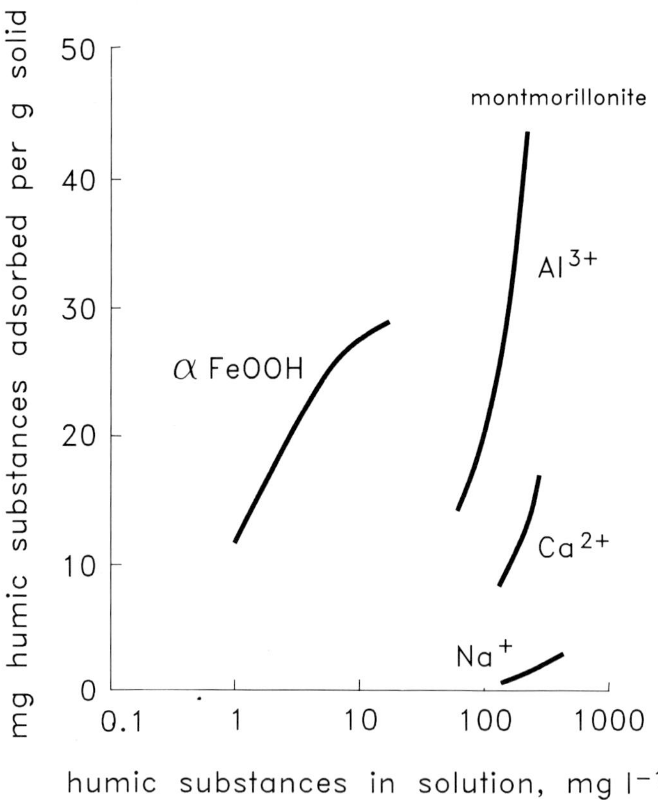

Fig. 3—Adsorption of humic substances by goethite ($\alpha FeOOH$) and by montmorillonite. The curve for $\alpha FeOOH$ (specific surface area 15 $m^2 g^{-1}$) is based on the data of Tipping (1981) for the adsorption of lakewater humics. The curves for montmorillonite (> 100 $m^2 g^{-1}$) are adapted with permission from Theng and Scharpenseel (1979) and refer to the adsorption of soil humic acid. The montmorillonite curves refer to solutions containing different cations as indicated; the curve for $\alpha FeOOH$ refers to 0.002 M $NaCl/HCO_3$. Results are for pH 7, except those for Al^{3+}-montmorillonite (pH 4).

concentration range 0.1–10 mg l^{-1}, whereas adsorption to montmorillonite takes place over the range 10–1000 mg l^{-1}. Weaker adsorption is consistent with the ease with which adsorbed humic acid can be removed from montmorillonite by washing (Theng 1979).

Further information on adsorption by the siloxane surface, and on other clay–polymer systems, can be found in Theng (1979). See also the books edited by Greenland and Hayes (1978, 1981).

ADSORPTION BY CARBONATES AND SULFATES

Wetzel and Otsuki (1974) showed that the precipitation and sedimentation of $CaCO_3$ removed "yellow organic acids" from the water of a marl lake, due to the adsorptive uptake of the acids by the precipitating carbonate. Cafe and Robb (1982) showed that polyacrylic acids (molecular weights 28,000 and 80,000 daltons) could adsorb to $BaSO_4$ and interpreted the adsorption data in terms of electrostatic attraction between positive Ba sites and the ionized groups of the polyacrylic acid; sulfate ions were also displaced by the adsorption of the acid. These findings suggest that natural organic acids may adsorb onto $BaSO_4$.

ADSORPTION BY ORGANIC SURFACES

Synthetic organic surfaces are known to interact with humic substances especially at low pH, and use is made of this in their isolation with XAD resins (Aiken 1985) and in water treatment with activated carbon (Weber et al. 1983). The adsorption in both cases is probably driven by the hydrophobic effect.

In the natural environment, organic surfaces include the outer walls of intact cells, and the surfaces of decomposition products such as plant litter and faecal pellets. Cell walls are composed largely of high molecular weight polysaccharides (Lehninger 1975) and these provide a hydrophilic surface of zero or negative charge. It is unlikely that humic substances or lower molecular weight organic acids would interact significantly with such a surface. However, other cell-wall components—lipids and proteins—have hydrophobic moieties that may undergo hydrophobic interactions with organic acids. Certain bacteria have a propensity to adhere to surfaces, and in at least some cases hydrophobic interactions are responsible. The apolar moieties exposed on the surfaces of these bacteria could conceivably interact with humic substances. Indirect evidence that such interactions can occur comes from work in which non-ionic detergents inhibited the hydrophobic attachment of bacteria to polystyrene (Paul and Jeffrey 1985). Decomposition products of plants and other organisms must have much less structural integrity than their living precursors, and are likely to present considerable hydrophobic surface for the possible adsorption of organic acids with apolar moieties.

Organic surfaces may possess positively charged groups, for example the protonated amino groups of proteins, and these could attract the negative charges of ionized organic acids electrostatically. In cases where the organic surface is negatively charged (probably the more usual situation), any adsorption by, for example, the hydrophobic effect, would probably require either specific (coordinative) or nonspecific coadsorption of cations to overcome electrostatic repulsion. In such cases adsorption is likely to increase as pH decreases.

An example of the adsorption of organic acids by intact cells comes from the work of Busch and Stumm (1968). These authors showed that several high molecular weight ($>10^5$) synthetic acidic polymers (polystyrene sulfonate, polyglutamic acid, polygalacturonic acid) could attach to the bacterium *Aerobacter aerogenes*, bringing about flocculation by bridging between cells. Hydrogen bonding, anion exchange, and interaction with surface cations were proposed as possible adsorption mechanisms. This may be a case where a large number of weak interactions, possible because of the large molecular size of the adsorbate, are responsible for adsorption.

PRECIPITATION AND SELF-ASSOCIATION

The solubility and aggregation behavior of humic substances can be rationalized at least to some extent by considering the interplay between their hydrophobic moieties on the one hand, and ionizable functional groups on the other, along the lines of Tanford's (1980) analysis of the solubility properties of straight-chain detergents, etc. Thus the precipitation of humic acid at low pH reflects the supremacy of hydrophobic interactions when solvation of the acid functional groups is largely abolished by their protonation. The higher solubility of uncharged fulvic acids—a feature exploited in their isolation—means that this material is less hydrophobic than the humic acids. Complexation with metal cations at neutral pH can also lead to the precipitation of humic substances.

As well as "straightforward" precipitation, there is evidence for the formation of relatively small, dispersed aggregates of humic substances in a process akin to micellization (Wershaw 1986).

Quantitative treatment of precipitation and self-association requires knowledge of the physical chemistry of humic substances in solution, and must take into account the heterogeneity—and therefore the range of solubilities—of the material.

ACCUMULATION AT THE AIR–WATER INTERFACE

Similar considerations apply to this phenomenon as to precipitation and self-association. The hydrophobic effect is the major driving force, apolar moieties of humic substances being removed from contact with the bulk aqueous phase while the hydrophilic parts remain in the aqueous phase. Substantial decreases in the surface tension of water can result; sedimentary fulvic acid has a surface activity similar to that of the well-known detergent sodium dodecyl (lauryl) sulfate (Hayase and Tsubota 1983). About 50% of the organic carbon of the sea surface microlayer is thought to be humic material (Sieburth 1983).

ENVIRONMENTAL CONSEQUENCES

The adsorption of organic acids by natural surfaces can have three kinds of effect. First, it may affect the solid phase by promoting or inhibiting dissolution or precipitation, or, when the solid is colloidal, by influencing aggregation. Second, it may alter the ways in which both adsorbent and adsorbate interact with other components such as metal ions, organic pollutants, and inorganic anions. The third kind of effect, and the one relevant to this Dahlem Workshop, is on the behavior of the organic acids themselves. In this context, the most obvious results of adsorption are the immobilization of organic acids on stationary surfaces of soils and sediments, and the control of solid-phase and solution concentrations of organic acids by interactions in suspension. Differential adsorption may fractionate organic acids. In lakes and oceans the gravity-driven transport of organic acids may occur when they are scavenged by precipitating particles (Wetzel and Otsuki 1974; Tipping and Woof 1983). Adsorbed organic acids may be more resistant to microbial attack than those in solution (Tate and Theng 1980). On the other hand surface adsorption may catalyze reactions among organic acids (Wang et al. 1983). It can be concluded that there is considerable scope for interactions at surfaces to influence the properties and concentrations of organic acids in aquatic ecosystems.

Acknowledgements. I thank J. Long for preparing the typescript, T.I. Furnass for assistance with the diagrams, and the Natural Environment Research Council for financial support.

REFERENCES

Aiken, G.R. 1985. Isolation and concentration techniques for aquatic humic substances. In: Humic Substances in Soil, Sediment, and Water, ed. G.R. Aiken, D.M. McKnight, R.L. Wershaw, and P. MacCarthy, pp. 363–385. New York: Wiley.

Aiken, G.R., D.M. McKnight, R.L. Wershaw, and P. MacCarthy, eds. 1985. Humic Substances in Soil, Sediment, and Water. New York: Wiley.

Birchall, S., D.J. Greenland, and M.H.B. Hayes. 1981. Adsorption of organic molecules. In: The Chemistry of Soil Processes, ed. D.J. Greenland and M.H.B. Hayes, pp. 221–400. Chichester: Wiley.

Brown, G., A.C.D. Newman, J.H. Rayner, and A.H. Weir. 1978. The structures and chemistry of soil clay minerals. In: The Chemistry of Soil Constituents, ed. D.J. Greenland and M.H.B. Hayes, pp. 29–178. Chichester: Wiley.

Busch, P.L., and W. Stumm. 1968. Chemical interactions in the aggregation of bacteria. Bioflocculation in waste treatment. *Env. Sci. Tech.* 2:49–53.

Cafe, M.C., and I.D. Robb. 1982. The adsorption of polyelectrolytes on barium sulphate crystals. *J. Coll. Int. Sci.* 86:411–421.

Davis, J.A. 1980. Adsorption of natural organic matter from freshwater environments by aluminium oxide. In: Contaminants and Sediments, ed. R.A. Baker, vol. 2, pp. 279–304. Ann Arbor: Ann Arbor Science.

Davis, J.A. 1982. Adsorption of natural dissolved organic matter at the oxide-water interface. *Geochim. Cosmo. Acta* 46:2381–2393.

Davis, J.A., and R. Gloor. 1981. Adsorption of dissolved organics in lake water by aluminium oxide. Effect of molecular weight. *Env. Sci. Tech.* 15:1223–1229.

Davis, J.A., and J.O. Leckie. 1978. Effect of adsorbed complexing ligands on trace metal uptake by hydrous oxides. *Env. Sci. Tech.* 12:1309–1315.

Greenland, D.J., and M.H.B. Hayes, eds. 1978. The Chemistry of Soil Constituents. Chichester: Wiley.

Greenland, D.J., and M.H.B. Hayes, eds. 1981. The Chemistry of Soil Processes. Chichester: Wiley.

Greenland, D.J., and C.J.B. Mott. 1978. Surfaces of soil particles. In: The Chemistry of Soil Processes, ed. D.J. Greenland and M.H.B. Hayes, pp. 328–353. Chichester: Wiley.

Hayase, K., and H. Tsubota. 1983. Sedimentary humic acid and fulvic acid as surface active substances. *Geochim. Cosmo. Acta* 47:947–952.

Kummert, R., and W. Stumm. 1980. The surface complexation of organic acids on hydrous γ-Al_2O_3. *J. Coll. Int. Sci.* 75:373–385.

Lehninger, A.L. 1975. Biochemistry. New York: Worth.

Mortland, M.M. 1970. Clay-organic complexes and interactions. *Adv. Agron.* 22:75–117.

Norde, W. 1980. Adsorption of proteins at solid surfaces. In: Adhesion and Adsorption of Polymers, Part B, ed. L.-H. Lee, pp. 801–825. New York: Plenum.

Paul, J.H., and W.H. Jeffrey. 1985. The effect of surfactants on the attachment of estuarine and marine bacteria to surfaces. *Can. J. Microbiol.* 31:224–228.

Schnitzer, M., and H. Kodama. 1966. Montmorillonite: effect of pH on its adsorption of a soil humic compound. *Science* 153:70–71.

Sieburth, J.M. 1983. Microbiological and organic-chemical processes in the surface and mixed layers. In: Air-Sea Exchange of Gases and Particles, ed. P.S. Liss and W.G.N. Slinn, pp. 121–172. Dordrecht: Reidel.

Tanford, C. 1980. The Hydrophobic Effect. New York: Wiley.

Tate, K.R., and B.K.G. Theng. 1980. Organic matter and its interactions with inorganic soil constituents. In: Soils with Variable Charge, ed. B.K.G. Theng, pp. 225–249. Lower Hutt, NZ: New Zealand Soc. of Soil Sci.

Theng, B.K.G. 1979. Formation and Properties of Clay-Polymer Complexes. Amsterdam: Elsevier.

Theng, B.K.G., and H.W. Scharpenseel. 1975. The adsorption of ^{14}C-labelled humic acid by montmorillonite. In: Proc. Int. Clay Conf., Mexico City, 1975, ed. S.W. Bailey, pp. 643–653. Barking, UK: Applied Science.

Tipping, E. 1981. The adsorption of aquatic humic substances by iron oxides. *Geochim. Cosmo. Acta* 45:191–199.

Tipping, E., D.W. Thompson, and C. Woof. 1989. Iron oxide particulates formed by the oxygenation of natural and model lakewaters containing Fe(II). *Arch. Hydrobiol.* 115:59–70.

Tipping, E., and C. Woof. 1983. Elevated concentrations of humic substances in a seasonally anoxic hypolimnion: evidence for co-accumulation with iron. *Arch. Hydrobiol.* 98:137–145.

Ulrich, H.-J., W. Stumm, and B. Cosović. 1988. Adsorption of aliphatic fatty acids on aquatic interfaces. Comparison between two model surfaces: the mercury electrode and γ-Al_2O_3 colloids. *Env. Sci. Tech.* 22:37–41.

Wang, T.S.C., M.-C. Wang, and P.M. Huang. 1983. Catalytic synthesis of humic substances by using aluminas as catalysts. *Soil Sci.* 136:226–230.

Weber, W.J., T.C. Voice, and A. Jodellah. 1983. Adsorption of humic substances: the effects of heterogeneity and system characteristics. *J. Am. Wat. Wks. Assn.* 5:612–619.

Wershaw, R.L. 1986. A new model for humic materials and their interactions with hydrophobic organic chemicals in soil-water or sediment-water systems. In: Transport and Transformation of Organic Contaminants, ed. D.L. Macalady. *J. Cont. Hydrol.* 1:29–45.

Wetzel, R.G., and A. Otsuki. 1974. Allochthonous organic carbon of a marl lake. *Arch. Hydrobiol.* 73:31–56.

Standing, left to right:
Simon Visser, Christian Steinberg, Uwe Münster, Robert Wetzel, Peter Behmel, Ed Tipping

Seated, left to right:
Bob Petersen, Jr., Olav Magnus Skulberg, David Francko, Diane McKnight, Egil Gjessing Not shown: Peter Werner

Organic Acids in Aquatic Ecosystems
eds. E.M. Perdue and E.T. Gjessing, pp. 223–243
John Wiley & Sons Ltd

Group Report
How Do Organic Acids Interact with Solutes, Surfaces, and Organisms?

D.M. McKnight, Rapporteur
P. Behmel
D.A. Francko
E.T. Gjessing
U. Münster
R.C. Petersen, Jr.
O.M. Skulberg
C.E.W. Steinberg
E. Tipping
S.A. Visser
P.W. Werner
R.G. Wetzel

INTRODUCTION

An aquatic ecosystem is comprised of interacting aquatic organisms, which range in size from bacteria to fish, and their chemical and physical environment. An obvious aspect of aquatic ecosystems is that organisms are always in intimate contact with water; a not so obvious consequence, however, is that dilute concentrations of dissolved solutes can significantly influence interactions in aquatic ecosystems. Organic acids in natural waters include a multitude of compounds; some, such as algal toxins, are exceedingly dilute and yet may have specific effects. Others, such as humic substances, are comparable in concentration to major ions and may have general or indirect effects. Organic acids can interact with different species of organisms, with other solutes such as cations and metals, and with particulate surfaces. The extent and intensity of these interactions will depend upon the chemical characteristics of the organic acid and upon its concentration relative to the abundance (or concentration) of a given organism, solute, or surface.

Organic acids are an important fraction (up to 90%) of the total pool of dissolved organic material. Here, we consider interactions that influence the temporal and spatial variability of organic acids in aquatic ecosystems by first reviewing critical aspects of these interactions, and then discussing those interactions relevant to water quality concerns, specifically eutrophication, acidification, and drinking water treatment.

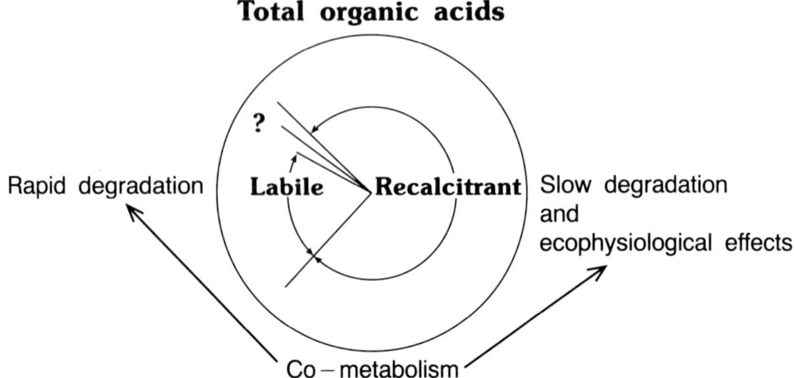

Fig. 1—Schematic diagram showing (*a*) that the total pool of dissolved organic acids is comprised of labile and recalcitrant species and (*b*) that the rates of degradation can be interrelated and dependent upon other aspects of the aquatic ecosystem.

BIOLOGICAL INTERACTIONS

Generally, there are two possible and nonexclusive ways by which organic acids directly affect aquatic organisms. One is by providing a substrate or a specifically required nutrient for growth and metabolism and the other is by influencing the physiological processes of an organism. The significance or potential for such biological interactions varies greatly among different organic acids and communities of organisms.

Organic Acids as Substrates

From an ecological perspective all dissolved organic acids are detrital (nonliving) organic carbon (Steinberg and Münster 1985). Uptake of organic acids as energy sources is a detrital pathway of carbon and energy transfer in an aquatic ecosystem, in contrast to a grazing (herbivorous) pathway (Wetzel et al. 1972).

Labile Organic Acids. Organic compounds that are rapidly used as energy-yielding substrates for heterotrophic growth are described as labile. Lower molecular weight organic acids (less than 500 daltons) may be important labile organic acids, although they comprise a lesser fraction of the total pool of dissolved organic acids, as represented in Fig. 1. In some environments, organic acids of higher molecular weight (up to 1000 daltons) also show considerable biodegradability (Hitsch and Werner 1988). Because of their labile nature, the contribution from degradation of these organic acids to the overall flux of organic carbon and energy through the detrital pathway may be much greater than their proportional concentration in the total organic pool. Measurements of flux or uptake rates can be derived from the measurement of the uptake of a radiolabeled low molecular weight organic acid (e.g., ^{14}C- or ^{3}H-labeled amino acids, carbohydrates, or fatty acids) and measurement of its *in situ* concentration. Because of their rapid turnover rates, *in situ* concentrations for these compounds are mostly in the nanomolar range and are quite difficult to measure. A further challenge is the numerous organic acid species in the environment. They all may contribute to the detrital carbon flux, but their individual contribution to the total carbon flux is still not easily measured. Further, we do not know if there are synergistic effects which vary with changing environmental conditions.

Consequently, the major limitations in aquatic ecosystems on the uptake of labile organic acids for microbial growth are not well known. Possible limiting factors are the chemical characteristics and speciation of a given organic acid, its concentration, and the availability of other nutrients (such as N and P) and cofactors.

Recalcitrant organic acids. As shown in Fig. 1, a greater fraction of the total dissolved organic acid pool is comprised of higher molecular weight organic acids, which are mainly, but not entirely, comprised of aquatic humic substances. Aquatic humic substances are further classified as fulvic acid (soluble at all pH values) and humic acid (soluble above pH 2); fulvic acids generally account for 40–60% of the dissolved organic acid pool. The number average molecular weights of isolated aquatic fulvic acids are usually between 500 and 1000 daltons (Aiken and Malcolm 1987); but it is also possible that these compounds could exist in natural waters as dimers, trimers, and aggregates of one kind or another.

These higher molecular weight compounds are not rapidly degraded; however, studies of organic carbon budgets for lacustrine systems show that these compounds are degraded over annual time scales (Wetzel 1983). Fulvic acids comprise a large portion of both dissolved and particulate organic material, therefore slow degradation of these compounds must also contribute substantially to the overall detrital flux of organic carbon. Because their degradation is known to occur, the descriptor "recalcitrant" may be a more appropriate term than "refractory" for the classification of aquatic humic substances.[1]

Studies in polyhumic lakes have shown that the biomass of bacteria plankton is positively correlated with humic material content, DOC, and

[1] **recalcitrant**: adj.: difficult to handle or operate. **refractory**: adj.: difficult to fuse, corrode, or draw out; esp. capable of enduring high temperature. (From Webster's New Collegiate Dictionary).

water color (Hessen 1985; Tranvik 1988). Because of their low inorganic nutrient content, polyhumic waters are thought to be less productive (Wetzel 1983) and are therefore also called dystrophic (from the Greek for bad nutrition). However in many dystrophic lakes a high ratio between respiration and primary production is found, possibly indicating that more carbon is available for the microheterotrophs than is supplied from autochthonous primary production (Salonen et al. 1983; Salonen 1984).

A characteristic of many aquatic bacteria is their small size, with DNA comprising a greater portion of cell biomass (Simon and Azam 1989). Bacterial isolates from humic-rich waters carry higher potential for degradation of polyphenols compared to isolates from clear water lakes (Tranvik and Hofle 1987). This higher potential for degradation of polyphenolic humic matter may be encoded in extrachromosomal DNA (plasmids). Bacterial isolates from polyhumic waters have higher quantities of plasmids, and many isolates have significant resistance to antibiotics and heavy metals (Schütt 1988a, 1988b, 1989).

The relative importance of different compounds in detrital pathways may change with time. Heterotrophic bacteria may respond quickly to short-duration pulses of labile organic compounds and be maintained between intervals by slow degradation of humic substances.

Although recalcitrant organic acids may not be energy sources that can be utilized efficiently by microorganisms, these organic acids comprise a large portion of dissolved organic nitrogen and dissolved organic phosphorus, and are potential sources of these critical nutrients. In this context, it may be that only a portion of a humic molecule is utilized microbially. Some N or P assimilation may occur, with concomitant CO_2 respiration, and the residual humic molecule may remain in solution and may recondense to form a new larger molecule. Under such a sequence, the turnover times for different moieties in humic molecules may be different. This process can also occur for smaller molecules, e.g., amino acids (Palenik and Morel 1989).

The extent to which biodegradation is accomplished by intracellular uptake of intact humic molecules is not known; however, it cannot be excluded as a possibility. Cell membranes cannot be represented as plastic sheets with pores; they are active transport systems that are adapted to external conditions and internal requirements. The size of aquatic fulvic acids does not preclude uptake. For example, the protein ferritine with a molecular weight of 2×10^6 daltons can be taken up intact by plant cells (Visser 1986). Another example is the uptake of Dextran (about 10^5 daltons) by protozoans (Sherr 1988). Single-celled organisms can have versatile ingestion mechanisms such as pinocytosis or phagocytosis. As an extreme example, an unpigmented (nonphotosynthetic) chrysophyte, *Spumella vulgaris*, with a normal diameter of 10 μm, can ingest into a food vacuole a diatom

Synedra acus, with a silica frustule length of 100 μm (Fig. 2). Such mechanisms may allow uptake of dissolved humic substances and even colloidal aggregates of humic substances.

Even though fulvic acids and other high molecular weight organic acids may only be used very inefficiently as sole substrates for maintenance and production by bacteria, co-metabolism of these compounds may account in part for their degradation. Co-metabolism is a process whereby these compounds are metabolized with other simple organic compounds (e.g., Horvath 1972). Aquatic bacteria rapidly increase activity in response to certain humic substances if present in combination with simple substrates, such as benzoic acid (e.g., De Haan 1977). Very little is known about these microbial degradation mechanisms and their environmental regulation, even though respiration and other metabolic measures clearly indicate that bacteria (and likely fungae) are responsible for most *in situ* decomposition.

Ecophysiological Effects of Organic Acids

Ecophysiology refers to the physiological "fitness" of an organism within its own ecosystem. Ecophysiological responses can be caused by both low molecular weight organic acids and high molecular weight organic acids, including humic acids. Examples of low molecular weight compounds having such effects are toxins and other allelochemic compounds released by aquatic organisms; these compounds may have effects at very low concentrations. Toxins produced by algae in eutrophic waters will be discussed subsequently. An example of a nontoxic effect is the effect of the regulatory molecule cyclic AMP on phytoplankton metabolism at concentrations as low as 5 nmol l^{-1} (Francko 1984a, 1984b).

Three general pathways for ecophysiological effects of humic substances can be identified. These involve: (*a*) changing the bioavailability of nutrients or toxicants by changing their chemical speciation; (*b*) sorption of humic molecules on cellular membranes; and (*c*) intracellular uptake of humic molecules (without subsequent degradation). Influences on the chemical speciation of nutrients and/or toxicants primarily occur through complexation reactions or indirectly through influences on the pH of the aquatic medium. Chemical species involved may be either metals, hydrophobic organic micropollutants, or anions, through such processes as formation of humic–iron–phosphate complexes (Francko, this volume). These interactions are further discussed in the broader context of chemical and physical interactions occurring in the environment.

The surface active properties of humic substances suggest that they may affect the structure, fluidity, or permeability of cell membranes, where humic molecules first encounter a cell (Visser 1982). Although such effects have not been studied to any extent, several mechanisms can be speculated

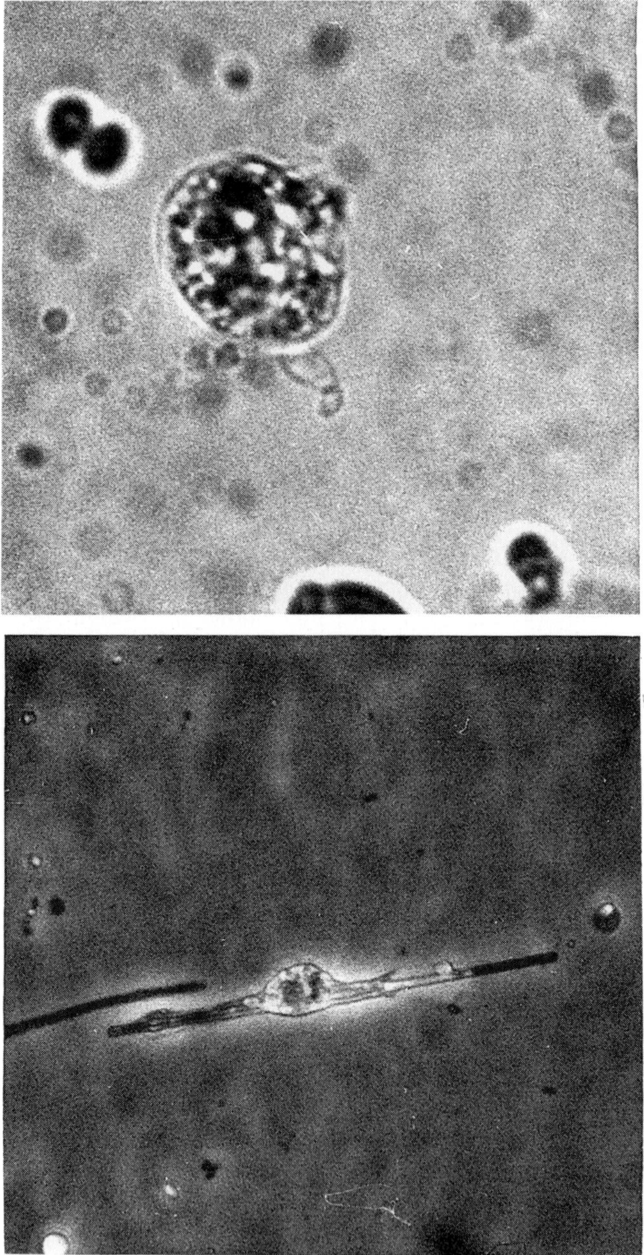

Fig. 2—Photomicrographs showing (A) a normal cell of *Spumella vulgaris*, a nonpigmented chrysophyte (magnification: 300 times); and (B) a *Spumella vulgaris* cell after ingestion of a diatom, *Synedra acus* (100 μm in length), into the food vacuole (magnification: approx. 50 times).

to exist. Many cells are surrounded by an extracellular layer of mucilage, composed of mucopolysaccharides, and humic substances could sorb and interact with these materials. The absorption of humic molecules at the cell membrane may protect the cell wall material from enzymatic attack (Behmel 1986; Münster et al. 1989). Humic acids may also affect the charge density in the electrical zone of influence of microbial cells, which often have a negative charge. In this context, the charge and size of these humic molecules are significant questions. Humic molecules could potentially act as transport compounds, facilitating the transport across the cell membrane of inorganic or organic nutrients. In polyacrylamide gradient gel electrophoresis, humic substances did not follow the retardation mechanisms commonly found for proteins, which suggests that a flexible structure may allow humic substances to move through cell membranes under certain conditions (Münster 1985; Steinberg and Münster 1985). Intracellular desorption of nutrients from humic carrier compounds may occur through changes in electrical potential or redox state. Although these possible membrane processes are potentially significant, only a few examples have been demonstrated to occur (Francko 1989). Some available techniques (Kaplan et al. 1982) have not yet been applied to the study of organic acid effects. Other potential physiological effects of extracellular or intracellular humic molecules are increases or decreases in enzymatic activities. Inhibition of enzymes could occur by competitive inhibition or by noncompetitive processes, such as changes in the tertiary structure of the enzymes. Activation of enzymes is equally possible, perhaps as a result of increasing the accessibility of the active site. Surfactants are known to aid in the induction of enzymes, and humic substances are known to have surfactant properties (Visser 1982, 1986). An indication of the potential importance of such processes is the observation that bacteria which exhibited denitrifying capabilities when isolated from the environment on standard culture medium became starch decomposers when cultured in the presence of humic substances (Visser 1985, 1988). These results suggest that humic substances may mediate the metabolic activities of microorganisms, allowing for differences in metabolism with changes in environment (e.g., movement through the soil column, transport from soil to aquatic systems).

The concentration of humic substances can be critical in determining the direction and intensity of an ecophysiological influence. At low concentrations, beneficial effects may occur, but at some break point within the concentration range of neutral pH waters, other toxicological effects may dominate (Petersen, this volume). Variations in the ability of organisms to exist in humic-rich waters are indicated by the dominance of flagellates in the phytoplankton of humic (dystrophic) waters (Carlough and Meyer 1989).

Chemical and Physical Interactions

Chemical and physical interactions involving organic acids in natural waters are greatly dependent upon concentration. The importance of concentration is a consequence of the law of mass action as expressed by mass action equations governing chemical processes. In contrast to biological interactions, organic acids present in dilute concentrations (ng–μg/l) will therefore not generally have a significant effect on the chemical and physical environment. The only exceptions are compounds which have high affinities for particular metals or ligands. For example, siderophores have stability constants for ferric iron–siderophore complexes of about 10^{30}.

The primary importance of concentration can be illustrated by considering the sorption of fulvic acids by hydrous metal oxides. If the concentration of oxides is low (less than 1.0 mg/l), then their surface properties will be largely determined by the sorption of fulvic acid, but the sorption will not substantially affect the distribution or chemical properties of the dissolved fulvic acids at 2 mg/l. Conversely, if a natural water is in contact with large quantities of hydrous metal oxides (e.g., B horizon in soils), then the concentration, distribution, and chemical characteristics of the dissolved fulvic acids will be changed.

The primary importance of concentration has two general consequences for understanding the processes causing spatial and temporal variability in organic acids. The first is that classes of organic acids that are most abundant in the total organic pool (Fig. 1) are likely to be most important in the context of altering the chemical/physical environment. Aquatic fulvic acids account for a large proportion of the organic acid pool; another large proportion is comprised of more hydrophilic organic acids that may have some similar properties to fulvic acids (Aiken, pers. comm.). The second consequence is that because (*a*) the range in concentrations of fulvic acids in natural waters is more than an order of magnitude, from <1 to above 30 mg C/l, and (*b*) our ability to measure fulvic acid concentration is limited, especially for binding site concentration, models that represent the major reactions with some simplifying assumptions should be able to reproduce or predict observations to some reasonable extent. If this approach is generally successful, then more detailed elaborations can be added to the models to test hypotheses that explain deviations between field and model results.

Complexation reactions. Complexation properties of organic acids are dependent primarily upon the number of complexation sites. At a first approximation, trace metal binding properties of different humic isolates are similar. Polyelectrolyte effects also influence complexation, but the ratio of molecular size to acidic group content is reasonably constant and the effect does not change substantially for humic substances. The metals that have been studied most are copper and aluminum, and the results from different laboratory studies of metal complexation by fulvic acids are similar (Buffle 1988).

The most abundant functional groups involved in metal binding are carboxylic acid groups. Evidence for this comes from spectroscopic studies showing copper–carboxylate charge transfer bands in the range of 240 nm (Piotrowicz et al. 1984). For transition metals, less abundant complexation sites may be nitrogen- and sulfur-containing functional groups (Frimmel et al. 1984; McKnight et al. 1988). These sites may be important at the concentration ratios of fulvic acids and dissolved metals found in natural waters.

Adsorption reactions. As illustrated in Fig. 3, adsorption reactions become increasingly important as the charge of organic acids decreases and their molecular weight and hydrophobic properties increase. The charge of an organic acid depends upon the number of acidic functional groups, the pH, and the concentration of metal ions. At the pH of most natural waters (pH 5–8), most carboxylic acid groups will be ionized (Petersen, this volume). With a decrease in pH below 5 there will be an increase in the proportion of un-ionized organic acids. The charge of organic acids also decreases with increasing metal ion concentrations; for example adsorption of humic substances by oxide surfaces is increased by additions of calcium (Tipping, this volume).

Sorption of organic acids by amphoteric and basic oxide surfaces is a ligand exchange process. Sorption per unit particle surface area is greater for these oxide surfaces than for aluminosilicates and calcium carbonates. The importance of different sorbing surfaces will depend upon the relative abundance of oxide, aluminosilicate, and calcium carbonate surfaces in an aquatic environment. The mechanisms of sorption are discussed by Tipping (this volume).

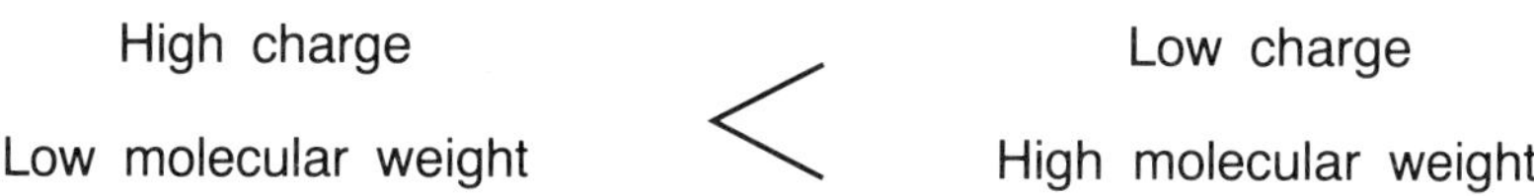

Fig. 3—Dependence of adsorption on charge and molecular weight of organic acids.

Photochemical reactions. Photochemical processes may affect organic acids in two ways: (*a*) by changing redox state of associated metals and bound constituents (e.g., Fe) and (*b*) through organic-acid-mediated formation of strong oxidants, which react with organic acids and other constituents. These processes are discussed by Francko (this volume). The importance of photodegradation of humic substances relative to slow microbial degradation is unknown for either clear water or humic-rich freshwaters. Hydroxyl radicals and organic radicals produced during photolysis could have toxic effects on organisms or may have other indirect ecophysiological effects.

Interactions controlling sources of organic acids to aquatic ecosystems. The diagram in Fig. 4 presents a simple illustration of two source pathways of organic acids from soils; in several zones chemical interactions can, in addition to microbial processes, control the fractionation and transport of organic acids. Fractionation occurs first between the solid phase organic material and the organic acids dissolved in the soil interstitial water. Only the weakly sorbed organic acids will enter the solution phase. When the upper (O/A) soil horizons are not saturated, the hydrologic flow path is through the mineral soil or B horizon, where large quantities of organic acids are sorbed by Al and Fe oxide solid phases. Another fractionation of organic acids occurs in this mineral soil or B horizon, and the soil interstitial

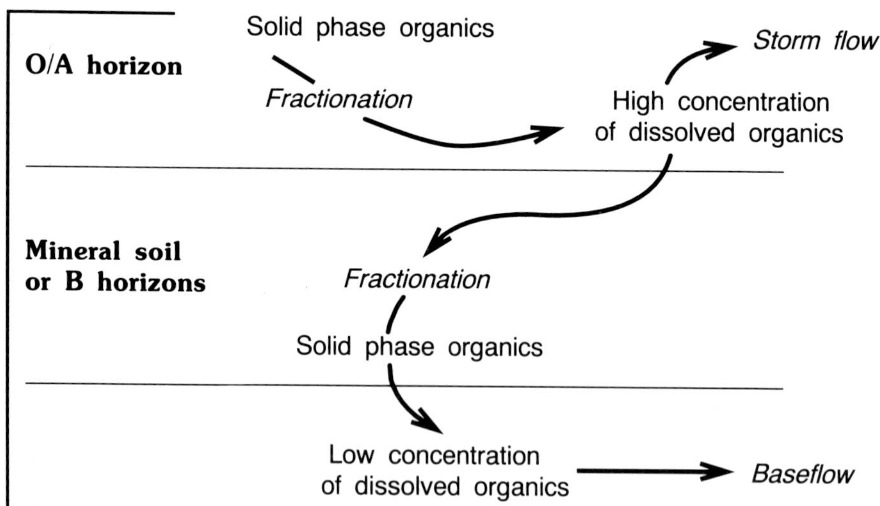

Fig. 4—Schematic diagram showing two source pathways for organic acids from soils.

water below this soil horizon could therefore have low concentrations of organic acids and a different distribution among the organic acid pool. These processes explain the lower organic acid concentrations in baseflow discharge to streams (Cronan, this volume). The changes in hydrologic flow path during periods of high infiltration rate (storm events and snowmelt) may result in a bypass of the mineral soil or B horizon, such that the interstitial waters from the O/A horizon enter a stream or lake more directly. In such events, the concentration of organic acids will reflect the initial concentrations in the O/A horizon water, and the extent of dilution by rainfall or snowmelt. Furthermore, the distribution and chemical characteristics of organic acids in the water will not have been influenced by the second fractionation at the mineral soil or B horizon. Depending upon particular soil types and profiles, geomorphology, and hydrology, these underlying processes can lead to different patterns of organic acid concentrations in response to storms and snowmelt.

INTERACTIONS OCCURRING IN AQUATIC ECOSYSTEMS

Small shallow freshwater lakes can be used as appropriate examples to test hypotheses about interactions in aquatic ecosystems. The common occurrence of small shallow lakes is illustrated in Fig. 5, which shows the occurrence of lakes of different surface areas and mean depths. Large lakes, with long residence times and phytoplankton as major sources of organic material, occur with a lesser geographical frequency than shallow lakes. In most shallow lakes, light penetrates to the lake bottom, and littoral zone primary

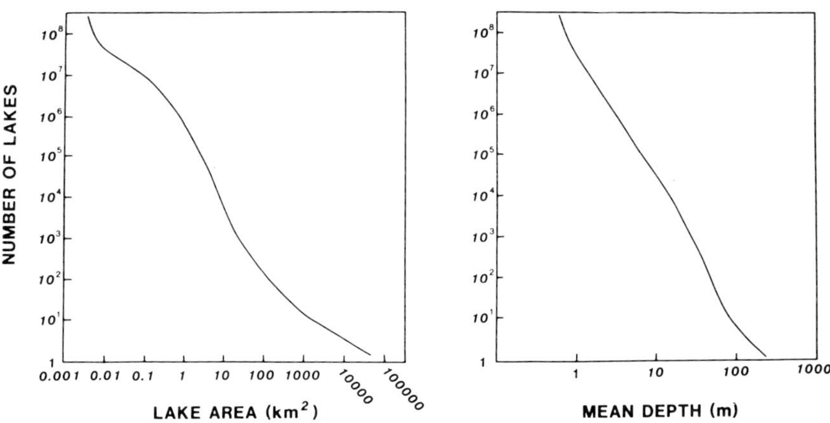

Fig. 5—Relation of number of lakes and reservoirs of the world versus surface area and mean depth (from Wetzel 1989).

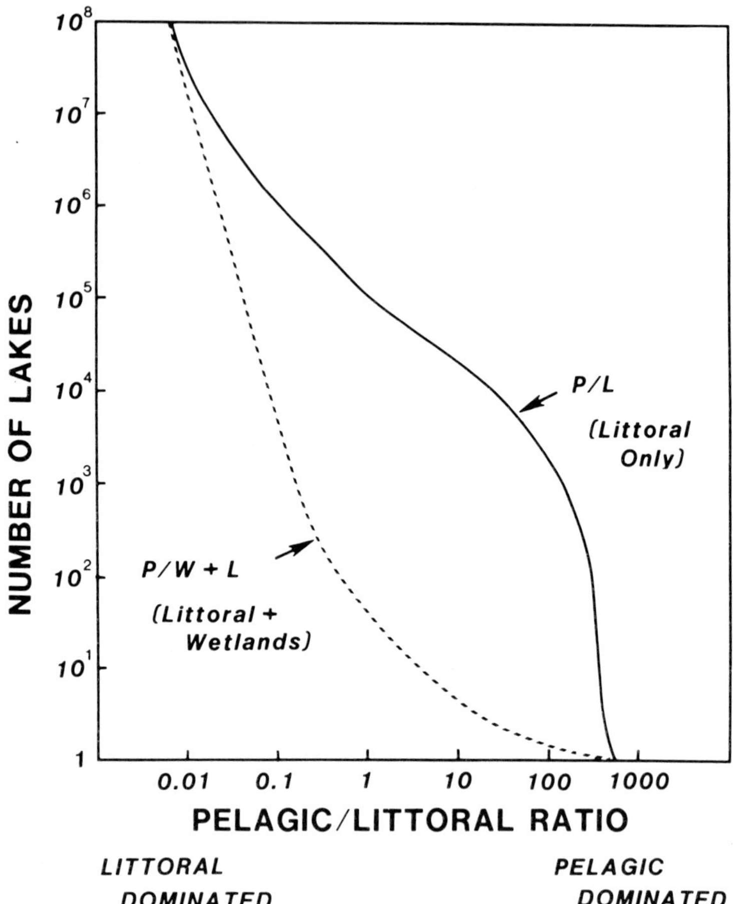

Fig. 6—Relation of the number of lakes and reservoirs of the world versus the ratio of pelagic zone (insufficient light reaching sediments to support attached photosynthetic growth) to littoral zone. A comparison is made among those lake ecosystems in which only the area of productivity of the littoral zone, L (water edge to depth where sufficient light is available to support benthic photosynthetic growth), is compared to the area or productivity of the pelagic zone, P (——). The more realistic relationship is also presented in which the area or productivity of the wetlands, W (portion of the lake ecosystem in which the soils are water saturated), plus that of the littoral zone is compared to that of the pelagic zone (----). From Wetzel (1989).

production is greater than pelagic zone primary production (Fig. 6). Periphyton developing on littoral surfaces can contribute significantly to primary production. The macrophyte surface area for periphyton growth was 24 m^2 per m^2 of surface area of littoral zone in one lake (Burkholder and Wetzel 1989). These shallow lakes are often surrounded by wetland areas.

In the context of the chemical properties of autochthonous organic acids, the dominance of littoral vascular plants is significant. Wetland and littoral plants have structural tissue composed of lignins and celluloses. Upon senescence, this structural tissue is deposited in lake sediments, where decomposition and production of dissolved organic acids mostly occurs under anaerobic reducing conditions. Soluble organic material from littoral plants can include labile and recalcitrant material; in a study of degradation of ^{14}C-labeled cattails, 50% of the soluble material was degraded within 48 hours and 80% was degraded within 100 days (Wetzel 1989). In these shallow lakes, transition zones between anoxic and oxic conditions will be important zones for sorption of organic acids by oxide and other sorptive surfaces. Biological processes related to activities of benthic organisms living at the sediment water interface may further influence the fluxes and composition of organic acids. These interface processes will also be influenced by seasonal changes and seasonal cycles of littoral and wetland plant growth, as well as meteorological events.

Ecosystem Level Interactions

Given that organisms have existed in aquatic ecosystems over evolutionary time scales, the development of complex and unexplored organic-acid-mediated ecosystem linkages seems probable. Such linkages could be manifested in detrital and herbivorous pathways of organic carbon cycling and energy flow.

Small shallow lakes dominated by littoral plants provide a useful example for consideration of trophic level linkages mediated by dissolved organic acids. The attenuation of light by colored fulvic acids can affect nutrient/light availability interactions of autotrophs. More highly colored systems will be shifted from dominance by planktonic production to dominance by littoral production because attached and emergent plants maintain their position in the light profile to avoid light limitation. These plants contain a greater content of structural material than phytoplankton, which in turn enhances production of recalcitrant colored organic acids during decomposition. This feedback loop may accelerate the progression of such a lake system from a pelagic to a wetland system (Wetzel 1979). These effects on littoral vs. planktonic primary production will also cascade upwards to higher trophic levels (e.g., planktonic herbivores such as daphnids and other cladocerans vs. benthic shredders and scrapers in littoral-dominated ecosystems). Under markedly acidified conditions in soft waters, dissolved organic acids can be reduced, presumably by condensation to particulate organic aggregates. Light transparency can increase under these specialized conditions and result in increased depth development of submersed macrophytes and periphyton.

Organic acids might also exert ecosystem-level controls if their presence affects organisms in higher trophic levels. One example of this would be indirect effects from controlling bioavailability of toxicants or nutrients to bacteria, phytoplankton, or macrophytes. Changes in energy processing and nutrient flux among assemblages of these organisms would cascade through both detrital and herbivorous food webs.

INTERACTIONS RELEVANT TO IMPORTANT WATER QUALITY ISSUES

Three current water quality problems include aspects that involve interactions with dissolved organic acids; the issues considered were eutrophication of lakes, acidification of lakes and streams by acidic deposition, and treatment of drinking water. In these water quality issues, interactions that were discussed in the previous sections in a theoretical (and sometimes speculative) manner take on practical significance. In the production of algal toxins in eutrophied lakes, allelochemic organic acids acting at low concentrations are used as an important example. For effects due to acidification, metal complexation and membrane sorption are important processes. The biological availability of recalcitrant organic acids is a key unknown in controlling microbial regrowth in distribution systems of public drinking waters.

Algal Toxins in Eutrophied Waters

During summer in eutrophied inland waters, an abundance of blue-green algae may reach very high cell densities (Paerl 1988). One major aspect of this problem that involves organic acids is the production of toxins by the dominating organisms (Skulberg et al. 1984). This situation is illustrated by the diagram in Fig. 7. Production of algal toxins is also a major problem during red tide episodes in marine and estuarine waters.

Although algal toxins encompass several classes of organic compounds, only a few toxins have been identified, synthesized, or toxicologically determined. Among the most intensively studied are hepatotoxic cyclopeptides and neurotoxic alkaloids produced by selected strains of cyanophyte species (Codd and Poon 1988). These toxins cause acute and chronic oral toxicity in animals (including cattle and humans). Most common bloom-forming cyanophytes are represented with toxic strains, e.g., *Oscillatoria, Microcystis, Anabaena*, and *Aphanizomenon*, and altogether 20 toxin-producing species have so far been identified. On a regional basis almost 50% of blooms of cyanophytes are composed predominantly of toxin-producing strains (Berg et al. 1986). Toxic cyanophyte blooms within geographical areas are intermittent and unpredictable. Toxin production is apparently an obligate process in toxin-producing strains, but the factors

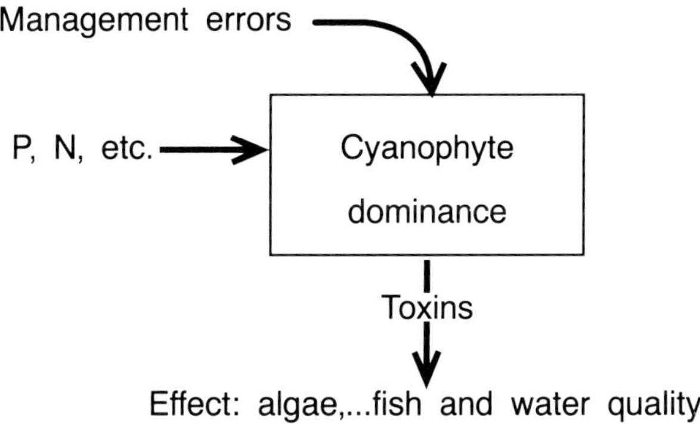

Fig. 7—Schematic diagram representing eutrophication effects on algal toxin production.

causing dominance of these strains in nature are not known (Skulberg and Skulberg 1985). The contents of toxins in living cells are rather high—up to 2 mg per gram of cells on a dry weight basis. The toxins are released into the water during degradation of cellular material and, under natural circumstances, may persist in the water for 3–4 weeks (Berg et al. 1987).

Extensive nuisance blooms causing severe fish kills in Scandinavian coastal waters during the early summer of 1988 were due to the prymnesiophyte flagellate *Chrysochromulina polylepis* (Underdal et al. 1989). The toxins involved—with chemical properties similar to those of acidic polar lipids—affected a broad range of aquatic organisms, including other algae, echinoderms, and fish. One toxin with hemolytic and ichthyotoxic properties (a digalactocyl-monacyl-glycerolipid) has been identified, suggesting that polyunsaturated fatty acids and hemolysins contribute to the fish kills and other deleterious effects of toxic algae (Skulberg, pers. comm.).

Acidification of Inland Waters

Many freshwaters are close to saturation with respect to calcium carbonate and are not very susceptible to acidification. However, there are geographically large areas of predominantly noncalcereous terrain, where acidification of natural waters from acidic deposition has proceeded over the past several decades. In these acidified natural waters decreases in the abundance of fish, crustaceans, and algae have occurred.

One of the toxic agents in acidified systems is dissolved free aluminum (Al^{3+} and hydrolysis products). Much variation in toxic effects between different streams and lakes has been explained by complexation of aluminum by organic acids (Nilssen 1982; Raddum and Saether 1981; Gjessing et al. 1989), and processes that determine relative concentrations of organic acids and aluminum are of paramount importance. However, there are some differences in toxic effects that remain unexplained after consideration of Al chemistry. These remaining toxic effects may be related to the characteristics of dissolved humic substances in acidified waters. Humic substances have a greater toxicity to aquatic invertebrates at lower pH values (pH 5 vs. pH 6 and 7) (Petersen, this volume). This pH-dependent effect may be caused by greater lipid solubility at low pH and changes in the partitioning of humic substances into cell membranes. Increased photoreduction of peroxides and other toxic photoproducts at low pH are another possible effect. In toxicological tests using salmon yearlings and *Daphnia*, humic substances purified by distilled water dialysis were more toxic from acidified waters than nonacidified waters (Gjessing, this volume). Sulfur interactions with humic substances may be involved. Another possible change in humic substances in acidified waters could be their molecular configuration or tertiary structure. The probability of such changes is an open question. Differences in predicted interactions arise from different values for the molecular weight of aquatic fulvic acids and different conceptual models for their structure and configuration. Several competing chemical processes may be acting, e.g., hydrogen bonding to water versus organic compounds, charge repulsion, bridging by metal ions, micelle formation, etc.

An important question is whether or not humic-rich systems are more or less susceptible to deleterious effects from acidic deposition. It can be argued that humic-enriched systems would be less vulnerable to acidification because of the abundance of anaerobic environments where sulfate and nitrate reduction by microbial populations would readily occur. These reduction processes decrease the concentrations of strong acids (SO_4^{2-} and NO_3^-) and increase alkalinity. If more acidic conditions were to develop, aluminum complexation reactions and potential increases in humic toxicity would be involved.

Bacterial Regrowth in Drinking Water Systems

One feature of chlorination of drinking waters is that the bactericidal properties remain as the treated drinking water moves through the distribution system. In other water treatment practices such as ozonation, bactericidal properties are not maintained and the potential exists for regrowth of bacteria within the distribution system, especially in the formation of biofilms. Ozonation may degrade otherwise recalcitrant dissolved organic acids. The resulting acids may be the main substrates for bacterial regrowth. Although the labile fraction of these organic acids may have low concentrations, the high flow rates in the distribution system result in a high flux and may enhance bacterial utilization compared to other dilute environments.

The treatment of drinking water may increase the potential of recalcitrant organic acids to serve as substrates for bacterial growth. This effect has been observed in ozonation (Werner and Hambsch 1986). Further study of these relationships would benefit from improved techniques for characterizing the bioavailability of recalcitrant organic acids and the effects of water treatment.

FUTURE RESEARCH

In considering future research needs for increasing understanding of organic acids and their environmental interactions, a modern version of a nursery rhyme comes to mind:

> Probable possible, my black hen,
> She lays eggs in the relative when.
> She doesn't lay eggs in the positive now,
> Because she is unable to postulate how. (Winsor 1958)

Because of the inherent complexities of aquatic ecosystems and inherent heterogeneities of organic acids in natural waters, research progress in this field may be limited by our ability to "postulate how"—i.e., to design definitive experiments, in either laboratory or field settings. In the context of the biodegradation of fulvic acids and other higher molecular weight organic acids, it is recommended that research must be directed to the co-metabolic degradation of organic acids of different sources under dominant aerobic and anaerobic environmental conditions of standing and running waters. Experimental approaches for this research could be based on the use of dual radioactive labels for the humic substrate and for the simple labile substrate. For the humic substrate, labeling of known functional groups could yield insight into microbial mechanisms of degradation. Another important question is whether or not intact humic molecules and other higher molecular weight organic acids are taken up by microbial cells, and if so, which of these organic acids and which microorganisms. Experimental resolution of this question requires specific reagents for determination of intact humic substances in living cells. Fluorescent dyes appear to be unsatisfactory, but other reagents have yet to be developed.

Another important area of research is the production and extracellular release of organic acids with growth-promoting or -inhibiting properties. The problems caused by toxic algal blooms will stimulate toxicological research related to organic acids in natural waters.

To progress in our understanding of chemical and physical interactions and ecosystem level interactions, we conclude that integrated field and laboratory studies are *essential*. The hypothesis that siderophores control iron availability to blue-green algae is an example of a hypothesis, developed largely from laboratory studies, that awaits testing by field measurements. Further, studies are needed to define the physicochemical nature of humic substances and other organic acids under different natural conditions, including such processes as aggregation. To quantify organic acid behavior in natural waters, strategies are required to simplify the highly complex chemistry of organic acids and develop usable predictive models. The question to be approached is: Can we explain what we observe in natural waters with such models?

For ecosystem level interactions, we need to develop testable hypotheses and follow through with field and laboratory experiments. The importance of pluralistic approaches in aquatic ecology (McIntosh 1987) cannot be overemphasized. Processes to be studied at the ecosystem level include the action of organic acids as toxicants, regulatory molecules, and energy sources. Field experiments, for example, could be based upon radiolabeling sources of organic acids (including biomarker compounds) in a whole lake system, and following the transformations which occur. Possible classes of biomarker compounds are fatty acids and lipopolysaccharides as tracers for cyanophytes. A more specific but important question that can be examined in laboratory and field experiments is: What are the effects of mineral acids on humic substances and to what extent do these effects increase (or decrease) the ecological impact of acidification? Greater understanding of the effect of pH on surface activity properties of humic substances would be helpful in this context.

REFERENCES

Aiken, G.R., and R.L. Malcolm. 1987. Molecular weight of aquatic fulvic acids by vapor pressure osmometry. *Geochim. Cosmo. Acta* 51:2177–2184.

Behmel, P. 1986. Die Sorption von Huminsäuren durch Bakterienzellwände, Analyse der Bindungsverhältnisse und Schutz gegen enzymatische Auflösung. In: Festschrift zum 65. Geburtstag von Prof. Dr. W. Ziechmann.

Berg, K., O.M. Skulberg, and R. Skulberg. 1987. Effects of decaying toxic blue-green algae on water quality—a laboratory study. *Arch. Hydrobiol.* 108(4):549–563.

Berg, K., O.M. Skulberg, B. Underdal, and T. Willén. 1986. Observations of toxic blue-green algae (Cyanobacteria) in some Scandinavian lakes. *Acta Vet. Scan.* 27:440–452.

Buffle, J. 1988. Complexation Reactions in Aquatic Systems: An Analytical Approach. Chichester: Wiley.

Burkholder, J.M., and R.G. Wetzel. 1989. Epiphytic microalgae on natural substrata in a hardwater lake: seasonal dynamics of community structure, biomass, and ATP content. *Arch. Hydrobiol. Supp.* 83:1–56.

Carlough, L.A., and J. Meyer. 1989. Protozoans in two southeastern blackwater rivers and their importance to trophic transfer. *Limnol. Ocean.* 34:163–177.

Codd, G.A., and G.K. Poon. 1988. Cyanobacterial toxins. In: Biochemistry of the Algae and Cyanobacteria, ed. L.I. Rogers and J.G. Gallon, pp. 283–296. Oxford: Clarendon.

De Haan, H. 1977. Effect of benzoate on microbial decomposition of fulvic acids in Tjeukemeer (the Netherlands). *Limnol. Ocean.* 22:38–44.

Francko, D.A. 1984a. Phytoplankton metabolism and uptake nucleotides. I. Nucleotide-induced perturbation of carbon assimilation. *Arch. Hydrobiol.* 100:341–354.

Francko, D.A. 1984b. Phytoplankton metabolism and cyclic nucleotides II. Nucleotide-induced perturbation of alkaline phosphatase activity. *Arch. Hydrobiol.* 100:409–421.

Francko, D.A. 1989. Modulation of photosynthetic carbon assimilation in *Selenastrum capricornutum* (Chlorophyceae) by CAMP: an electrogenic mechanism. *J. Phycology* 25:305–313.

Frimmel, F.H., A. Immerz, and H. Niedermann. 1984. Complexation capacities of humic substances isolated from freshwater with respect to Copper(II), Mercury(II) and Iron(II, III). In: Complexation of Trace Metals in Natural Waters, ed. C.J.M. Carmer and J.C. Duinker, pp. 329–343. The Hague: Nijhoff/Junk.

Gjessing, E.T., G. Riise, R.C. Petersen, and E. Andruchow. 1989. Bioavailability of aluminum in the presence of humic substances at low and moderate pH. *Sci. Total Env.* 81/82:683–690.

Hessen, D.O. 1985. The relation between bacterial carbon and dissolved humic compounds in oligotrophic lakes. *FEMS Microbiol. Ecol.* 31:215–223.

Hitsch, E., and P. Werner. 1988. Probleme bei der Trinkwasseraufbereitung aus Karstwasser. *Vom Wasser* 70:223–243.

Horvath, R.S. 1972. Microbial co-metabolism and the degradation of organic compounds in nature. *Bacteriol. Rev.* 36:146–155.

Kaplan, A., D. Zenvirth, C. Reinhold, and J.A. Berry. 1982. Involvement of a primary electrogenic pump in the mechanism of HCO_3^- uptake by the cyanobacterium *Anabaena Variabilis*. *Plant Physiol.* 69:978–982.

McIntosh, R.P. 1987. Pluralism in ecology. *Ann. Rev. Ecol. Syst.* 18:321–342.

McKnight, D.M., K.A. Thorn, R.L. Wershaw, J.M. Bracewell, and G.W. Robertson. 1988. Rapid changes in dissolved humic substances in Spirit Lake and South Fork Castle Lake, Washington. *Limnol. Ocean.* 33(6):1527–1541.

Münster, U. 1985. Investigations about structure, distribution and dynamics of different organic substrates in the DOM of lake Plußsee. *Arch. Hydrobiol. Supp.* 70:429–480.

Münster, U., P.E. Plon, and J.N. Lammi. 1989. Evaluation of the measurements of extracellular enzyme activities in a polyhumic lake by means of studies with 4-methylumbelliferyl-substrates. *Arch. Hydrobiol.* 17.

Nilssen, J.P. 1982. Acidification in southern Norway: seasonal variation of aluminum in lake waters. *Hydrobiol.* 94:217–221.

Paerl, H.W. 1988. Growth and reproductive strategies of freshwater blue-green algae (cyanobacteria). In: Growth and Reproductive Strategies of Freshwater Phytoplankton, ed. C.D. Sandgren, pp. 261–315. Cambridge: Cambridge Univ. Press.

Palenik, B., and F.M.M. Morel. 1989. Amino acid utilization by marine phytoplankton: a novel mechanism. *Limnol. Ocean.*, in press.

Piotrowicz, S.R., G.R. Harvey, D.A. Boran, C.P. Weisel, and M. Springer-Young.

1984. Cadmium, copper, and zinc interactions with marine humus as a function of ligand structure. *Marine Chem.* 14:333–346.
Raddum, G., and O. Saether. 1981. Chironomid communities in Norwegian lakes with different degrees of acidification. *Verh. Int. Verein. Limnol.* 21:399–405.
Salonen, K. 1984. Peculiarities in the limnology of small polyhumic lakes. *Lammi Notes* 11:5–7.
Salonen, K., K. Kononen, and L. Arvola. 1983. Respiration of plankton in two small, polyhumic lakes. *Hydrobiol.* 101:65–70.
Schütt, C. 1988a. Ecology of aquatic organisms. I. Habitat-specific plasmid patterns in natural bacterial populations of fresh-water lakes. *Verh. Int. Verein. Limnol.* 23:1821–1824.
Schütt, C. 1988b. Plasmid-DNA in natural bacterial populations of four brownwater lakes (South Sweden). *Arch. Hydrobiol. Beih.* 31:133–139.
Schütt, C. 1989. Plasmids in the bacterial assemblage of a dystrophic lake: evidence for plasmid-encoded nickel resistance. *Microb. Ecol.* 17:49–62.
Sherr, E.B. 1988. Direct use of a high molecular weight polysaccharide by heterotrophic flagellates. *Nature* 335:348–351.
Simon, M., and F. Azam. 1989. Protein content and protein synthesis rates of planktonic marine bacteria. *Mar. Ecol. Prog. Ser.* 51:201–213.
Skulberg, O.M., G.A. Codd, and W.W. Carmichael. 1984. Toxic blue-green algal blooms in Europe: a growing problem. *Ambio* 13:244–247.
Skulberg, O.M., and R. Skulberg. 1985. Planktic species of *Oscillatoria* (Cyanophyceae) from Norway—characterization and classification. *Arch. Hydrobiol. Supp.* 71(1/2): 157–174.
Steinberg, C., and U. Münster. 1985. Geochemistry and ecological role of humic substances in lakewater. In: Humic Substances in Soil, Sediment and Water, ed. G.R. Aiken, D.M. McKnight, R.L. Wershaw, and P. MacCarthy, pp. 105–146. New York: Wiley.
Tranvik, L.J. 1988. Availability of dissolved organic carbon for planktonic bacteria in oligotrophic lakes of differing humic content. *Microb. Ecol.* 16:311–322.
Tranvik, L.J., and M.G. Hofle. 1987. Bacterial growth in mixed cultures on dissolved organic carbon from humic and clear waters. *Appl. Env. Microbiol.* 53(3):482–488.
Underdal, B., O.M. Skulberg, E. Dahl, and T. Aune. 1989. Disastrous bloom of *Chrysochromulina polylepis* (Prymnesiophyceae) in Norwegian coastal waters (1988)—mortality in marine biota. *Ambio* 18(5):265–270.
Visser, S.A. 1982. Surface active phenomena by humic substances of aquatic origin. *Rev. Fran. Sci. Eau* 1:285–295.
Visser, S.A. 1985. Physiological action of humic substances on microbial cells. *Soil Biol. Biochem.* 17:457–462.
Visser, S.A. 1986. Effects of humic substances on plant growth. In: Humic Substances, Effect on Soil and Plants, pp. 89–135. Rome: Reda.
Visser, S.A. 1988. Absorption of humic substances by plant cells and their transport within plant tissues. In: Proc. Int. Symp. Humus et Planta IX, p. 232. Prague.
Werner, P., and B. Hambsch. 1986. Investigations on the growth of bacteria in drinking water. Intern. workshop on water c, Mulhouse, France, 9–10 April 1986.
Wetzel, R.G. 1979. The role of the littoral zone and detritus in lake metabolism. *Arch. Hydrobiol. Beih.* 13:145–161.
Wetzel, R.G. 1983. Limnology, 2nd ed. Philadelphia: Saunders.
Wetzel, R.G. 1989. Land-water interfaces: metabolic and limnological regulators. Baldi Memorial Lecture. *Verh. Int. Verein. Limnol.* 24, in press.

Wetzel, R.G., P.H. Rich, M.C. Miller, and H.L. Allen. 1972. Metabolism of dissolved and particulate detrital carbon in a temperate hard-water lake. *Mem. Inst. Ital. Idrobiol.* 29 (Supp.):185–243.
Winsor, F. 1958. The Space Child's Mother Goose. New York: Simon & Schuster.

Organic Acids in Aquatic Ecosystems
eds. E.M. Perdue and E.T. Gjessing, pp. 245–260
John Wiley & Sons Ltd

Patterns of Organic Acid Transport from Forested Watersheds to Aquatic Ecosystems

C.S. Cronan

Department of Botany and Plant Pathology
University of Maine
Orono, Maine 04469, U.S.A.

Abstract. There are significant temporal and spatial variations in the concentrations of dissolved organic carbon (DOC) and organic acids in surface waters. These variations are strongly affected by patterns of DOC transport from terrestrial to aquatic ecosystems. Differences in DOC export from terrestrial systems are largely determined by (*a*) hydrologic characteristics of the watershed and (*b*) the organic chemistry of the separate hydrologic source compartments within the watershed. In forested watersheds, DOC and organic acidity may originate from one or more of the following source compartments: atmospheric deposition, the forest canopy, the forest floor, the mineral soil, riparian wetlands, or the groundwater zone. On average, DOC concentrations increase from 1–3 mg C l^{-1} in precipitation, to 8–12 mg C l^{-1} in canopy throughfall, to 20–40 mg C l^{-1} in forest floor leachate, and then decrease to 4–8 mg C l^{-1} in B horizon lateral flow, and to 1–3 mg C l^{-1} in groundwater. The net flux of organic acids to a given surface water will be a function of the variable organic chemistry in each of these source compartments and the variable hydrologic transfer fluxes from the compartments to the stream or lake.

INTRODUCTION

There are significant spatial and temporal variations in the concentrations of dissolved organic carbon (DOC) and organic acids in surface waters (Fig. 1). These differences are presumably determined by the interactive effects of multiple biogeochemical processes and by the balance between hydrogeochemical input and output fluxes to aquatic ecosystems. Because of the potentially important effects of organic acids on water quality, there is a need to identify and quantify the major environmental parameters and relationships controlling the patterns of organic acid chemistry in stream

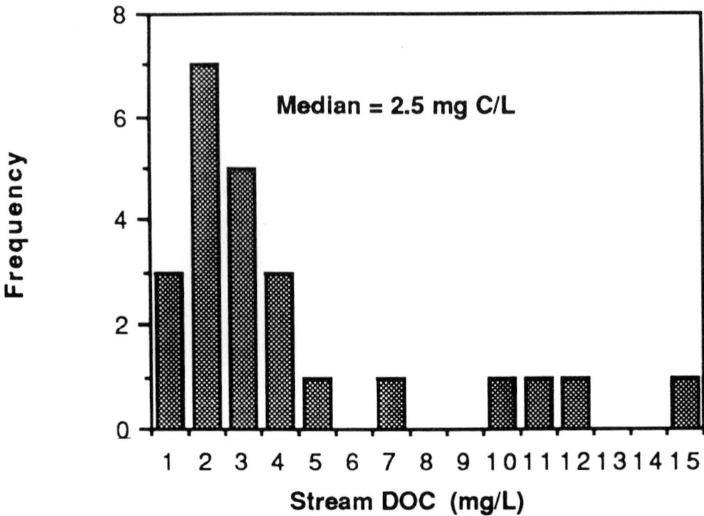

Fig. 1—Distribution of mean DOC concentrations for 24 streams draining small catchments in North America. The bars represent DOC intervals of 1 mg/l, beginning with 0–1 mg/l.

and lake ecosystems. In the discussion that follows, the term "organic acids" will refer to the entire molecular weight spectrum of aliphatic and aromatic organic compounds containing acidic carboxyl and hydroxyl functional groups.

This chapter presents an analysis of the patterns of organic acid transport from forest ecosystems to surface waters. In other sections of this book, there are related chapters concerning the contributions of organic acids from wetlands and aquatic environments. The release and transport of DOC and organic acids have been studied in a range of forest ecosystems, including temperate mixed hardwood forests (Bruckert and Jacquin 1969; McDowell and Wood 1984; David and Driscoll 1984; Cronan and Aiken 1985; Driscoll et al. 1985), warm temperate mixed deciduous forests (Hoffman et al. 1980; Meyer and Tate 1983), warm temperate swamp ecosystems (Brinson et al. 1980; Mulholland 1981), cool temperate coniferous forests (Ugolini et al. 1977; Dawson et al. 1978; Sollins and McCorison 1981; Bergkvist 1986), boreal coniferous forests (Ugolini et al. 1987), montane coniferous forests (Cronan et al. 1978; Graustein 1981; Litaor 1987; Ugolini et al. 1988), and tropical lowland forests (Frangi and Lugo 1985). In this chapter, the results from these investigations will be analyzed to determine how variations in surface water DOC and organic acidity may be influenced by inputs from forest ecosystems.

Specific focal questions for this chapter are: What are the major sources of organic acids in forest ecosystems? What is the range of variation in the concentrations of DOC and organic acids for each source compartment? How do the concentrations in each source compartment vary as a function of changes in vegetation type, season, climate, hydrology, soil type, and disturbance? How does the qualitative composition of organic acids vary by source compartment and in response to changes in environmental conditions? How are hydrologic fluxes partitioned among source compartments? What are the potential fluxes of DOC and organic acids from each source compartment to the aquatic ecosystem? Can we account for the organic acid content of a given surface water by using a mass balance model that combines inputs from various source compartments?

SOURCES OF DOC AND ORGANIC ACIDS

In order to understand the dynamic behavior of organic acids in natural waters, it is first necessary to identify the major sources and sinks for these organic solutes. For any given watershed, the potential sources include: atmospheric deposition, the forest canopy, the forest floor pool of decaying litter and humus, soil organic matter, plant roots and fungi, wetland peat deposits, aquatic sediments, aquatic detritus, and aquatic organisms. Organic acid inputs from these sources may reach surface waters directly or may be transported to streams or lakes as canopy throughfall, forest floor lateral flow, mineral soil lateral flow, or groundwater inputs.

If we view surface waters as a mixture of these various hydrologic inputs, an important concern is to determine how these source waters differ in terms of DOC and organic acid chemistry. As shown in Fig. 2, there is considerable inter-watershed variability in the DOC concentrations that have been reported for each of the hydrologic source compartments. Perhaps the largest range of differences in DOC concentrations can be found among forest floor leachates (Fig. 3). Yet, there are also relatively distinct differences in the mean DOC concentrations between compartments. On average, DOC concentrations increase from 1–3 mg C l^{-1} in precipitation, to 8–12 mg C l^{-1} in canopy throughfall, to 20–40 mg C l^{-1} in forest floor leachates, and then decrease to 4–8 mg C l^{-1} in B horizon lateral flow, and to 1–3 mg C l^{-1} in groundwater.

Although DOC is a useful indicator of organic acidity in natural waters, the composition of DOC may include variable amounts of acidic, neutral, and basic compounds. Thus, it is important that we develop an understanding of the chemical differences associated with the DOC of each ecosystem source compartment. Until now, there have only been fragmented efforts to characterize the organic chemistry of solutions collected from different ecosystem levels. In addition, there have been few attempts to follow the

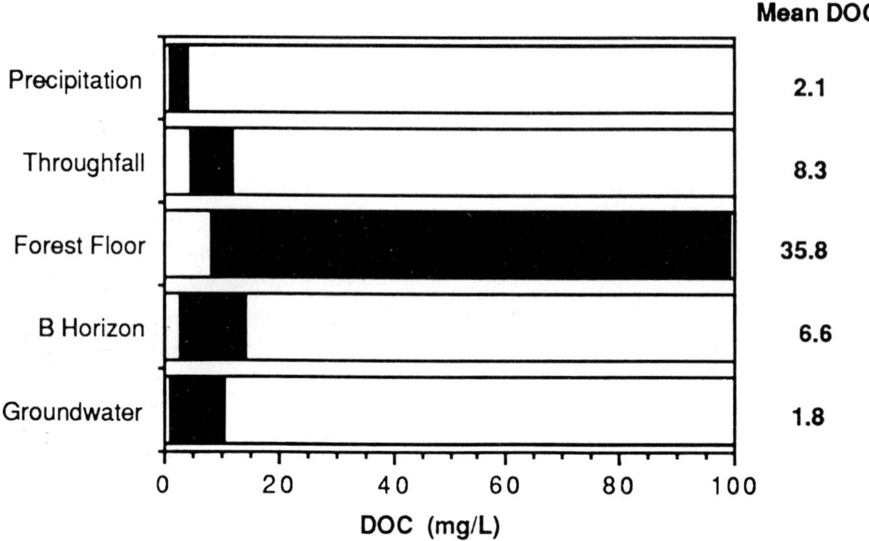

Fig. 2—Ranges of DOC observed in solutions collected at different ecosystem levels in forested watersheds in North America, Central America, Europe, and Asia.

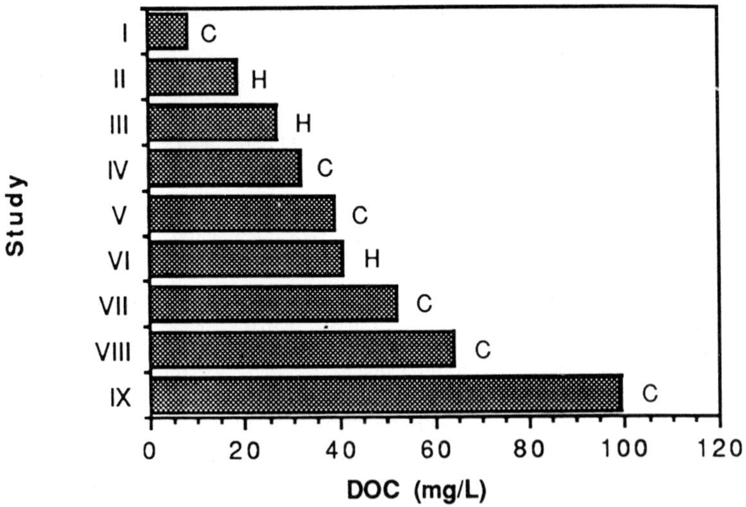

Fig. 3—Variations in DOC concentrations for forest floor leachates from nine different hardwood (H) and coniferous (C) forests in North America.

detailed changes in organic chemistry along flow gradients in a watershed ecosystem. Because of this, it is difficult to make comprehensive generalizations regarding the chemistry of organic solutes in the different terrestrial source compartments.

There have been several attempts to describe the organic acidity of precipitation and canopy throughfall inputs using chromatographic techniques. For example, Galloway et al. (1976) have shown that wet deposition may contain relatively high concentrations of simple aliphatic acids such as formic and acetic acids. It is probably reasonable to assume that 10–50% of the DOC in precipitation is comprised of these types of low molecular weight organic acids. Hoffman et al. (1980) have also examined some of the chemical properties of organic solutes in canopy throughfall from a southern hardwood forest.

A number of studies have identified specific organic acids that occur in forest soil solutions. These acids include the following aliphatic and aromatic simple acids: acetic, aconitic, adipic, butyric, caproic, citric, formic, fumaric, glycollic, lactic, maleic, malic, malonic, oxalic, propionic, succinic, valeric, benzoic, ferulic, p-hydroxybenzoic, p-coumaric, vanillic, protocatechuic, and salicylic (Kononova 1966; Bruckert 1970; Bruckert and Jacquin 1969; Graustein et al. 1977; Cronan et al. 1978; McColl and Pohlman 1986). In general, these low molecular weight acids are thought to decompose rapidly and to represent a relatively small percentage of the DOC.

Cronan and Aiken (1985) used XAD-8 fractionation to characterize the changes in major DOC fractions between different ecosystem levels in a mixed northern hardwood watershed in New York. They found that although DOC fractions in soil solutions were similar between hardwood and coniferous plots, there were consistent differences in DOC composition between different soil depths and ecosystem strata. O horizon leachates contained approximately 45% hydrophobic acids, 40% hydrophilic acids, and 15% neutrals plus bases. In comparison, B horizon solutions exhibited a shift toward a smaller percentage of hydrophobic acids (~25%) and a larger percentage of hydrophilic acids (~55%) and neutrals plus bases (20%). Surface waters exhibited another set of patterns related to mixing and in-lake changes in DOC composition. Other investigators have also reported a selective retention of hydrophobic acids (Vance and David 1989) or fulvic acids (Dawson et al. 1978) in the B horizon.

McDowell and Likens (1988) examined the changes in phenolics, aldehydes, carbohydrates, primary amines, and carboxylic acids between ecosystem compartments in a northern hardwood forest. They found that carboxylic acids represented approximately 11% of the total DOC in precipitation, 4% of the total DOC in throughfall, and <2% of the total DOC in seeps.

ENVIRONMENTAL VARIATIONS IN DOC

Given the distinct variations in DOC concentrations within forest ecosystems, there is a significant need to identify and quantify the environmental factors and relationships that control the abundance and composition of these organic solutes. Until that information is developed, our models for predicting the behavior of organic acids in natural systems will be severely constrained. At this point, a number of general, though not universal, relationships have been observed between variations in DOC and changes in environmental conditions in forest ecosystems. Several studies have shown that DOC concentrations in throughfall and forest floor leachates may be 50 to 100% higher in coniferous stands as compared to hardwood stands in the same region (David and Driscoll 1984; Cronan and Aiken 1985; Driscoll et al. 1988). Riha et al. (1986) have also shown pronounced differences in soil solution DOC sampled beneath different coniferous species. Hoffman et al. (1980) reported that DOC concentrations in throughfall may vary by a factor of 3 to 4 between growing and dormant seasons in deciduous forests. There may also be significant seasonal variations in forest floor leachate concentrations. For example, Cronan and Aiken (1985) found that O horizon DOC concentrations were approximately 20% higher in summer as compared to winter in a northern hardwood forest. McDowell and Likens (1988) reported that seasonal variations in the concentrations of DOC in forest floor leachates were inversely related to the amount of water flow. However, DOC concentrations in B horizon solutions in the same system exhibited no obvious seasonal pattern or relationship to hydrology.

David et al. (1989) presented laboratory results suggesting that the solution chemistry of canopy throughfall may influence the solubilization of organic matter from the O horizon. They found inverse relationships between forest floor DOC and both acidity and ionic strength. In addition, they reported a positive relationship between water contact time and DOC release from O horizon samples.

There is also strong evidence that climate and soil drainage conditions exert a crucial influence on DOC concentrations in surface soil horizons. On well-drained and moderately drained soils, there tends to be an inverse correlation between average soil temperature and DOC concentrations in surface soil leachates. Thus, DOC concentrations generally increase in cooler environments. In poorly drained areas, there is a tendency for DOC concentrations in surface horizons to reach very high levels, whether the site is a cool northern peatland or a warm southern blackwater swamp.

The few studies concerned with the effects of disturbance on DOC concentrations and export have not demonstrated consistent significant relationships (Sollins and McCorison 1981; Meyer and Tate 1983; McDowell and Likens 1988). Overall, there is much to be learned about the environmental relationships controlling variations in the release and transport of DOC and organic acids in terrestrial ecosystems.

A MASS-BALANCE MIXING MODEL FOR DOC EXPORT

As a first approximation, we can use a DOC mixing model to understand the variations in organic acid transport from forest ecosystems to surface waters. For any watershed ecosystem, the DOC flux between the terrestrial and aquatic systems will be a function of the variable DOC concentrations in the separate source compartments and the different hydrologic transfer fluxes from the compartments to the stream or lake (Fig. 4). The net DOC flux or weighted DOC concentration reaching the surface water can be calculated by specifying the DOC concentration for each compartment and by routing appropriate percentages of the surface runoff through the different hydrologic source compartments.

As an example, Table 1 illustrates a simulation for a northern hardwood forest watershed in summer. The mean DOC concentrations for the five source compartments are shown and it is assumed that the watershed runoff is derived from different proportional contributions from direct precipitation inputs, direct throughfall inputs, lateral flow from the forest floor in riparian variable source areas, lateral flow from the B horizon, and from groundwater. In this case, the predicted weighted DOC concentration in stream water is 3.57 mg C l^{-1}. Additional simulation results presented in Table 2 show that, depending upon assumptions about the DOC and hydrologic contributions from each compartment, the weighted DOC in runoff can be made to vary from about 2.45 mg C l^{-1} in a groundwater-dominated system to approximately 8 to 12 mg C l^{-1} during the springmelt period when water flow originates primarily from the O, E, and B horizons. For all these cases, it must be remembered that the model does not include wetland inputs or a consideration of in-stream or in-lake processes affecting DOC.

Figures 5a–5d illustrate a simulation of the monthly changes in runoff, DOC export, and weighted stream DOC concentration for hardwood versus coniferous forest ecosystems in the northeastern U.S. The major assumptions incorporated into the simulation were as follows: (*a*) hydrologic routing shifted to upper soil layers during the spring and fall high flow periods and to lower soil layers during midsummer; (*b*) throughfall DOC dropped from 8 mg C l^{-1} during the growing season to 2 mg C l^{-1} during the dormant season; (*c*) forest floor DOC dropped 20% during the winter–spring period; and (*d*) DOC was 50% higher in the coniferous versus hardwood forest floor. In the case of the hardwood system, the estimated annual carbon export to the stream was approximately 3.6 g C m^{-2}. By comparison, carbon export for the coniferous system was roughly 25% higher at 4.5 g C m^{-2}. For reference, the carbon export for a groundwater-dominated system would

TABLE 1. A spreadsheet mixing model for DOC transport to surface waters. For each ecosystem source compartment, the baseline mean DOC concentration is given and the percentage of runoff derived from that compartment is indicated. The state modifier allows one to increase the forest floor DOC for coniferous systems or to change the throughfall DOC between growing and dormant seasons. The net DOC column shows the calculated mass of DOC derived from each source compartment. The final entry in the net DOC column is the weighted mean DOC in stream water for the specified conditions.

Source	State	Baseline DOC mg/l	State Modifier	Hydrologic %	Net DOC mg/l	% Contribution
Rain	HDW Summer	1.5	1	0.01	0.015	0.4
Throughfall		8	1	0.01	0.08	2.2
Forest Floor		25	1	0.03	0.75	21
B Horizon		6	1	0.1	0.6	16.8
Groundwater		2.5	1	0.85	2.125	59.5
Stream DOC					3.57	

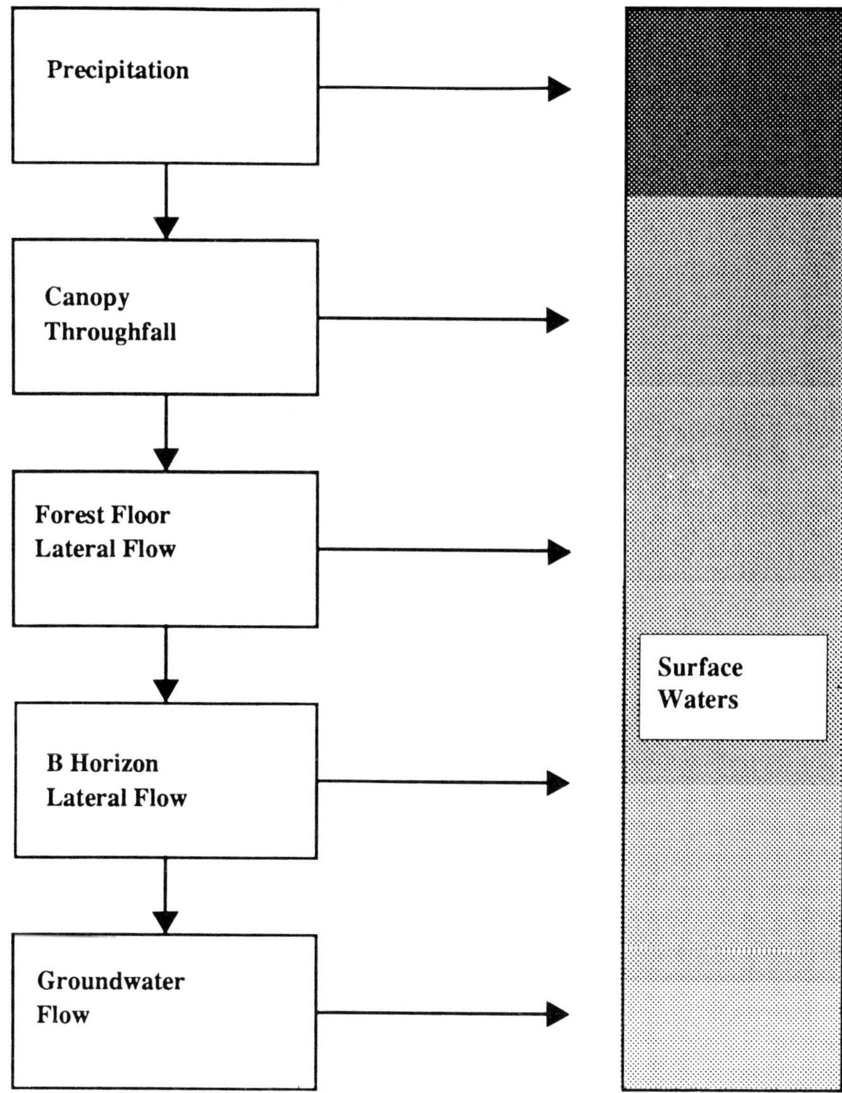

Fig. 4—The major DOC and hydrologic source compartments that may contribute organic acids to surface waters in forested watersheds. This box model does not include wetland and aquatic internal sources. Several of the compartments may also function as sinks for DOC.

TABLE 2. Four additional simulations of DOC patterns for the forested watershed described in Table 1. In the first panel, the system is calibrated as a groundwater-dominated watershed. The second panel shows a hardwood system during a storm event when 25% of the flow originates in the O and B horizons of variable source areas. The third panel shows a hardwood system during snowmelt when the water table rises and vertical infiltration decreases, forcing the majority of flow through the O and B horizons. The final panel shows the same snowmelt scenario in a coniferous watershed.

Source	State	Baseline DOC mg/l	State Modifier	Hydrologic %	Net DOC mg/l	% Contribution
Rain	HDW Summer	1.5	1	0.05	0.075	3
Throughfall		8	1	0	0	0
Forest Floor		25	1	0	0	0
B Horizon		6	1	0	0	0
Groundwater		2.5	1	0.95	2.375	96.9
Stream DOC					2.45	

Source	State	Baseline DOC mg/l	State Modifier	Hydrologic %	Net DOC mg/l	% Contribution
Rain	HDW Summer	1.5	1	0	0	0
Throughfall		8	1	0	0	0
Forest Floor		25	1	0.05	1.25	28.9
B Horizon		6	1	0.2	1.2	27.7
Groundwater		2.5	1	0.75	1.875	43.3
Stream DOC					4.325	

Source	State	Baseline DOC mg/l	State Modifier	Hydrològic %	Net DOC mg/l	% Contribution
Rain	HDW	1.5	1	0.03	0.045	0.5
Throughfall	Snowmelt	8	0.25	0.02	0.04	0.4
Forest Floor		25	0.8	0.2	4	49.6
B Horizon		6	1	0.6	3.6	44.6
Groundwater		2 5	1	0.15	0.375	4.6
Stream DOC					8.06	

Source	State	Baseline DOC mg/l	State Modifier	Hydrologic %	Net DOC mg/l	% Contribution
Rain	CON	1.5	1	0.03	0.045	0.3
Throughfall	Snowmelt	8	0.25	0.02	0.04	0.3
Forest Floor		25	1.5	0.2	7.5	64.8
B Horizon		6	1	0.6	3.6	31.1
Groundwater		2.5	1	0.15	0.375	3.2
Stream DOC					11.56	

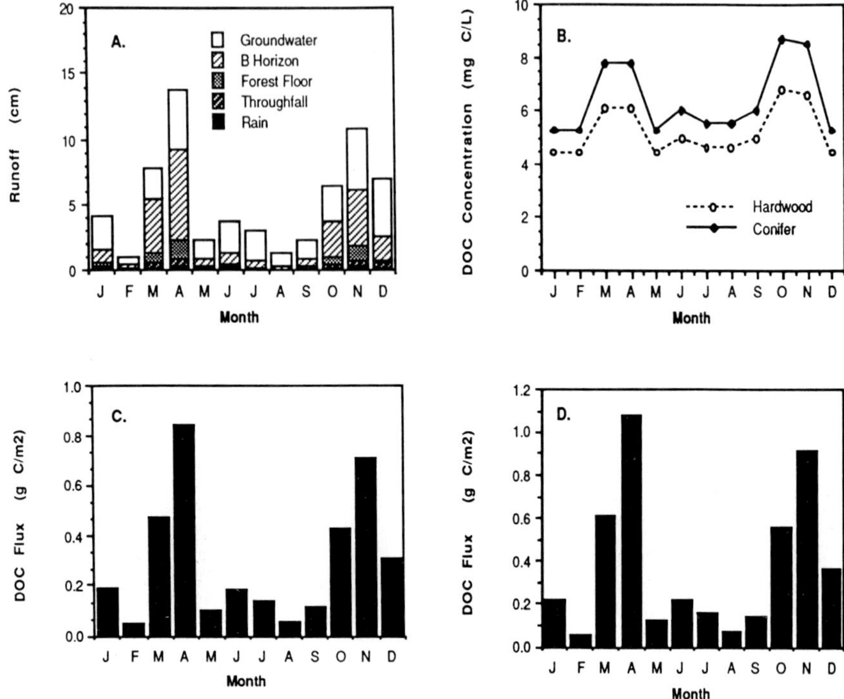

Fig. 5—Predicted monthly patterns of runoff (A), weighted mean DOC concentration in stream water (B), and DOC export for hardwood (C) and coniferous (D) watersheds in the northeastern U.S. The simulation was run using the mixing model in Table 1 and runoff data from a small watershed in New York. For each month the estimated amount of runoff derived from each ecosystem compartment is shown in panel A. The DOC flux estimate in panel C is for a hardwood forest, while panel D shows the simulated DOC export patterns for a coniferous forest.

have been about 1.3 g C m^{-2} under these calibration conditions. These results indicate the potentially large influence of hydrology and vegetative cover on patterns of DOC export to streams.

One can also take the simple mixing model one step further by incorporating DOC fractionation data into the simulation. Cronan and Aiken (1985) reported DOC fractionation data for O horizon and B horizon soil solutions and stream water in three forested catchments in New York. For the inlet stream to Sagamore Lake, they measured a mean DOC concentration of 8 mg C l^{-1} and found that the relative proportions of hydrophobic acids:hydrophilic acids:neutrals plus bases fell along a mixing line between the composition of the O and B horizon soil solutions (Fig. 6).

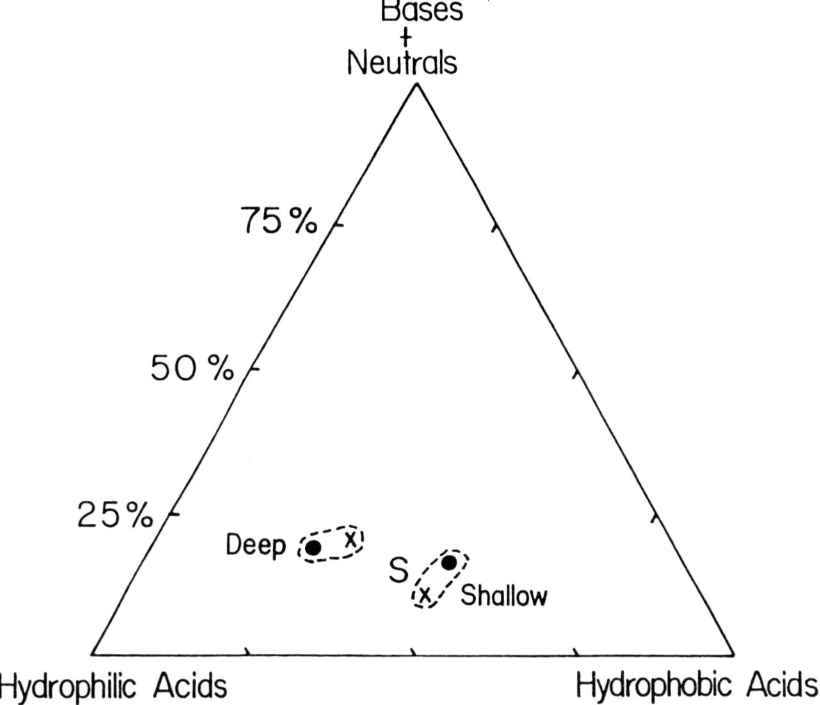

Fig. 6—Ternary diagram showing the relative percentages of hydrophobic acids, hydrophilic acids, and neutrals plus bases in O horizon solutions (shallow), B horizon solutions (deep), and stream water (S) for the Sagamore mixed hardwood watershed in New York.

As shown in Table 3, the DOC mixing model accounted for both the average DOC concentration in stream water at Sagamore watershed as well as the relative distribution of DOC fractions in the stream.

Overall, these simulations demonstrate that much of the considerable variation in streamwater DOC export can be accounted for by simple differences in hydrologic flow and DOC concentration between ecosystem source compartments. For this type of model to be effective, it will be necessary to have an accurate means of predicting hydrologic flow paths and an improved understanding of the variations in DOC concentrations and composition within each compartment. Ultimately, models such as these should be expanded into dynamic simulation models capable of including wetland processes, rate-dependent biological and geochemical processes, and detailed information concerning the organic chemistry of source and receptor waters.

TABLE 3. An extension of the mixing model to include the distribution of DOC fractions in the source and receptor waters at Sagamore watershed, New York. The DOC fractions in the source compartments are those reported by Cronan and Aiken (1985). The field data and the model output for this system both indicate that the inlet stream contains about 8 mg C l^{-1} and that the DOC fractions include about 40% hydrophobic acids, 45% hydrophilic acids, and 15% neutrals plus bases.

Source	State	Baseline DOC mg/l	State Modifier	Hydrologic %	Net DOC mg/l	% Contribution	% Hydrophobic	% Hydrophilic	% Neutrals
Rain	HDW Summer	1.5	1	0	0	0	0	0	0
Throughfall		8	1	0	0	0	0	0	0
Forest Floor		25	1	0.2	5	62.1	0.46	0.41	0.13
B Horizon		6	1	0.3	1.8	22.3	0.27	0.53	0.2
Groundwater		2.5	1	0.5	1.25	15.5	0.4	0.45	0.15
							40	44	14
Stream DOC					8.05				

Acknowledgements. This research was supported by grant BSR-8807826 from the National Science Foundation. The author thanks M.B. David for his helpful review comments.

REFERENCES

Bergkvist, B. 1986. Leaching of metals from a spruce forest soil as influenced by experimental acidification. *Wat. Air Soil Poll.* 31:901–916.

Brinson, M.B., H.D. Bradshaw, R.N. Holmes, and J.B. Elkins. 1980. Litterfall, stemflow, and throughfall nutrient fluxes in an alluvial swamp forest. *Ecology* 61:827–835.

Bruckert, S. 1970. Influence des composes organiques solubles sur la pedogenese en milieu acide. *Ann. Agron.* 21:421–452.

Bruckert, S., and F. Jacquin. 1969. Interaction entre la mobilité de plusieurs acides organiques et de divers cations dans un sol a mull et dans un sol a mor. *Soil Biol. Biochem.* 1:275–294.

Cronan, C.S., and G.R. Aiken. 1985. Chemistry and transport of soluble humic substances in forested watersheds of the Adirondack Park, New York. *Geochim. Cosmo. Acta* 49:1697–1705.

Cronan, C.S., W.A. Reiners, R.C. Reynolds, and G.E. Lang. 1978. Forest floor leaching: contributions from mineral, organic, and carbonic acids in New Hampshire subalpine forests. *Science* 200:309–311.

David, M.B., and C.T. Driscoll. 1984. Aluminum speciation and equilibria in soil solutions of a Haplorthod in the Adirondack Mountains (New York, USA). *Geoderma* 33:297–318.

David, M.B., G.F. Vance, J.M. Rissing, and F.J. Stevenson. 1989. Organic carbon fractions in extracts of O and B horizons from a New England Spodosol: effects of acid treatment. *J. Envir. Qual.* 18:212–217.

Dawson, H.J., F.C. Ugolini, B.F. Hrutfiord, and J. Zachara. 1978. Role of soluble organics in the soil processes of a Podzol, Central Cascades, Washington. *Soil Sci.* 126:290–296.

Driscoll, C.T., R.D. Fuller, and D.M. Simone. 1988. Longitudinal variations in trace metal concentrations in a northern forested ecosystem. *J. Envir. Qual.* 17:101–107.

Driscoll, C.T., N. van Breemen, and J. Mulder. 1985. Aluminum chemistry in a forested spodosol, New Hampshire. *Soil Sci. Soc. Am. J.* 49:437–443.

Frangi, J.L., and A.E. Lugo. 1985. Ecosystem dynamics of a subtropical floodplain forest. *Ecol. Monogr.* 55:351–369.

Galloway, J.N., G.E. Likens, and E.S. Edgerton. 1976. Acid precipitation in northeastern United States: pH and acidity. *Science* 194:722–724.

Graustein, W.C. 1981. The effects of forest vegetation on solute acquisition and chemical weathering: a study of the Tesuque Watersheds near Santa Fe, New Mexico. Ph.D. diss., Yale University, New Haven.

Graustein, W.C., K. Cromack, and P. Sollins. 1977. Calcium oxalate: occurrence in soils and effect on nutrient and geochemical cycles. *Science* 198:1252–1254.

Hoffman, W.A., S.E. Lindberg, and R.R. Turner. 1980. Some observations of organic constituents in rain above and below a forest canopy. *Env. Sci. Tech.* 14:999–1002.

Kononova, M.M. 1966. Soil Organic Matter: Its Nature, Its Role in Soil Formation and in Soil Fertility. New York: Pergamon.

Litaor, M.I. 1987. Aluminum chemistry: fractionation, speciation, and mineral equilibria of soil interstitial waters of an alpine watershed, Front Range, Colorado. *Geochim. Cosmo. Acta* 51:1285–1295.

McColl, J.G., and A.A. Pohlman. 1986. Soluble organic acids and their chelating influence on Al and other metal dissolution from forest soils. *Wat. Air Soil Poll.* 31:917–927.

McDowell, W.H., and G.E. Likens. 1988. Origin, composition, and flux of dissolved organic carbon in the Hubbard Brook Valley. *Ecol. Monogr.* 58:177–195.

McDowell, W. H., and T. Wood. 1984. Podzolization: soil processes control dissolved organic carbon concentrations in stream water. *Soil Sci.* 137:23–32.

Meyer, J.L., and C.M. Tate. 1983. The effects of watershed disturbance on dissolved organic carbon dynamics of a stream. *Ecology* 64:33–44.

Mulholland, P.J. 1981. Organic carbon flow in a swamp-stream ecosystem. *Ecol. Monogr.* 51:307–322.

Riha, S.J., G. Senesac, and E. Pallant. 1986. Effects of forest vegetation on spatial variability of surface mineral soil pH, soluble aluminum and carbon. *Wat. Air Soil Poll.* 31:929–940.

Sollins, P., and F.M. McCorison. 1981. Nitrogen and carbon solution chemistry of an old growth coniferous forest watershed before and after cutting. *Wat. Res. Res.* 17:1409–1418.

Ugolini, F.C., R. Dahlgren, S. Shoji, and T. Ito. 1988. An example of andosolization and podzolization as revealed by soil solution studies, southern Hakkoda, northeastern Japan. *Soil Sci.* 145:111–124.

Ugolini, F.C., R. Minden, H. Dawson, and J. Zachara. 1977. An example of soil processes in the *Abies amabilis* zone of central Cascades, Washington. *Soil Sci.* 124:291–302.

Ugolini, F.C., M.G. Stoner, and D.J. Marrett. 1987. Arctic pedogenesis. I. Evidence for contemporary podzolization. *Soil Sci.* 144:90–100.

Vance, G.F., and M.B. David. 1989. Effect of acid treatment on dissolved organic carbon retention by a spodic horizon. *Soil Sci. Soc. Am. J.* 53:1242–1247.

Organic Acids in Aquatic Ecosystems
eds. E.M. Perdue and E.T. Gjessing, pp. 261–279
John Wiley & Sons Ltd

Spatial and Temporal Scales of Dissolved Organic Carbon in Streams and Rivers

J.R. Sedell[1] and C.N. Dahm[2]

[1]*USDA Forest Service*
Pacific Northwest Research Station
FSL Laboratory, 3200 Jefferson Way
Corvallis, OR 97331, U.S.A.

[2]*Department of Biology*
University of New Mexico
Albuquerque, NM 87131, U.S.A.

Abstract. Regional variation in riverine ecosystem dissolved organic carbon (DOC) concentration generally ranges from 0.5–40 mg C/l. Downstream concentrations of DOC within a catchment show no predictable general trends. Where anthropogenic activity increases as a river becomes larger, a general increase in DOC is apparent. In less disturbed river basins there does not appear to be a strong link between stream size and DOC concentration. The amount of swamps and wetlands within a basin strongly influences the amount of DOC. Small streams are more variable in DOC concentrations than larger streams and rivers. Historically, river floodplains and small streams retained more water and supported greater primary production which contributed more DOC to riverine ecosystems.

INTRODUCTION

Dissolved organic carbon (DOC) in freshwater ecosystems, normally the largest fraction of organic matter present, has long been a topic of interest to geochemists, limnologists, and stream ecologists (Thurman 1985). Quantification of specific components, structural characteristics, utilization as an energy resource by microbes, interaction with metals, and effects on the light spectrum are topics that have received considerable attention. Much, however, remains to be learned about the chemical structure of DOC in natural waters and the roles that DOC plays in the functioning of aquatic ecosystems. Recently, one role played by the largest component of DOC, the organic acids, has come under closer scrutiny. The role of organic acids as a source of acidity in natural waters has generally been considered to be relatively constant temporally. It has been suggested, however, that DOC is not temporally constant over the long term in areas receiving strong acid deposition and that DOC concentrations decrease as acidification proceeds.

Support for the hypothesis of decreasing organic acids in surface waters draining catchments with increasing, strong acid deposition has been presented from a lake survey of the eastern United States (Sullivan et al. 1988). The concentration of DOC decreased in the northeastern and upper Midwest sections of the United States across a gradient of increasing wet sulfur deposition. Paleolimnological evidence has also been put forth to argue that lake acidification is accompanied by a loss of organic matter (Davis et al. 1985). In other areas, notably eastern Canada and much of Scandinavia, a link between acidification of catchments and a decline in DOC has not been found. Historical changes in the amount and type of DOC within a drainage basin are difficult to infer or quantify and unequivocal linkage of long-term changes in DOC to acid deposition is still to be proven. In particular, distinguishing between DOC differences which are linked to varying land use patterns, vegetation type, soil type, and/or agricultural activities from those due to strong acid deposition can be problematic.

In this chapter, we will focus on sources of DOC, spatial variation in DOC concentration, and the temporal variability in concentration of DOC in streams and rivers in a variety of lotic environments worldwide. In addition, we will highlight historical changes in streams, rivers, riparian zones, and catchments which are very likely to have influenced the type and amount of DOC in surface waters over temporal scales comparable to those attributed to long-term acidification.

SPATIAL PATTERNS OF DOC IN STREAMS AND RIVERS

Meybeck (1982) estimated DOC transport by world rivers to the oceans. An average DOC concentration of 5.75 mg C/l was calculated from available data for large rivers. This is equivalent to a transport of 2150 kg of carbon as DOC annually from each square kilometer of land surface. This estimate of DOC concentration and transport was based upon data from only a few major rivers. These values, however, have held up well to a growing data set for rivers in a wide variety of climatic zones. A global average value for DOC of approximately 5 mg C/l remains a good estimate.

Regional variation in the concentration of DOC can range from less than 1 mg C/l to upwards of 30 mg C/l. For example, in North America, DOC concentrations in excess of 10 mg C/l commonly occur in streams and rivers draining regions with extensive wetlands such as swamps of the Atlantic Coastal Plain (Mulholland and Kuenzler 1979) or northern bogs (McKnight et al. 1985). Meyer (1986) has also studied blackwater streams and rivers in the southeastern United States where mean monthly concentrations of

DOC in the Ogeechee River ranged from 6–17 mg C/l. A small tributary to the Ogeechee, Black Creek, had an annual average DOC concentration of 30.8 mg C/l with occasional mean monthly concentrations exceeding 40 mg C/l. Streams draining forested upland catchments of the southeastern United States contained much lower concentrations of DOC. Mean annual discharge-weighted concentration of DOC for four small streams at Coweeta in North Carolina ranged from 0.60 to 2.33 mg C/l (Tate and Meyer 1983). Tate and Meyer (1983) also summarized mean discharge-weighted concentration of DOC from 14 montane catchments throughout North America. All 14 sites fell below the mean worldwide river average of 5 mg C/l with a range from 0.56–3.9 mg C/l. Other forested areas, such as boreal forest streams, have somewhat higher concentrations of DOC on average with typical DOC concentrations from 5–15 mg C/l.

Mulholland and Watts (1982) synthesized existing data from rivers and streams throughout North America to come up with an estimate of organic carbon transport to the oceans. Although their data are presented as total organic carbon (TOC), the bulk of the material in transport for most systems was probably in the DOC fraction. A range of TOC concentrations from 1.6–21.7 mg C/l was found for 1977 and 1978. Regional variation in annual export was primarily attributed to differences in annual runoff, but the concentration of DOC in various streams and rivers was not strongly correlated to discharge alone. In general, regional variations in DOC concentration can span the range from about 0.5–40 mg C/l.

The concentration of DOC also changes substantially as water is routed through a basin. McDowell and Likens (1988) have summarized the concentration data for DOC for precipitation, throughfall, soil solution, streamside seeps, and stream water in the Hubbard Brook Valley of New Hampshire. Precipitation contained on average 1 mg C/l. Throughfall concentrations averaged 12 and 34 mg C/l during two years of sampling. Upper eluvial soil horizons had an average DOC concentration of 28 and 38 mg C/l during two years of measurement. The upper B soil horizon had a DOC concentration of 6 mg C/l and the B horizon at 30 cm was 3 mg C/l. Seeps feeding Bear Brook had an average DOC concentration of 1.7 mg C/l while the stream average was 1.8 mg C/l. A dynamic range of DOC concentrations of about 40 times precipitation averages occurred within various zones of the catchment, but the average concentration of DOC in the stream was only 0.8 mg C/l above the precipitation average.

Spatial variation in DOC along the length of the Amazon River has been reported by Richey et al. (1980). During both rising and high water, the concentration of DOC was relatively uniform, averaging 4.2 and 6.5 mg C/l, respectively. The range of DOC values along a 2000 km transect during rising water was from 3.4–6.0. The range along a 3400 km transect during high water was from 3.9–9.9. Waters from the Rio Negro, a lowland region with an expansive floodplain, were normally about a factor of two higher than those in the mainstem Amazon. Overall, the water in the Amazon ranged from 3.4–9.9 mg C/l during these two cruises. No discernible downstream pattern was seen and a somewhat smaller range of DOC concentrations was found relative to North America or Europe. Downstream concentrations of DOC within a catchment have been observed to increase, decrease, or show no predictable changes dependent on the specific study. A general trend has not emerged. Where anthropogenic activity increases as streams and rivers grow larger, a general increase in DOC has been observed. In less disturbed environments, there does not appear to be a strong link between stream order and DOC concentration. Streams and rivers need not be accumulators of soluble organic material as they progress to the sea.

Streams are connected with their watersheds primarily through their interaction with the riparian component of the watershed. The nature of the river systems dictates that the form of this coupling is different between small streams and large rivers.

Small streams are linked to the landscape by virtue of their small size, relatively large surface area/volume ratio, and their great abundance relative to larger water courses. Changes in the hydrologic coupling and any degradation most often occur through removal of the riparian forest. Changes in the riparian forest affect throughfall, litter fall, and retention of organic inputs.

Large streams are often linked to the landscape by virtue of extensive floodplains and complex channel patterns. Because of their greater width and relatively short length, direct riparian inputs can be minimal under low flow conditions. Hydrologic decoupling and degradation occur through channelization or flow regulation, confining the river to a single channel and denying it annual or more frequent access to its floodplain. Many riparian litter and tree inputs that were referred to for small streams occur in an analogous way on the floodplain of large rivers. Large rivers go to the forest, rather than wait for forest inputs.

Rivers and streams have many different geomorphic reaches. Those reaches have varying abilities to influence DOC concentration. For example, gorges on highly constrained reaches of rivers do not interact with a floodplain, have high velocities, and have relatively small natural organic inputs into the reach from the edges. In a braided section, the river system is characterized by multiple channels, bars, and unstable islands, and the expected organic inputs and biomass from the floodplain are medium. In a meandering section of a river, a great diversity of abandoned channels, cut off meanders (oxbow lakes), side-arm sloughs, and marshes allows a mosaic of floodplain wetlands usually well connected hydrologically with the main channel and with the aquifer. Because of the geomorphic stability and varied riparian forests, there is a high production of biomass.

Unconstrained reaches which have a high valley width/channel width ratio are more hydrologically retentive regardless of stream size. Both subsurface flow and surface area of water are greater in these areas. The primary production of the stream and floodplain vegetation is also high in these areas. These metabolic hotspots along a stream are characterized by lower gradients and often aggrading channels indicating reduced stream power.

TEMPORAL VARIATION IN DOC CONCENTRATION

Large Rivers

Annual variation in concentration of DOC in large rivers tends to be dampened by the hydrologic regime (Welcomme 1985). Spatial variations within the basin are often integrated into values which change only minimally during the year. Occasionally, however, hydrologic processes linking the expansive riparian areas to the main channels can contribute significantly to the concentration of DOC within the river. This can occur either during initial inundation of the riparian zones or during the falling limb of the hydrograph when stored water along the main channels drains back into the river. Examples of annual patterns of DOC concentrations in four large rivers are shown in Fig. 1. The four rivers are the Columbia, Ganges, Gambia, and Amazon and they drain large portions of the North American, Asian, African, and South American continents. Sampling locations and details of sampling methods and chemical analyses are given in Richey et al. (1980), Dahm et al. (1981), Lesack et al. (1984), and Ittekkot et al. (1985).

The Columbia River has a remarkably constant concentration of DOC throughout an annual cycle. A small increase in DOC concentration occurred in the late spring when runoff was peaking, but total variability was small. The entire range of DOC measured in 1973 and 1974 was from 1.81–2.47 mg C/l. The extensive network of large dams on the lower reaches of the river is a likely cause for the limited variation in the concentration of DOC on an annual basis. The other factor influencing the low DOC would be the limited floodplain area within this canyon-dominated river basin.

The Ganges River showed a much wider range of DOC values than the Columbia River (1.3–9.3 mg C/l). Oxbow lakes, ponds, and topographic depressions in the lower reaches of the river are hypothesized as major sources for increased DOC such as occurred in July 1981 (Ittekkot et al. 1985). When discharge is low, biogeochemical processes occurring within the wetted margins of the river, often under anaerobic conditions, result in

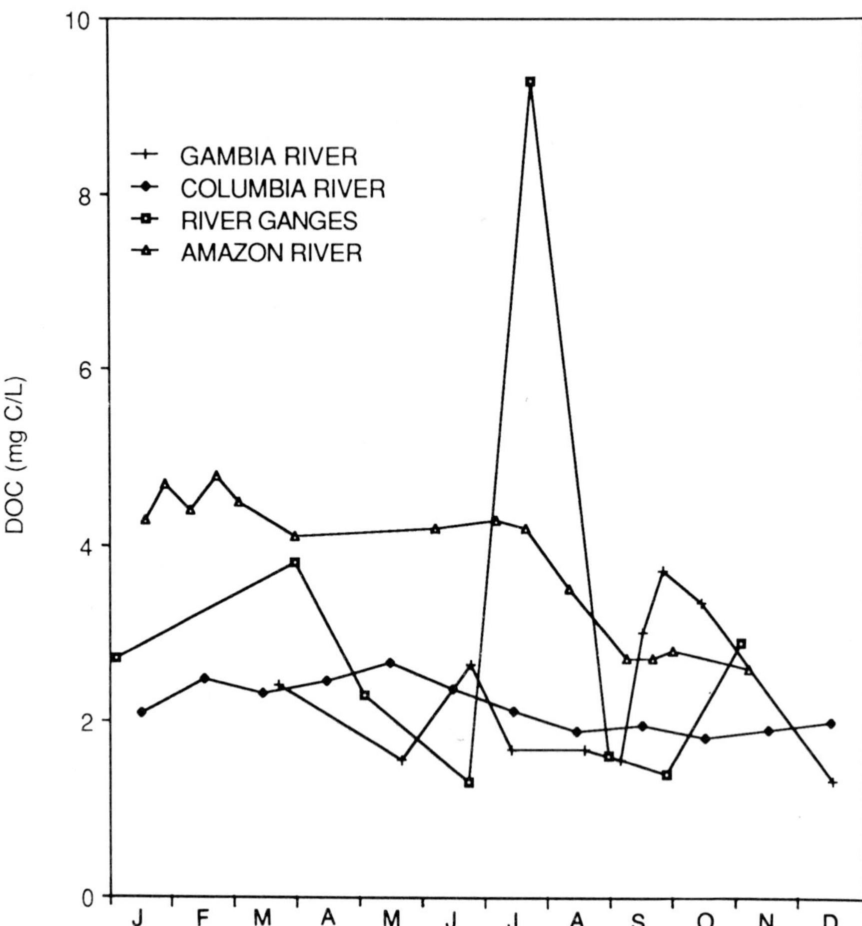

Fig. 1—Riverine concentration of DOC from the Amazon, Ganges, Gambia, and Columbia rivers over an annual cycle.

increased concentrations of DOC in those waters. With increasing water discharge from the river, the accumulated DOC is flushed out into the main channel through thorough mixing of river water with the previously isolated floodplain features. Chemical characterization of the DOC during various times of the year supports the conclusion that the wetlands adjacent to the main channel are major contributors of DOC to the river during periods of rising and high water.

The Gambia River had a somewhat smaller dynamic range of DOC concentrations than the Ganges, but much more variability than the Columbia River. The concentration range for DOC on an annual basis was 1.3–3.7 mg C/l in 1980–1981. A period of elevated DOC concentrations occurred from August to October while the minimum value occurred in December (Lesack et al. 1984). The concentration of DOC displayed counterclockwise hysteresis with rising and falling discharge (Fig. 2). Concentration of DOC peaked in late September (3.7 mg C/l) when discharge had fallen to about one-half the maximum level. Little change in DOC concentration was measured during the rising limb of the hydrograph. Minimum DOC

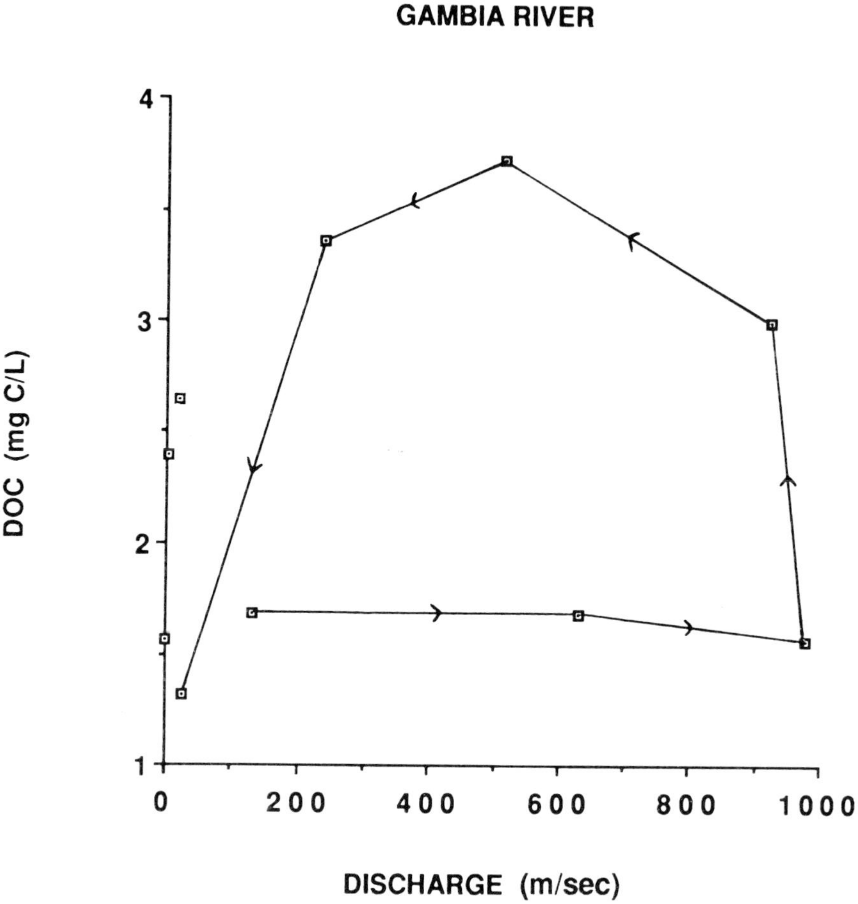

Fig. 2—DOC concentrations in the Gambia River display a counterclockwise hysteresis with rising and falling discharges due to draining DOC-enriched wetlands.

concentration was found to occur near lowest discharge rates for the river. In this river, the draining of DOC-enriched wetlands, marshes, and backwaters in the margins of the river, after peak discharge, as the likely source for the increased DOC and counterclockwise hysteresis with discharge.

The annual pattern for DOC concentrations in the Amazon River in 1983 upstream of the Rio Negro is shown in Fig. 1, and the relationship between riverine discharge and DOC is shown in Fig. 3 (unpublished data provided by A.H. Devol and J.E. Richey, University of Washington, Seattle, WA). The patterns are very compatible with those reported by Furch (1985).

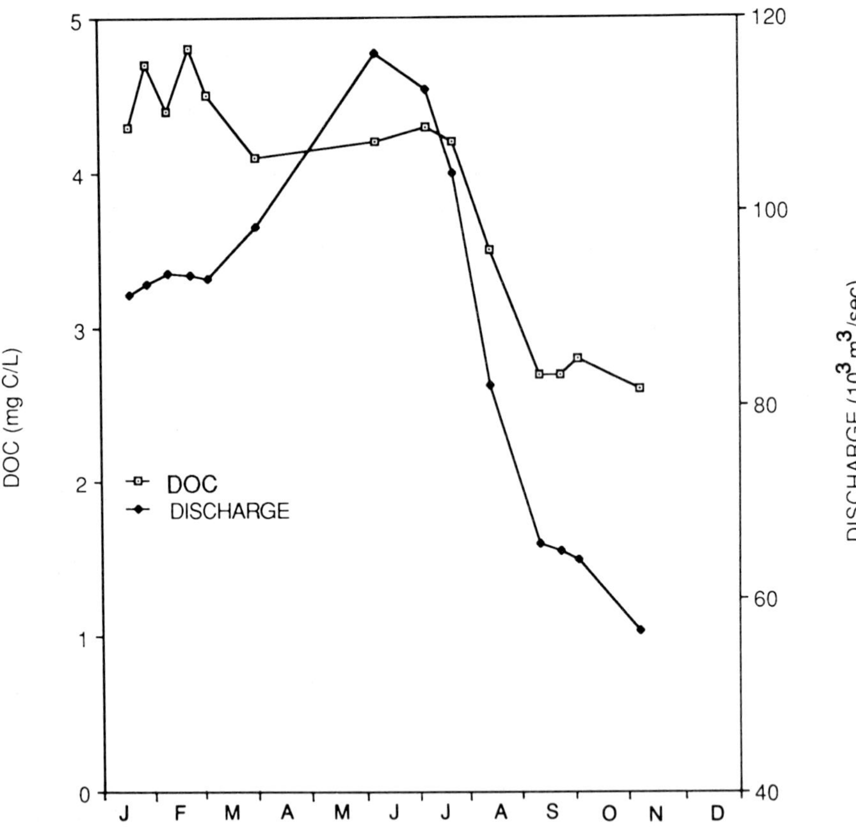

Fig. 3—Discharge and DOC concentration of the Amazon River over an annual cycle in 1983 from a station just above the confluence of the Rio Negro.

Concentrations of DOC are above 4.0 mg C/l during the rising limb and high water times of the hydrograph. As the river recedes, DOC concentrations decline to values consistently less than 3.0 mg C/l during the low water phase. Hysteresis in this section of the Amazon is clockwise with increases in DOC during rising discharge and decreases in DOC as discharge declines.

Intermediate-sized Rivers

We define intermediate-sized rivers on the basis of catchment areas between 10^2–10^5 km^2. Examination of the hydrographs for intermediate-sized rivers indicates discharge spikes which are more accentuated than those of large rivers and which persist for days to weeks. The variation of DOC on an annual basis shows a wider dynamic range than for big rivers but a smaller range than for little streams and small rivers. Mean DOC levels in this river size reflect both water storage in adjacent wetlands and primary production on the floodplain.

Good examples of the annual variation of DOC in intermediate-sized rivers are found in the northeastern and southeastern U.S. In the northeastern U.S., Fisher (1977), studying the Fort River (107 km^2), and Klotz and Matson (1978), investigating the Shetucket River (1330 km^2), found that DOC levels varied inversely with discharge and ranged from 1.7–14.8 and 2.1–18.3 mg C/l, respectively. Generally, the summer and fall low flows had high DOC levels. Both of these rivers are moderately urbanized and have extensive woodlots in the catchment area.

In the southern U.S., DOC levels can average from between 2–4 mg C/l in semiarid areas to between 10–27.6 mg C/l in the productive bottomland hardwood forests of the Southeastern and Gulf Coast. Aside from the climate, which drives higher or lower rates of primary production, river floodplain interactions play a role in the annual DOC variation. For example, the Brazos River (90,000 km^2) drains a semiarid region that is dewatered, channelized, dammed, and has had much of its riparian vegetation removed. The mean annual DOC concentration is about 3.3 mg C/l, and its annual variability is 1.7–7.7 mg C/l (Malcolm and Durum 1976). The DOC levels on an annual basis reflect a clockwise hysteresis with DOC increasing with discharge to a maximum peak point, then declining before peak discharge. This river is highly eutrophic, and there is a suggestion that DOC concentrations are low because of enhanced bacterial activity, but a limited source region of terrestrial vegetation adjacent to the channel may be a likely reason.The Sopchoppy River (264 km^2) in a rich bottomland hardwood forest varies annually between 6.2–52.0 mg C/l and has a mean of 27 mg C/l (Malcolm and Durum 1976). This river exhibits an annual

DOC pattern that has an almost direct relationship with discharge. This indicates to us a wide, rich floodplain with very little levee development and hence a higher hydrologic connectivity between the river channel and floodplain.

We use these data to illustrate the comparison with big river DOC, showing accentuated annual variability and high terragenic primary production results in higher concentrations of DOC.

Small Streams and Rivers

Streams also show distinctive variability in the concentration of DOC on the scale of hours to a few days in addition to the weekly and seasonal patterns common to rivers. Storms are important regulators of DOC concentration in streams. In general, DOC concentrations increase during the rising limb of a hydrograph. The relative response of the concentration of DOC during a storm is also linked to antecedent precipitation within the basin, season, and hydrology of the catchment during storms. Mechanisms which have been postulated to explain the observed increase in DOC during storms include (*a*) channel flushing and elongation, (*b*) changes in the flow path of water through the soil, (*c*) input of throughfall directly entering the stream, and (*d*) flushing of the hyporheic zone into the stream.

Figure 4 shows an example of the response of DOC concentrations within a first-order stream in western Oregon to a series of small storms over a two-week period in September 1977. The storms followed a time of dry weather when the concentration of DOC was generally 2–3 mg C/l. An initial gradual increase in discharge to approximately four times baseline was accompanied by an increase in DOC to 7–9 mg C/l. A more intense storm increased discharge more than an order of magnitude from baseline and a maximum concentration of DOC was measured during the rising limb of the hydrograph during this storm. The concentration of DOC reached 24.6 mg C/l and then decreased to 4.47 mg C/l on the falling limb of the hydrograph after the storm ended. Increases in the concentration of DOC up to one order of magnitude above background values can occur in association with storm flows.

Another factor which has been shown to cause daily fluctuations in the concentration of DOC in small streams is algal primary production (Kuserk et al. 1984). Early morning minima were followed by a mid-afternoon peak in a second-order stream in a pasture with a verdant streambed community. Diel increases in DOC of 35–66% were measured at six stations along the stream. Photochemical reactions might also produce diel patterns in DOC concentrations in some streams, but such a linkage has yet to be shown conclusively.

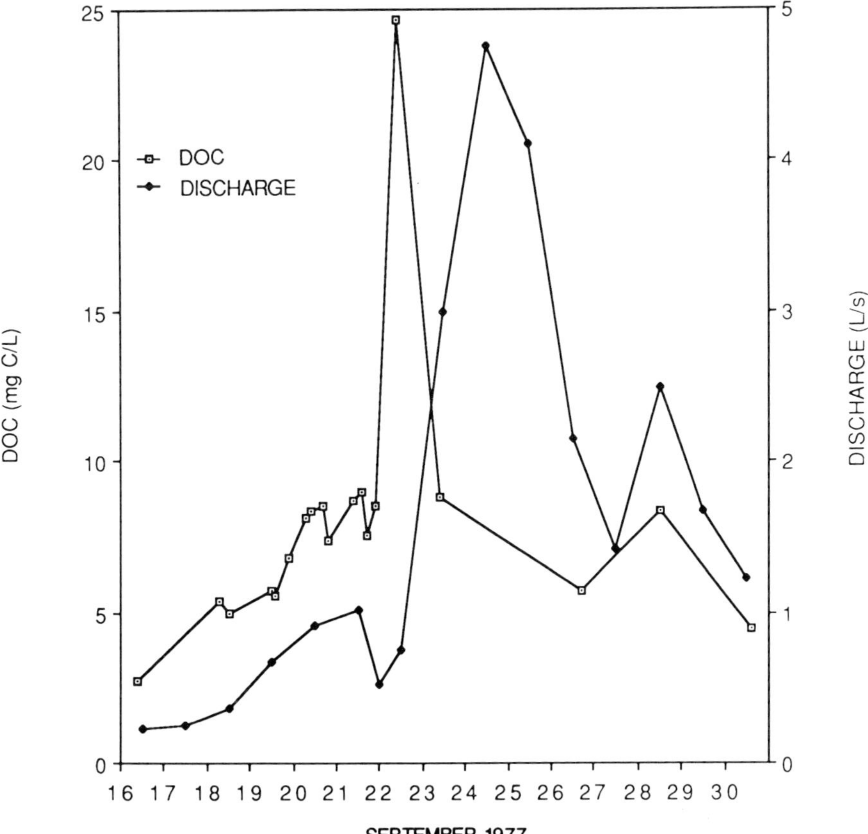

Fig. 4—Discharge and DOC concentration in a small stream in western Oregon over a two-week period during which a series of storms added precipitation to the catchment.

Historical Changes in Hydrologic Retention and Floodplain Vegetation

The river system and its associated floodplain vegetation are shaped by the physical energy of water moving downhill and the sediment load. Schafer (1973) superbly illustrates these concepts for the Rhine River in Germany where the size, shape, location, and migration rates of oxbow lakes along the floodplain respond to changes in sediment loads and discharge.

Sedell and Froggatt (1984) documented such a change for 25 km of the Willamette River, Oregon. Between 1854 and 1967, the Willamette River became increasingly isolated from its floodplain as a result of channelization and agricultural modification of the riparian forests. In 1854, the riparian forest was 1.5–3 km wide and in contact with >250 km of river edge on oxbow lakes and cutoff sloughs. By 1967 the length of river edge was systematically decreased to 64 km, a reduction of 74 percent, and the riparian forest was restricted to the adjacent bank. Much of the change in riparian forest interaction was completed by 1910 owing to snag removal and river navigation improvements (levees). These changes have resulted in a severe reduction in the ability of alluvial reaches to retain sediments and organics, the quantity of organic inputs, and primary production on the floodplain.

Hesse et al. (1989) dramatically illustrate this point on the lower Missouri River from Sioux City, Iowa, to the mouth at St. Louis. Channelization of this reach of the Missouri River directly eliminated over 192,000 ha of aquatic habitat and wetlands from the active erosional zone of the river. Channelization, along with the flood protection provided by mainstem and tributary reservoirs, fostered agricultural, urban, and industrial encroachment on 95% (728,460 ha) of the floodplain. This extensive development dramatically changed the composition of the natural plant communities that formerly colonized the floodplain and reduced available supplies of organic material by at least 65%. Native floodplain vegetation plus increased water retention on the floodplain will produce more DOC (tea bag effect). Native plant communities contribute large amounts of DOC (Moore 1987) from throughfall, stemflow, and root decomposition as well as 20–25% of the litter fall leaching out as DOC. Although agricultural production on the floodplain can reach 9 t/ha/yr of biomass (Ovington et al. 1963), much of this production is physically removed from the cropped field (floodplain) in the form of grain, forage, or domestic livestock. Native plant communities, which are periodically flooded, can contribute biomass from 6 t/ha/yr for grasslands to 16 t/ha/yr for cattail marshes to the ecosystem.

The total of worldwide wetland areas has drastically decreased in the last 40 years. The conterminous U.S. has only 46% of the original 8.7×10^7 ha of wetlands remaining (Tiner 1984). In a twenty-year period between the mid-1950s and mid-1970s, there was a net 3.6×10^6 ha loss (for every ha of wetland gained about 5.5 ha were lost). Tiner (1984) further estimates that about 2.0×10^5 ha of wetlands continue to be lost annually to agriculture and estuarine filling.

Gains in wetlands in the U.S. have come in the form of open water habitat from reservoirs, farm ponds, and coastal subsidence and total 0.8×10^6 ha. These open water gains were both modest in area and biological productivity as compared to the 4.45×10^6 ha losses in the productive forested and emergent wetlands during the same period. As an illustration, the lower Mississippi River floodplain originally included over 9.7×10^6 ha of bottomland forested wetlands. By 1937, only 4.7×10^6 or <50% of these remained. In the 1980s, there are less than 2.1×10^6 ha remaining—roughly 20% of the original acreage (Hefner and Brown 1984; Tiner 1984).

Figure 5a provides a cross-sectional view of a degraded small stream system. In this example, the stream has cut down through previously deposited alluvium. As a result, the channel and associated vegetation have changed dramatically. Species typical of wetland conditions have largely disappeared and the channel continues to erode laterally. There is little subsurface storage of water and the stream is characterized by intermittent flow. In contrast, Fig. 5b illustrates a previously eroded channel that supports a diversity of riparian vegetation and has undergone recovery. The vegetation provides relative stability to stream banks and causes deposition of sediment; over time the channel has undergone aggradation. Such aggradation is often a natural consequence of allowing streamside vegetation that may have been modified by historical grazing, logging, agriculture, or other management practices an opportunity to again function and exert its influence on flow conditions, characteristics of the channel, and DOC concentrations. A consequence of this aggradation process is that the water table will similarly rise. In some cases, a formerly intermittent stream may flow perennially.

The DOC concentration, retention, and production both on the surface and in the saturated zone are much higher in Fig. 5b than in Fig. 5a because of greater primary production and fine root biomass (Fig. 5c). The subsurface riparian soils have concentrations 4–10 times that of stream water (Moore 1987; C. Dahm, unpublished). For large rivers, the lowering of the water table in river valleys in both Europe and North America is well documented by Decamps et al. (1988). This lowering of the water table below the rooting zones is a result of gravel mining, trapping of sediments by upstream dams, and channelization. The retention of water in the area, as well as the carbon inputs, contributes to higher DOC (Naiman et al. 1988).

The landscape processes of deforestation, combined with floodplain levee systems, tend to isolate the river from the floodplain and accelerate the runoff of storm events. Figure 6 illustrates the relationship between increasing area of freely flooding floodplain and stage/discharge ratios expressed as degree of deviation from "pristine" (100%) conditions. As illustrated by Belt (1975) for the Mississippi River, long-term mean stage/discharge ratios increase as freely flooding floodplains decrease in area or become unavailable to floodwaters. Simultaneously, the annual variation about the long-term mean stage/discharge ratio increases with the decrease in freely flooding bottomlands, resulting in significant hydrologic changes in the river and in the adjacent bottomland hardwood wetlands. Such changes in hydrologic conveyance do not occur in response to elimination of a single site from

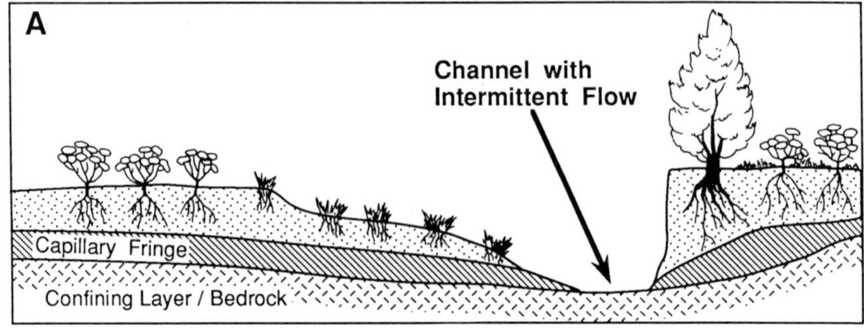

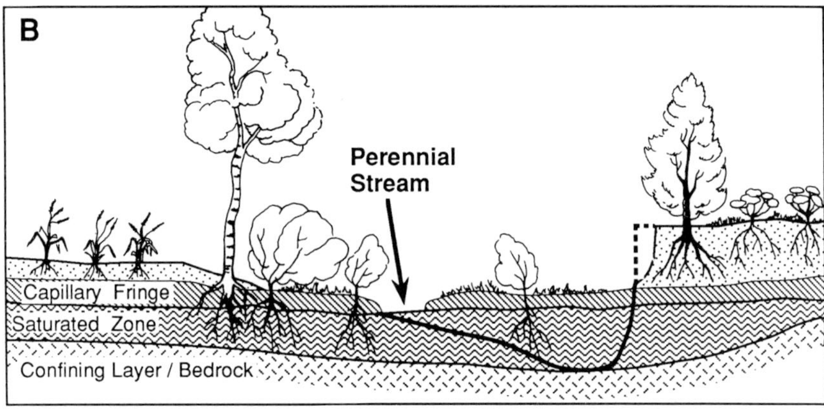

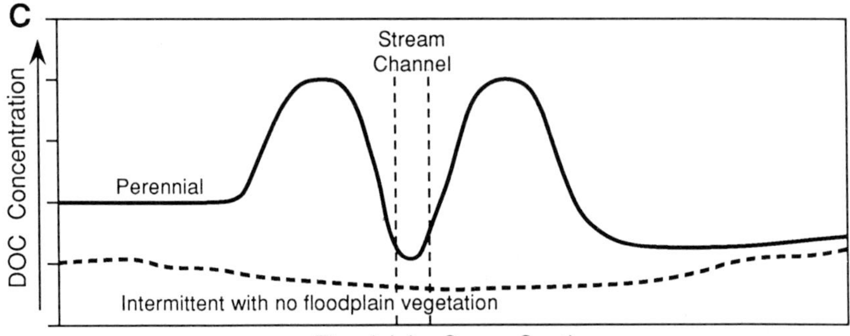

Fig. 5—General characteristics and DOC functions of riparian areas associated with rangeland streams (adapted from Elmore and Beschta 1987).

the bottomlands, but emerge as a result of incremental decreases in freely flooding bottomland. This type of alteration in the hydrologic regime of a channel network results from landscape-level processes (Lee and Gosselink 1988).

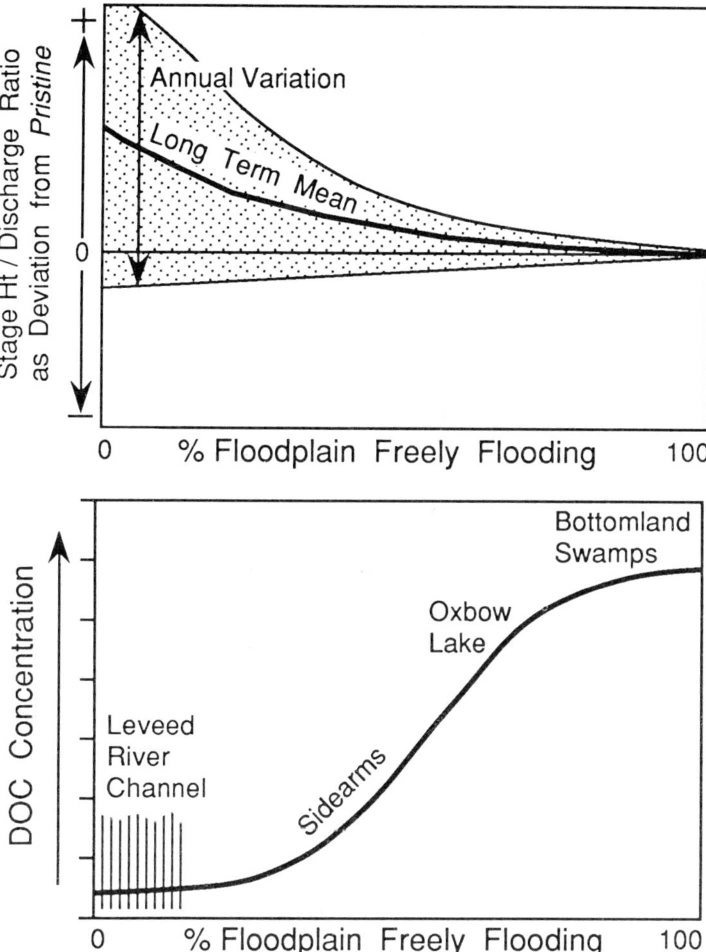

Fig. 6—Increased hydrologic conveyance from disconnecting the floodplain from the river channel (adapted from Lee and Gosselink 1988).

Changes in DOC concentration are hypothesized to vary directly with change in hydrologic conveyance within systems which have not been drastically urbanized or receive significant human effluents (Fig. 6b). The different floodplain features display a trend in DOC concentrations which increase with distance and connectivity with the main channels or aerial extent of the floodable floodplain or wetlands.

Not only is the floodplain becoming less productive in terms of primary production, but also the retention of water on the land is greatly reduced. This results in a potential reduction in DOC.

DISCUSSION – POSSIBLE ISSUES FOR CONSIDERATION

Concentrations of DOC in natural waters range over approximately two orders of magnitude. The causes behind this variability are important to the energetics, buffering characteristics, and nutrient dynamics of these ecosystems. The causal mechanisms which control the concentration and quality of DOC in various aquatic ecosystems are important to elucidate. Hypotheses that predict declining concentrations of DOC in regions receiving strong acid deposition need to be evaluated mechanistically in addition to the correlative procedures used to date. Therefore, we present the following questions as possible topics for further discussion.

1. What are the sources and routing of DOC in different types of surface waters?
2. What factors predispose some natural waters to contain high or low concentrations of DOC?
3. What is the relative importance of anthropogenic changes to surface water runoff in altering the quantity and characteristics of DOC?
4. What is the effect on the aquatic ecosystem of changing from a complex deciduous and coniferous leaf litter source to an algal-macrophyte-dominated DOC source?
5. What is the effect on the aquatic ecosystem of DOC derived from anaerobic processes as opposed to aerobic processes?
6. How do structural differences in the types of organic acids affect availability of limiting nutrients?

Question 1 includes such topics as (*a*) the relative contribution of soil organic matter versus autochthonous sources of DOC to the soluble organic carbon pool; (*b*) the importance of riparian, wetland, hyporheic, and main channel contributions to total DOC; and (*c*) spatial and temporal variability in the source areas for DOC within a catchment. Historically, there has been a general consensus among organic geochemists that DOC in surface waters is derived mainly from soil organic matter. The variability of the concentration of DOC in time and space and the generally higher concentrations of DOC in waters which drain regions with extensive wetlands, riparian zones, and gallery forests suggest a major contribution from the floodplains of streams and rivers. Is the geomorphology of the catchment, particularly along the drainage network, a strong determinant of the average DOC concentration? Which river ecosystems derive the majority of their DOC from upland terrestrial environments and which ecosystems are primarily dependent on carbon sources within the permanently saturated wetland areas of the floodplain?

Question 2 focuses on what the major sources and sinks of DOC are within a catchment. Two major potential sources of DOC for most surface waters are the upper organic-rich horizons of upland soils and organic-rich depositional zones within the floodplain. Two major potential sinks for DOC are the lower horizons of upland soils and bacterial metabolism. The interplay of these sources and sinks, coupled with the hydrologic pathways of water in the basin, produces the array of DOC concentrations measured worldwide.

Question 3 encompasses the wide range of impacts that human activities can potentially exert on DOC dynamics within a catchment over temporal scales ranging from hours to millennia. Anthropogenic activities may result in changes which could either increase or decrease DOC concentrations in the water. Possible mechanisms which are postulated to decrease DOC include (*a*) structures which accelerate the rate of runoff (e.g., levees, channelized streams and rivers, large organic debris removal, drainage canals and ditches, etc.); (*b*) lowered water table through wetland removal, enhanced erosion, dredging, etc.; (*c*) vegetation reduction within the riparian corridor; (*d*) large reservoirs; and (*e*) acid deposition throughout the basin. Human activities which can lead to a possible increase in DOC concentrations within streams and rivers include: (*a*) eutrophication associated with urban, industrial, or agricultural loading of nutrients; (*b*) increased rice production throughout much of the world; (*c*) riparian zone protection and reestablishment; (*d*) wetland zone protection and reestablishment; and (*e*) increased channel complexity from such agents as large organic debris and beaver activity. A wide variety of human impacts, both enhancing or restricting DOC inputs into streams and rivers, are often occurring within a basin simultaneously. Differentiating between the major effects of human intervention within a catchment is critical for assigning an unequivocal dominant mechanism to explain a directional shift in DOC concentration through time.

Questions 4–6 address the fact that vast areas of DOC-producing features have been disconnected and removed from influencing rivers. The sources of organic acids have changed along with the processes from which DOC is derived. In general there is greater in-channel production of algae and less structurally complex organic material entering streams. Algal organics would have more carbohydrates and proteins. Leaf litter has more polymeric forms, such as phenolics, terpenes, and alkaloids. We, as well as M. Perdue (pers. comm.), have suggested that those systems with extensive floodplains and adjacent wetlands have a large proportion of DOC derived from anaerobic decomposition.

What are the structural differences in DOC derived from anaerobic processes as compared to more prevalent aerobic processes today? If there are major structural differences in DOC caused by both differences in organic sources as well as decomposition processes, what are the effects on limiting nutrient availability? There are suggestions that some DOC makes limiting nutrients more available and countersuggestions that DOC from wetlands may suppress blue-green algal blooms in adjacent water bodies.

The general question of temporal and spatial variations of organic acids at the ecosystem level has been discussed above in terms of DOC. Implicit in the discussion is the assumption that most of the DOC is in the form of organic acids and that the equivalents of carboxyl groups per gram of DOC are relatively constant. An additional topic for discussion, one for which a much smaller set of data is presently available, is the temporal and spatial variation of organic acids within the larger DOC pool. Can it be generally assumed that the concentration of organic acids closely follows the concentration of DOC in most aquatic ecosystems?

Acknowledgements. The National Science Foundation supports ecosystem-level research at Oregon State University (grants BSR-8414325 and BSR-8508356) and University of New Mexico (BSR-8616438) which we gratefully appreciate. J. Barnett, D. Coor, and L. Nelson prepared the figures. Special thanks to M. Meybeck for a thorough review.

REFERENCES

Belt, C.B., Jr. 1975. The 1973 flood and man's construction of the Mississippi River. *Science* 189:681–684.

Dahm, C.N., S.V. Gregory, and P.K. Park. 1981. Organic carbon transport in the Columbia River. *Est. Coast. Shelf Sci.* 13:645–658.

Davis, R.B., D.S. Anderson, and F. Berge. 1985. Palaeolimnological evidence that lake acidification is accompanied by loss of organic matter. *Nature* 316:436–438.

Decamps, H., M. Fortune, F. Gazell, and G. Pautau. 1988. Historical influence of man on the riparian dynamics in a fluvial landscape. *Landsc. Ecol.* 1:163–173.

Elmore, W., and R.L. Beschta. 1987. Riparian areas: perceptions in management. *Rangelands* 9(6):260–265.

Fisher, S.G. 1977. Organic matter processing by a stream segment ecosystem: Fort River, Massachusetts, U.S.A. *Int. Rev. Ges. Hydrobiol.* 62:701–727.

Furch, K. 1985. Dissolved carbon in a floodplain lake of the Amazon and in the river channel. *Mitt. Geol.-Paläont. Inst. Univ. Hamburg*:285–298.

Hefner, J.M., and J.D. Brown. 1984. Wetland trends in the southeastern United States. *Wetlands* 4:1–12.

Hesse, L.W., G.R. Chaffink, and J. Brabender. 1989. Missouri River mitigation: a system approach. *Fisheries* 14:11–15.

Ittekkot, V., S. Safiullah, B. Mycke, and R. Seifert. 1985. Seasonal variability and geochemical significance of organic matter in the River Ganges, Bangladesh. *Nature* 317:800–802.

Klotz, R.L., and E.A. Matson. 1978. Dissolved organic carbon fluxes in the Shetucket River of eastern Connecticut, U.S.A. *Freshwat. Biol.* 8:347–355.

Kuserk, F.T., L.A. Kaplan, and T.L. Bott. 1984. *In situ* measures of dissolved organic carbon flux in a rural stream. *Can. J. Fish. Aq. Sci.* 41:964–973.

Lee, L.C., and J.G. Gosselink. 1988. Cumulative impacts on wetlands: linking scientific assessments and regulatory alternatives. *Envir. Manag.* 12(5):519–602.

Lesack, L.F.W., R.E. Hecky, and J.M. Melack. 1984. Transport of carbon, nitrogen, phosphorus, and major solutes in the Gambia River, West Africa. *Limnol. Ocean.* 29:816–830.

Malcolm, R.L., and W.H. Durum. 1976. Organic carbon and nitrogen concentrations and annual organic carbon load of six selected rivers of the United States. USDI-Geological Survey, Water Supply Paper 1817-F, Washington, D.C.

McDowell, W.H., and G.E. Likens. 1988. Origin, composition, and flux of dissolved organic carbon in the Hubbard Brook Valley. *Ecol. Monogr.* 58:177–195.

McKnight, D., E.M. Thurman, R.L. Wershaw, and H. Hemond. 1985. Biogeochemistry of aquatic humic substances in Thoreau's Bog, Concord, Massachusetts. *Ecology* 66:1339–1352.

Meybeck, M. 1982. Carbon, nitrogen, and phosphorus transport by world rivers. *Am. J. Sci.* 282:401–450.

Meyer, J.L. 1986. Dissolved organic carbon dynamics in two subtropical blackwater rivers. *Arch. Hydrobiol.* 108:119–134.

Moore, T.R. 1987. Dissolved organic carbon in forested and cutover drainage basins, Westland, New Zealand. *Int. Ass. Sci. Hydrol. Publ.* 167:481–487.

Mulholland, P.J., and E.J. Kuenzler. 1979. Organic carbon export from upland and forested wetland watersheds. *Limnol. Ocean.* 24:960–966.

Mulholland, P.J., and J.A. Watts. 1982. Transport of organic carbon to the oceans by rivers of North America: a synthesis of existing data. *Tellus* 34:176–186.

Naiman, R.J., H. Decamps, J. Pastor, and C.A. Johnston. 1988. The potential importance of boundaries to fluvial ecosystems. *J. N. Am. Benthol. Soc.* 7(4):289–306.

Ovington, J.D., D. Heitkamp, and D.B. Lawrence. 1963. Plant biomass and productivity of prairie, savanna, oakwood and maize field ecosystems in central Minnesota. *Ecology* 44:52–63.

Richey, J.E., J.T. Brock, R.J. Naiman, R.C. Wissmar, and R.F. Stallard. 1980. Organic carbon: oxidation and transport in the Amazon River. *Science* 207:1348–1351.

Schafer, W. 1973. Alkrhein Verbund am Nördlichen Oberrhein. *Cour. Forsch. Senckenberg* 7:1–63.

Sedell, J.R., and J.L. Froggatt. 1984. Importance of streamside vegetation to large rivers: the isolation of the Willamette River, Oregon, U.S.A., from its floodplain. *Verh. Int. Verein. Limnol.* 22:1828–1834.

Sullivan, T.J., J.M. Eilers, M.R. Church, D.J. Blick, K.N. Eshleman, D.H. Landers, and M.S. De Hann. 1988. Atmospheric wet sulphate deposition and lakewater chemistry. *Nature* 331:607–609.

Tate, C.M., and J.L. Meyer. 1983. The influence of hydrologic conditions and successional state on dissolved organic carbon export from forested watersheds. *Ecology* 64:25–32.

Thurman, E.M. 1985. Organic Geochemistry of Natural Waters. Dordrecht:Nijhoff.

Tiner, R.W. 1984. Wetlands of the United States: current status and recent trends. National Wetlands Inventory 1984-439-855-814/10870. Washington, D.C.:U.S. Fish and Wildlife Service.

Welcomme, R.L. 1985. River Fisheries. Fisheries Technical Paper 262. Rome: FAO.

Organic Acids in Aquatic Ecosystems
eds. E.M. Perdue and E.T. Gjessing, pp. 281–299
John Wiley & Sons Ltd

Production and Utilization of Dissolved Organic Carbon in Riverine Ecosystems

J.L. Meyer

Zoology Department and Institute of Ecology
University of Georgia
Athens, GA 30602, U.S.A.

Abstract. Major sources of dissolved organic carbon (DOC) in streams and rivers are found within riparian zones and stream channels. In addition to DOC excreted by primary producers in the channel, DOC is rapidly leached from terrestrial leaves falling into streams. A more important DOC source appears to be DOC leached from material stored in the streambed. Leaching of this material is facilitated by biological activity and, in some cases, may occur under anaerobic conditions. The amount of DOC produced by these kinds of sources has been reduced by human activities that reduce channel storage and decouple rivers and floodplains.

DOC is removed from the water column of lotic ecosystems by biotic and abiotic processes at rates ranging from <10–550 mg C $m^{-2}h^{-1}$. Abiotic removal processes include sorption, photooxidation, and particle formation. Biotic utilization of DOC is largely bacterial and varies with the chemical nature of DOC and the bacterial community. Epilithic microbial communities are important sites for DOC uptake in many streams. The extent of contact between water and sediments is a critical determinant of rates of DOC utilization in rivers. Consequences of DOC utilization include alteration of biogeochemical cycling of other elements and increased secondary production in the ecosystem.

Three areas for future research are highlighted: (*a*) ecologically relevant chemical analyses of natural DOC that can be used routinely to characterize biological availability of DOC in aquatic ecosystems; (*b*) information on the rates and biotic availability of products of very slow leaching of stored organic matter; (*c*) hydrologic studies including work on water flow paths and residence times in the watershed to clarify DOC sources and work on water movement in streambed sediments to quantify exchanges of DOC between the sediments and water column.

INTRODUCTION

In global geochemistry, rivers are viewed as linkages between major storage reservoirs on land and in oceans. Although rivers do serve as transporters, their role as modifiers and retainers of materials is more significant in global processes than is often acknowledged and ecologically much more interesting. It is this aspect of dissolved organic carbon (DOC) in riverine ecosystems that is addressed in this chapter, namely, what are major sources and sinks for DOC in rivers? Production and utilization of DOC in rivers will influence the amount of organic carbon transported to oceans, provide a trophic resource for lotic organisms, and impact cycling and transport of other elements. Research on production and utilization of dissolved organic matter in riverine ecosystems has generally dealt with total DOC rather than singling out organic acids; hence in this chapter I discuss production and utilization of DOC, of which organic acids are a major component. The proportion of riverine DOC that is made up of organic acids varies from 49–92% (Aiken 1985), although in average river water, organic acids comprise about 90% of DOC (Thurman 1985).

SOURCES OF DOC

Terrestrial and aquatic primary production is the ultimate source of riverine DOC. Although a major fraction of primary production in a watershed may occur on upland soils, production occurring in areas near stream channels is a particularly important DOC source for streams. On the uplands, DOC leached from vegetation by rainwater enters upper soil horizons, where additional leaching of stored organic matter results in a large flux of DOC to mineral horizons. These horizons appear to be particularly effective at removing DOC from solution (e.g., by sorption on iron and aluminum oxides [McDowell and Likens 1988]); hence water that enters the stream via subsurface seeps is low in DOC concentration. In situations where mineral horizons are bypassed (e.g., during intense rainstorms when lateral flow through surface soil horizons becomes important) or where the capacity of the mineral soils to sorb organic matter is low, water entering streams from uplands has a higher DOC concentration. The critical factor here is the lower, mineral soil horizons, which can decouple upland DOC sources from streams. DOC coming from the uplands, which has percolated through a well-developed soil, sets a low, baseline DOC concentration in stream water. This idea has been more fully developed by Cronan (this volume).

The key sources of spatial and temporal variation in DOC concentration in pristine streams are the organically rich streamside and floodplain soils as well as organic matter stored and produced in the channel. These sources are emphasized in this chapter (see also Hemond, this volume). Increasing concentrations of DOC have been observed along channels of headwater streams and larger rivers (e.g., Fig. 1), although other patterns have also been observed (e.g., Sedell and Dahm, this volume). The increases documented in Fig. 1 are primarily a result of increases in the higher

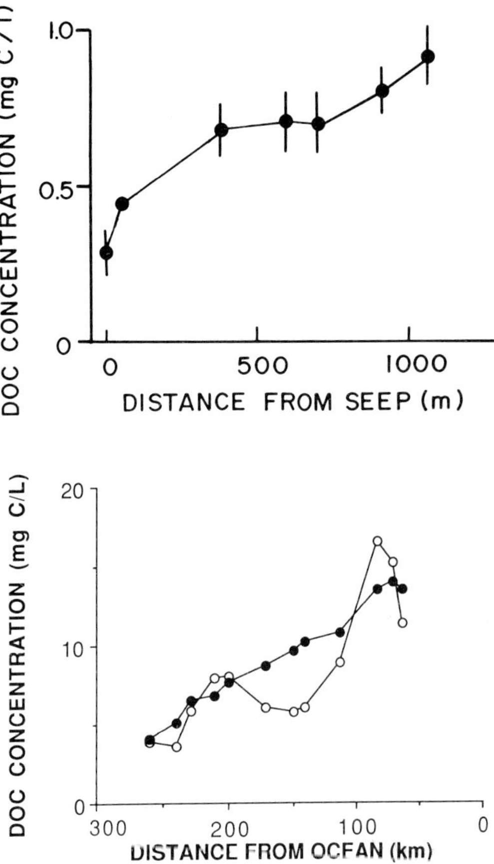

Fig. 1—Changes in DOC concentration along rivers. A. Increase in DOC concentration along Hugh White Creek, a headwater stream draining an undisturbed forest in the southern Appalachians. Values are annual means ± 95% confidence limits (reprinted from Meyer, Tate et al. 1988). B. Increase in DOC concentration along the Ogeechee River, Georgia, at a time of low discharge (open circles) and high discharge (closed circles) (J.L. Meyer and R.T. Edwards, unpubl. data).

molecular weight fractions of DOC. Increases are observed despite the fact that biotic and abiotic processes remove DOC from the water column and in the absence of variation in DOC concentration in groundwater entering the stream; this implies that there are sources of DOC in or near the channel. The common observation of rapidly increasing DOC concentrations during storms also points to the importance of DOC sources that are in or near the channel, as will be discussed below.

Leaching of Stored Organic Matter

Generation of DOC from terrestrial detritus can be viewed in two phases: a rapid flush of soluble organic compounds from fresh detritus within the first day of exposure to water (5–30% of detrital carbon [Petersen and Cummins 1974]) and a slow release of DOC during detrital decomposition. We know more about the former than the latter. The initial leachate is often rich in carbohydrates; for example, sugar maple leaf leachate contained 13% monomeric carbohydrates, 10% polymeric carbohydrates, and 32% phenolics (McDowell 1985). In regions with extensive deciduous vegetation, the initial leaching phase results in a pulse of labile DOC into the system during autumn. The absence of a striking autumn increase in DOC concentration in most streams despite this input (e.g., McDowell 1985) gives testimony to the ability of stream ecosystems to utilize this DOC. The fraction of organic carbon in leaf litter that winds up as DOC is usually calculated as the amount leached in this initial period, yet leaching continues at low rates for much longer periods. Bacterial and fungal growth in litter is at the expense of DOC which has been released by exoenzymes. The organic carbon represented by microbial biomass and respiration is a reflection of the amount of DOC produced. Hence the fraction of leaf litter inputs that become DOC is clearly much greater than the 5–30% leached initially.

An unknown but potentially large fraction of riverine DOC comes from slow leaching of stored organic matter. To evaluate this DOC source, we must consider major sites of organic matter storage in streams. There are three important sites of organic matter storage that contribute DOC to stream ecosystems: depositional zones behind dams created by woody debris, sediments in the wetted stream channel, and litter and soils in floodplains and riparian zones. Debris dams are obvious sites of organic matter storage in streams in forested regions; both woody debris and the particulate organic matter it traps (e.g., leaf litter) release some of their carbon as DOC. Much of this organic matter is recalcitrant; yet even that will leach DOC, albeit slowly. For example, radiolabeled lignocellulose DOC was leached from leaf litter at a rate of 0.5% of initial labeled lignocellulose per day after 13 weeks of incubation (Moran and Hodson 1989). DOC from these recalcitrant sources is likely to be richer in fulvic and humic acids and hence have a higher content of acidic functional groups. Although the rate at which DOC is leached from recalcitrant material in streams is very low, the amount of material stored in stream channels can be so great that this can be a major source of DOC. For example, a leaching rate of only 0.1% per year of woody debris stored in an Oregon headwater stream would account for measured DOC export in that stream (Cummins et al. 1983).

Although debris dams in small streams are the sites of organic matter storage that first catch the eye, most organic matter stored in streams is in the form of fine particulate organic matter (FPOM) in streambed sediments. This material will also leach DOC, and elevated levels of DOC are commonly found in streambed interstitial water (e.g., Crocker and Meyer 1987). Interstitial DOC concentration is a function of sediment organic matter content, microbial production and utilization of DOC, and water exchange (Crocker and Meyer 1987). Stream biota play a role in accelerating this leaching process. The importance of fungi and bacteria in solubilization of organic matter has long been recognized, but larger organisms also play a role via their feeding activities. Leaf-shredding invertebrates are common in streams with accumulations of leaf detritus. They ingest large quantities of detritus, have low assimilation rates, and hence produce large quantities of DOC and leachable FPOM (Meyer and O'Hop 1983). Where these organisms are abundant, this can be a significant source of DOC (Meyer and O'Hop 1983). For example, when macroinvertebrate biomass was experimentally reduced in a headwater Appalachian stream, less DOC was generated in the experimental stream than in an unaltered reference stream (J.B. Wallace, pers. comm.).

Streamside riparian zones, including floodplains, are significant sources of DOC for streams. Variable source areas that are important sources of water during stormflow also provide inputs of DOC to the stream. Rainstorms flush water from isolated pools along the channel perimeter and leach DOC from organic-rich streamside source areas, such as organic matter stored in areas of channel expansion. Storm discharge commonly has higher DOC content; reports from 11 streams in latitudes ranging from 6 to 68° indicate that DOC concentrations increase during storms (Meyer, McDowell et al. 1988). If there were no additional source areas mobilized during storms, we would expect to see either no change in concentration or, more likely, a decrease in concentration resulting from dilution by a larger volume of water. Higher DOC concentration coupled with larger volumes of water mean that storm runoff is an important component of DOC transport. For example, when increases in DOC concentration during storms were not included in transport calculations, annual DOC transport was 35–60% of that calculated when storm-induced changes in DOC concentrations were included (Meyer and Tate 1983). Detailed analyses of chemical composition of DOC during storms are limited, but the proportion of phenolics, monomeric carbohydrates, and polymeric carbohydrates has been shown to increase during rainstorms in a forest stream (McDowell and Likens 1988).

The recognition that areas of organic matter accumulation in the streambed and riparian zone can be anaerobic (Dahm et al. 1987) leads to intriguing questions on the potential contributions of organic acids generated under anaerobic conditions. The quantity and quality of DOC would be influenced by the extent to which streambed debris accumulations and streamside source areas are anaerobic. Anaerobic metabolic pathways commonly yield products such as acetate, proprionate, and butyrate that are labile, although polymeric material is also being solubilized to compounds that are relatively inert and hence accumulate in the anaerobic region. Waters from these anaerobic stream areas are enriched in DOC as well as other nutrients (Dahm et al. 1987). When waterflow through these zones increases, they would become a DOC source to the stream. Anaerobic areas in the streambed are associated with organic matter accumulations behind retention structures such as woody debris and beaver dams; these retention structures have been greatly reduced by human activity (Dahm et al. 1987). Smaller debris accumulations will probably also be anaerobic in their interior, particularly when they occur in areas of reduced flow such as organic matter buried in sediments. Streamside pockets of organic matter in waterlogged conditions are also likely to be sites of anaerobic metabolism. If these kinds of areas are common in a riverbed and its riparian zone, then anaerobically produced DOC will be of importance. We are lacking critical field data from a wide range of riverine ecosystems that allow us to assess the extent of these anaerobic zones. A historical assessment suggests they are less common today than prior to human alteration of sites of organic matter storage (Dahm et al. 1987).

The significance of streamside wetlands as a source for streamwater DOC is widely recognized. In low-gradient blackwater rivers of the U.S. Coastal Plain, DOC is high in concentration (10–20 mg C/l) and has its origin in swampy floodplains. Soils from the Ogeechee River floodplain in Georgia leached DOC during simulated flooding in the laboratory (Meyer 1986). When we use these data to calculate the amount leached from the floodplains during a rise in water level of 50 cm, the amount leached from the floodplain is adequate to account for 90% of DOC transported in the river over an 11-day period when floodplain water levels rose 54 cm (Meyer 1986). Clearly DOC leached from soils of productive floodplain forests is a significant source of DOC to rivers. The historical impact of man in decoupling rivers and their floodplains has reduced the amount of DOC entering streams from this source. It is likely that the amount of DOC transported to the ocean by pristine river systems with extensive floodplains, large amounts of stored woody debris, and large beaver populations was considerably greater than that observed today.

Autochthonous Production of DOC

In unshaded streams, DOC generated by primary producers in the stream channel contributes to the total DOC load of the stream. DOC is produced by living as well as decomposing algae and macrophytes. Algal epiphytes on macrophyte leaves increase rates of DOC excretion in some macrophytes (Hough and Wetzel 1975). Under baseflow conditions in well-lit streams algal-generated DOC resulted in a diel variation in DOC concentration with late-afternoon maxima 21–44% greater than the daily minima (Kaplan and Bott 1982). Rates of DOC production range from 11.7 mg DOC $m^{-2}h^{-1}$ for a few hours of maximal production from stones colonized heavily by *Ulothrix* (Kaplan and Bott 1982) to 130 mg DOC $m^{-2}d^{-1}$ in a bed of *Lemna minor* (Baker and Farr 1987). Glutamate is a common form of DOC released by phytoplankton, although not by macrophytes (Hough and Wetzel 1975); however, macrophyte-produced DOC is generally composed of low molecular weight compounds that are readily available to microbes (e.g., Baker and Farr 1987). Low molecular weight compounds released by stream algae can also play a role as cofactors, stimulating the decomposition of more refractory DOC, although additional refractory material is also produced during this process (Geller 1986). The contribution of autochthonous DOC to metabolism of lotic biota is probably greater than its contribution to DOC in transport. One reason for this is the lability of algal- and macrophyte-produced compounds. In addition, the close proximity of autotrophs and heterotrophs in epilithic assemblages ensures that much of the DOC produced is consumed within the epilithic community and does not escape to the water column (Lock et al. 1984).

UTILIZATION OF DOC IN LOTIC ECOSYSTEMS

Numerous experiments have demonstrated removal of DOC from stream water in natural streams (Table 1). DOC removal has also been measured in recirculating microcosms with sediments. Microcosms offer a useful technique for comparing uptake of DOC from different sources, but it is difficult to relate the generally higher uptake rates measured in microcosms to a rate of removal in the natural channel. Hence I have not included these data in Table 1. Measured uptake rates vary over two orders of magnitude, from <10 to 550 mg C $m^{-2}h^{-1}$, as a consequence of the quality and concentration of DOC added, streambed geomorphology, water depth and velocity, and biotic activity. DOC uptake increases with DOC concentration, and higher rates than reported in Table 1 have been measured with high concentrations of sucrose (Meyer, Tate et al. 1988). Uptake rates that are measured at high concentrations are more indicative of a stream channel's potential for DOC uptake than of uptake rates that are occurring under ambient conditions. Different rates of DOC uptake are measured in streams even when DOC source and concentration are the same. For example, uptake rates of added sucrose during summer in headwater streams in North Carolina and Oregon varied by an order of magnitude as a consequence of differences in channel geomorphology and biotic activity (Table 1). An areal measure of uptake (mg C $m^{-2}h^{-1}$) is particularly sensitive to water depth and velocity and to the measurement of streambed area, which is generally taken as mean width times length of the section;

TABLE 1. Uptake of added DOC in natural streams.

Location	DOC Source	Concentration (mg C/l)	Uptake Rate (mg C $m^{-2}h^{-1}$)	Reference
Pennsylvania	Ambient	3–8	9–16	Kuserk et al. (1984)
Pennsylvania	Bovine manure	12	51–237	Kuserk et al. (1984)
Pennsylvania	Jewel weed	–	147	Kuserk et al. (1984)
Ontario	Cedar leachate	10.5	250–550	Lush and Hynes (1978)
North Carolina	Sucrose	1	240	Meyer, Tate et al. (1988)
North Carolina	Sucrose	4	527	Munn (1989)
North Carolina	Leaf leachate	0.2–0.5	30–240	Meyer, Tate et al. (1988)
New Hampshire	Sugar maple	3.7–5.6	0–326	McDowell (1985)
New Hampshire	Yellow birch	5	75–291	McDowell (1985)
New Hampshire	Spruce leachate	5.5	115–180	McDowell (1985)
Oregon	Sucrose	4	38	Munn (1989)

hence variations in bed roughness in different streams contribute to reported variations in uptake rate. We have observed a 20-fold variation in DOC uptake rate across stream reaches that differ in geomorphology, with lowest rates in gravel reaches and highest rates in reaches with debris dams; these differences are greater than could be explained solely by variation in water residence time in the two reaches (Munn 1989). Given the wide range in stream geomorphology, DOC concentrations, and sources used, it is notable that the variation in uptake rate reported in Table 1 is not even greater.

What are the processes underlying the observed uptake? For sake of clarity, I have separated them into abiotic and biotic processes, although both are occurring simultaneously. Estimates of the relative importance of biotic vs. abiotic pathways vary with the experimental technique and the nature of the streams being studied; for example, Dahm (1981) reports that 20% of sorption is abiotic, whereas McDowell (1985) estimates abiotic sorption to be 50%. It is important to recognize that the two are not necessarily competing processes: DOC sorbed by abiotic processes can be available for subsequent biotic uptake. Abiotic sorption of DOC onto sediments is a process analagous to the trapping of particulate organic matter behind debris dams. It is a retention process which can be followed by biotic utilization of the retained material.

Abiotic Processes

DOC can be rapidly removed from solution by chemical adsorption onto sediment particles, either in the streambed (e.g., McDowell 1985) or suspended in the water column (e.g., Ertel et al. 1986). For example, Rio Negro humic acids are selectively adsorbed onto fine suspended sediments immediately below the junction of this river and the more turbid Rio Solimoes; this results in a concentration of humic substances 70–83% of that expected (Ertel et al. 1986). The capacity of sediments to adsorb DOC has frequently been associated with the presence of iron and aluminum sesquioxides; a similar mechanism is responsible for retention of DOC in mineral soil horizons (e.g., McDowell 1985). Chemical sorption of DOC is generally rapid and is quickly saturated, whereas microbial utilization is commonly a slower process that continues for a longer time (e.g., Dahm 1981).

In lotic ecosystems the sites of abiotic and biotic uptake of DOC are primarily in the streambed, so the extent to which water comes in contact with the sediment becomes critical in determining the rates at which DOC is removed from the water column. Recent work investigating the rate at which surface waters enter sediments has shown remarkably rapid sediment penetration. In a headwater stream in the Appalachians, a conservative tracer added in the water column penetrated to depths of 10 cm within 5 minutes, and minimum rates of infiltration into the sediments ranged from 0.2 cm min^{-1} in fine sediments to 1 cm min^{-1} in cobble substrate (Munn and Meyer 1988). Studies in California streams have also documented the extensive hydrologic and chemical interactions between water in the stream channel and the surrounding streambed; added solutes accumulate in transient storage zones of slowly moving water in the streambed (Bencala et al. 1984). This kind of transient storage clearly prolongs the time of contact of DOC with the sediment, and increases the amount of DOC that can be removed from the water column. Exchange between surface and interstitial waters can be extremely patchy and is influenced by channel geomorphology and bed permeability. Using streambed temperature data to examine waterflow in a Michigan stream, White et al. (1987) found surface water to be influencing streambed temperatures to depths of 50 cm. Interaction with that large a volume of sediment would allow for considerable DOC uptake! Large objects such as boulders or woody debris in the channel as well as regular features such as riffles and pools influence the patterns of waterflow in the sediments (White et al. 1987) and hence the capacity of the sediments to sorb solutes like DOC.

In this chapter I have discussed sediments as both a source and a sink for DOC. Clearly the next question is: are they a net source or a net sink? Unfortunately there is not a simple answer to this question. The answer will depend on sediment chemical composition, organic matter content, biotic activity, and the type of DOC entering the system.

Exposure to direct sunlight will influence DOC quantity and quality. Exposure to ultraviolet radiation under aerobic conditions will result in photooxidation of the humic substances in natural waters and hence a reduction in DOC content or a change in DOC quality. The kinds of changes that can occur include oxidation and bleaching of aromatic rings and quinone formation (McKnight et al. 1988). The abiotic process of photooxidation generally renders DOC more accessible to bacterial degradation (e.g., Geller 1986) and is a phenomenon worthy of further study in natural systems.

The phenomenon of particle formation from DOC has both biotic and abiotic components and greatly influences the availability of this organic matter to the biota. Abiotic flocculation of DOC will occur as a consequence of increases in pH, calcium ion concentration, salinity, and turbulence (Wotton 1988). Bubbles traversing the water column as well as those generated at the surface will accumulate a surface microlayer of hydrophobic organic compounds (e.g., fatty acids), which can turn into particles when the bubble bursts (Wotton 1988). The importance of this phenomenon in nature will obviously vary considerably with the nature of the stream channel and streamwater DOC content. In a Swedish stream, particle generation from DOC was observed over a 60 m reach of stream which included a waterfall (Petersen 1986). The rate of DOC removal by flocculation was calculated as 62 mg C $m^{-2}h^{-1}$ (Petersen 1986), a rate in the mid-range of DOC removal rates measured by experimental additions of DOC (Table 1), although most of those were with more labile sources of DOC. In part, the rate of DOC flocculation by bubbles is limited by the relatively short lifetime of bubbles in the stream (seconds). The particles formed by flocculation are generally small, ranging from 4–10 μm in diameter (Petersen 1986). Particles of this size are readily available to a wide range of filter feeders and are probably a more nutritive food resource than similar-sized particles produced by the breakdown of particulate detritus (Wotton 1988). In high-gradient streams with considerable turbulence, this mechanism of DOC loss cannot be overlooked.

Foam produced in zones of turbulence is an obvious yet poorly studied feature of rivers. Although it is most apparent in rivers receiving anthropogenic surfactants, it is also found in pristine river systems, where it can trap particulate organic matter such as algae, fungal spores, and detritus. In marine systems, foam offers patches of high nutrient concentration, supports high populations of bacteria and protozoa (Tsyban 1971), and can meet a major fraction of the nutritional requirements of invertebrates such as amphipods (Barlocher et al. 1988). Its chemical and ecological significance in freshwaters is unknown.

Biotic Utilization of DOC

Bacterial uptake of DOC is the most important biotic sink for DOC in riverine ecosystems. The rate of uptake by bacteria is dependent on temperature, numbers and kinds of bacteria present, and chemical nature of the DOC. The effect of DOC quality on uptake has been most widely investigated. Numerous studies report bacterial uptake and growth on specific compounds such as glucose and leucine. These are of limited value in helping us understand the utilization of DOC in natural systems, primarily because such a small fraction of naturally occurring DOC is composed of these clearly identifiable compounds. Other studies have demonstrated different rates of microbial utilization depending on the source of DOC, specifically the leaf species from which it was leached (e.g., Dahm 1981). In general, leachates from nonvascular plants are most readily utilized, followed by macrophytes and deciduous species, with conifer leachates being utilized least. Others have examined the ability of bacteria to use different molecular weight fractions of DOC. Low molecular weight fractions are most available, with intermediate molecular weight fractions largely unavailable (Meyer et al. 1987) or even inhibitory, to bacterial metabolism (Ford and Lock 1987). The proportion of DOC used by bacteria varies widely with the nature of DOC and the experimental conditions. For example, in three days bacteria used 54% of low molecular weight DOC

(<1000 nominal molecular weight) and 21% of very high molecular weight DOC (>10,000 nominal molecular weight) from a blackwater river (Meyer et al. 1987); in four days of exponential growth, bacteria used from 15–23% of naturally occurring DOC from humic and clear lakes (Tranvik and Hofle 1987). The efficiency of conversion of DOC to bacterial tissue also varies, although bacterial growth efficiencies on complex natural DOC fall between 20 and 30%; these estimates include growth on DOC that is frequently considered refractory, such as that produced during lignocellulose decomposition (Meyer et al. 1987; Tranvik and Hofle 1987; Moran and Hodson 1989).

Despite the fact that differential abilities of bacterial species and strains to degrade various substrates are a major criterium used in bacterial taxonomy, we know relatively little about temporal or spatial variation in the ability of natural bacterial communities to degrade naturally occurring DOC. In a prairie stream, isolates of the natural bacterial community at an upstream grassland site were able to utilize the DOC produced from grasses at that site and unable to use DOC coming from gallery forests at a downstream site; isolates from the gallery forest site could use DOC from both sources (McArthur et al. 1985). This may be a consequence of physiological shifts in enzymatic capabilities of the bacteria or changes in bacterial species in the community. Whether the underlying mechanism is a physiological or genetic change in the bacteria, it is important to recognize that the capacity of a stream to utilize DOC will vary with the composition of its bacterial community and the nature of the DOC to which it is normally exposed.

Nonbacterial components of the biotic community can also utilize DOC, although the rates at which they remove DOC from the water will probably be considerably less than for bacteria, for reasons of surface/volume ratio if nothing else. One component of the DOC pool that seems particularly vulnerable to uptake by larger organisms is the high molecular weight fraction, which may actually be colloidal. Because of the operational nature of the definition of DOC (material passing a 0.45 μm pore-size filter), some of what we are calling dissolved may actually be colloidal (Lock et al. 1977). This highest molecular weight fraction can be taken up by organisms ranging in size from protozoa (Sherr 1988) to blackflies (Wotton 1988). Although this material may be significant in terms of the nutrition of the organisms involved, this pathway is unlikely to be a significant sink for DOC unless a large fraction of the DOC is in this highest molecular weight fraction. In a blackwater river we have found that only 11% of the DOC is >100,000 nominal molecular weight (Meyer 1986).

In headwater streams, biotic utilization of DOC is primarily in the benthos. The water column becomes progressively more important as one moves further downstream. In very large rivers, like the Amazon, water column metabolism of DOC is a major carbon sink. For example, Richey et al. (1980) estimate that annual water column respiration in the Amazon is equal to half of the total carbon exported. One of the centers of biotic activity in smaller streams is the epilithon, the film of autotrophic and heterotrophic microbes living on the surfaces of rocks and other stable substrates. In a study comparing the relative importance of different sites for DOC uptake in third- and fourth-order Welch streams, Micklebaugh et al. (1984) found that uptake by the epilithic community on surface stones and underlying gravel was more important than uptake by the community associated with the organic floc in the sediments. The development of the epilithic community can be linked to the DOC content of the water. For example, in a New Zealand stream (DOC concentration of 5 mg C/l), an organic layer of slime, bacteria, and fungi on rock surfaces grew in the dark, whereas a very limited epilithic community developed in another stream with lower DOC concentration (<0.5 mg C/l). Under conditions of full sunlight, the epilithic community in both streams was dominated by autotrophs (Rounick and Winterbourn 1983). These epilithic communities were active in removing added DOC (leaf leachates) from the water column. In a Pennsylvania stream, a supply rate of 28 mg natural DOC/h supported a bacterial biomass of 10 μg C/g sediment; bacterial biomass was reduced when the rate of DOC supply from the water column was reduced (Bott et al. 1984).

The rate of diffusion across the organic layer appears to be a major barrier to more rapid uptake of DOC by epilithon. The community is embedded in a biotically produced mucopolysaccharide matrix that limits the rate of exchange with surrounding water; this reduces exchange with the water column but also reduces leakage of DOC produced by autotrophs within the community (Lock et al. 1984). The initial step in utilization of water column DOC by the epilithic community is adsoption of DOC by the organic matrix (Ford and Lock 1987). The presence of high molecular weight DOC has been shown to reduce microbial use of lower molecular weight DOC, apparently by competing for adsorption sites on the organic layer (Ford and Lock 1987).

There is a tight linkage between geomorphology of a river channel and its capacity to remove DOC from the water column. Channel geomorphology will also affect which components of the system are most critical in the uptake process. For example, in headwater streams in both Oregon and North Carolina, debris dams were the sites of 81 and 60% (respectively) of all DOC uptake (Munn 1989). An understanding of channel geomorphology is the first step in addressing spatial variation in DOC utilization in rivers. Critical geomorphic features include presence of backwater and side channel areas where slower water velocity will allow for longer contact between water column and sediment, and hence greater DOC uptake. Factors that increase water penetration into the hyporheos (e.g., sediment permeability, hydraulic head) will also be associated with more rapid removal of DOC from the water column. An understanding of the flow paths of water in natural stream channels, in particular the time of contact between water and sediments, is a critical determinant of DOC utilization in streams.

Consequences of DOC Utilization

Apart from the obvious consequence of a reduction in DOC concentration and hence transport to the ocean, what is the impact of DOC utilization on ecosystem processes in rivers? One obvious impact is that utilization of DOC has an impact on the biogeochemical cycles of other elements. For example, adding humic substances to lake water alters phosphorus cycling either because of increased competition between algae and bacteria after the elimination of bacterial substrate limitation or because the humic acids sequester organophosphorus compounds (Stewart and Wetzel 1982). Similar processes would also alter cycling of other elements.

Humic substances are photoreactive, and hence a reduction in DOC content can alter the photic environment of a water body. A reduction in DOC can lead to less light attenuation and hence could have a stimulatory effect on primary productivity in systems where it is limited by light penetration. Dissolved humic substances are also involved in photosensitized reactions that transfer absorbed light energy to other organic chemicals in waters via formation of triplet states of the humic substances (Zepp et al. 1985). A reduction in content of humic substances will reduce the extent of these photosensitized transformations in the ecosystem. The impact of photochemical changes on bioavailability and toxicity of organic acids has been thoroughly reviewed by Francko (this volume).

Microbial attack on DOC will also alter the remaining DOC and affect its chemical reactivity. For example, in a study of long-term changes in fulvic acids after a massive injection of DOC into Spirit Lake by volcanic activity, McKnight et al. (1988) found a decrease in phenolic hydroxyl content, total sulfur content, and content of aliphatic and heterocyclic sulfur groups, and an increase in carbohydrate content and in the heterogeneity of the fulvic acids. Coupled to these changes, particularly the changes in sulfur-containing moieties, was a reduction in copper binding sites on the fulvic acids. The biotic and abiotic alteration of DOC discussed earlier can clearly have an influence on its chemical makeup and hence on its reactivity with other ions such as metals.

Utilization of DOC also has an impact on the stream food web. Bacterial biomass produced at the expense of DOC becomes food for higher trophic levels; hence DOC utilization is one of the bases of riverine food webs. As in planktonic systems, bacteria are consumed by protozoa, both in the water column (L. Carlough, unpubl. data) and in the benthos (Bott and Kaplan

1989). Benthic deposit-feeders such as chironomids and oligochaetes are common in many streambeds and can utilize bacterial biomass in the sediments. Bacteria in the water column are captured by filter-feeding insects such as larval blackflies. Many filter feeders produce mucus which can adsorb DOC as well as trap particulate material; the mucus is then ingested by the animal. Another valuable food resource for the stream biota that is derived in part from DOC is the epilithon—bacteria, algae, and other organisms in a mucopolysaccharide matrix attached to hard benthic substrates. Experimental studies have measured assimilation efficiencies from 18 to 74% in insects feeding on epilithon produced solely from water column DOC (Rounick and Winterbourn 1983). Filtering and gathering collectors that have consumed material derived from DOC are then consumed by higher trophic levels such as predacious insects and fish. Hence utilization of DOC increases secondary production in a riverine ecosystem.

FUTURE RESEARCH DIRECTIONS

Although there are myriads of exciting research questions in the area of DOC production and utilization in rivers, there is a particular need for ecologically relevant chemical and hydrologic studies. Interdisciplinary teams of chemists and ecologists need to investigate the factors regulating biotic utilization of DOC. Teams of hydrologists and ecologists should study water storage and movement and its effect on DOC utilization in streambed sediments.

If we can develop routine chemical methods to characterize the biological availability of natural DOC, we will have a powerful tool for predicting DOC utilization throughout a river basin. In the past chemists have tried to identify specific compounds in naturally occurring DOC, and biologists have studied the utilization of specific compounds because of their availability in radiolabeled form. Readily identifiable compounds comprise a small fraction of naturally occurring DOC, and many of those that are present have formed complexes with humic substances and therefore would not be detected by the normal analytical procedures. Hence a different, more holistic approach is needed, which characterizes all DOC present with respect to its biotic availability, not one that seeks to dissect the DOC into identifiable compounds. The methods should be applicable routinely, so that determinations could be made on a large number of samples in a relatively short time. One approach is to relate elemental or functional group content to the biotic availability of the DOC. In preliminary studies in our laboratory, we have found a significant relationship between bacterial growth and the oxygen:carbon and nitrogen:carbon ratios of DOC (J.L. Meyer and E.M. Perdue, unpubl. data). Other new approaches are needed. Perhaps the ratio of soluble lignin degradation products to nitrogen in DOC will be a useful indicator of its rate of utilization, just as the lignin:nitrogen ratio appears to be a good indicator of decomposition rate of particulate organic matter (Melillo et al. 1982).

Ecologically relevant hydrologic studies of the flow paths of water in the basin and in the channel are also needed. The major factor controlling the rate of DOC uptake in a channel is the amount of time the water spends in contact with the sediments. Even though uptake rates are low, if a long time is spent in contact with sediments, then the impact of this on total DOC utilization in the river will be considerable. Laboratory studies have measured a threefold variation in uptake rates as a function of DOC quality (e.g., Dahm 1981). That variation may be swamped in nature by the geomorphically induced variation between different river systems in the fraction of water in transport that is in contact with the sediments. Studies of this nature are also critical to our understanding of uptake of other elements in riverine systems.

The need for a better understanding of hydrologic flow paths extends beyond the stream channel and into the streamside and watershed source areas. The amount of time spent in contact with different soil horizons will control the amount of DOC leached from organic matter and the amount sorbed by the mineral soil, and hence have a major impact on the amount of DOC transported to the stream channel. Our attention in this work should be focused near the stream channel. We need to assess the importance of streamside source areas and determine if and when they become anaerobic, because that will influence both the generation of DOC and its chemical nature.

Coupled with these studies of waterflow, we need to better understand the very slow leaching of organic material buried in the streambed. Most organic matter is stored in the streambed as fine particles, yet we know virtually nothing about the rates at which DOC is generated from this material or its availability to the biota. This material is generally presumed to be refractory, but recent studies have demonstrated microbial utilization of supposedly refractory DOC. Although rates of leaching and utilization will be considerably lower than for freshly introduced litter, the vast quantities and long periods of time that this material is exposed to stream water and to benthic microbes make it a potentially important component of DOC dynamics in streams.

Acknowledgements. This research was supported by grants BSR 8514328 and BSR 8705744 from the National Science Foundation.

REFERENCES

Aiken, G.R. 1985. Isolation and concentration techniques for aquatic humic substances. In: Humic Substances in Soil, Sediment and Water, ed. G.R. Aiken,

D.M. McKnight, R.L. Wershaw, and P. MacCarthy, pp. 363–386. New York: Wiley.

Baker, J.H., and I.S. Farr. 1987. Importance of dissolved organic matter produced by duckweed (*Lemna minor*) in a southern English river. *Freshwat. Biol.* 17:325–330.

Barlocher, F., J. Gordon, and R.J. Ireland. 1988. Organic composition of sea foam and its digestion by *Corophium volutator* (Pallas). *J. Exp. Mar. Biol. Ecol.* 115:179–186.

Bencala, K.E., V.C. Kennedy, G.W. Zellweger, A.P. Jackman, and R.J. Avanzino. 1984. Interactions of solutes and streambed sediment. *Wat. Res. Res.* 20:1797–1814.

Bott, T.L., and L.A. Kaplan. 1989. Densities of benthic protozoa and nematodes in a Piedmont stream. *J. N. Am. Benthol. Soc.* 8:187–196.

Bott, T.L., L.A. Kaplan, and F.T. Kuserk. 1984. Benthic bacterial biomass supported by streamwater dissolved organic matter. *Microb. Ecol.* 10:335–344.

Crocker, M.T., and J.L. Meyer. 1987. Interstitial dissolved organic carbon in sediments of a southern Appalachian headwater stream. *J. N. Am. Benthol. Soc.* 6:159–167.

Cummins, K.W., J.R. Sedell, F.J. Swanson, G.W. Minshall, S.G. Fisher, C.E. Cushing, R.C. Petersen, and R.L. Vannote. 1983. Organic matter budgets for stream ecosystems: problems in their evaulation. In: Stream Ecology, ed. J.R. Barnes and G.W. Minshall, pp. 299–354. New York: Plenum.

Dahm, C.N. 1981. Pathways and mechanisms for removal of dissolved organic carbon from leaf leachate in streams. *Can. J. Fish. Aq. Sci.* 38:68–76.

Dahm, C.N., E.H. Trotter, and J.R. Sedell. 1987. Role of anaerobic zones and processes in stream ecosystem productivity. In: Chemical Quality of Water and the Hydrologic Cycle, ed. R.C. Averett and D.M. McKnight, pp. 157–178. Chelsea, MI: Lewis.

Ertel, J.R., J.I. Hedges, A.H. Devol, J.E. Richey, and M. Ribeiro. 1986. Dissolved humic substances of the Amazon River system. *Limnol. Ocean.* 31:739–754.

Ford, T.E., and M.A. Lock. 1987. Epilithic metabolism of dissolved organic carbon in boreal forest rivers. *FEMS Microbiol. Ecol.* 45:89–97.

Geller, A. 1986. Comparison of mechanisms enhancing biodegradability of refractory lake water constituents. *Limnol. Ocean.* 31:755–764.

Hough, R.A., and R.G. Wetzel. 1975. The release of dissolved organic carbon from submersed aquatic macrophytes: diel, seasonal and community relationships. *Verh. Int. Verein. Limnol.* 19:939–948.

Kaplan, L.A., and T.L. Bott. 1982. Diel fluctuations of DOC generated by algae in a piedmont stream. *Limnol. Ocean.* 27:1091–1100.

Kuserk, F.T., L.A. Kaplan, and T.L. Bott. 1984. *In situ* measures of dissolved organic carbon flux in a rural stream. *Can J. Fish. Aq. Sci.* 41:964–973.

Lock, M.A., R.R. Wallace, J.W. Costerton, R.M. Ventullo, and S.E. Charlton. 1984. River epilithon: toward a structural-functional model. *Oikos* 42:10–22.

Lock, M.A., P.M. Wallis, and H.B.N. Hynes. 1977. Colloidal organic carbon in running waters. *Oikos* 29:1–4.

Lush, D.L., and H.B.N. Hynes. 1978. The uptake of dissolved organic matter by a small spring stream. *Hydrobiol.* 60:271–275.

McArthur, J.V., G.R. Marzolf, and J.E. Urban. 1985. Response of bacteria isolated from a pristine prairie stream to concentration and source of soluble organic carbon. *Appl. Env. Microbiol.* 49:238–241.

McDowell, W.H. 1985. Kinetics and mechanisms of dissolved organic carbon retention in a headwater stream. *Biogeochem.* 1:329–352.

McDowell, W.H., and G.E. Likens. 1988. Origin, composition, and flux of dissolved organic carbon in the Hubbard Brook Valley. *Ecol. Monogr.* 58:177–195.

McKnight, D.M., K.A. Thorn, R.L. Wershaw, J.M. Bracewell, and G.W. Robertson. 1988. Rapid changes in dissolved humic substances in Spirit Lake and South Fork Castle Lake, Washington. *Limnol. Ocean.* 33:1527–1541.

Melillo, J.M., J.D. Aber, and J.F. Muratore. 1982. Nitrogen and lignin control of hardwood leaf litter decomposition dynamics. *Ecology* 63:621–626.

Meyer, J.L. 1986. Dissolved organic carbon dynamics in two subtropical blackwater rivers. *Arch. Hydrobiol.* 108:119–134.

Meyer, J.L., R.T. Edwards, and R. Risley. 1987. Bacterial growth on dissolved organic carbon from a blackwater river. *Microb. Ecol.* 13:13–29.

Meyer, J.L., W.H. McDowell, T.L. Bott, J.W. Elwood, C. Ishizaki, J.M. Melack, B.L. Peckarsky, B.J. Peterson, and P.A. Rublee. 1988. Elemental dynamics in streams. *J. N. Am. Benthol. Soc.* 7:410–432.

Meyer, J.L., and J. O'Hop. 1983. Leaf-shredding insects as a source of dissolved organic carbon in headwater streams. *Am. Midl. Nat.* 109:175–183.

Meyer, J.L., and C.M. Tate. 1983. The effects of watershed disturbance on dissolved organic carbon dynamics of a stream. *Ecology* 64:33–44.

Meyer, J.L., C.M. Tate, R.T. Edwards, and M.T. Crocker. 1988. The trophic significance of dissolved organic carbon in streams. In: Forest Hydrology and Ecology at Coweeta, ed. W.T. Swank and D.A. Crossley, pp. 269–278. New York: Springer.

Micklebaugh, S., M.A. Lock, and T.E. Ford. 1984. Spatial uptake of dissolved organic carbon in river beds. *Hydrobiol.* 108:115–119.

Moran, M.A., and R.E. Hodson. 1989. Formation and bacterial utilization of dissolved organic carbon derived from lignocellulose. *Limnol. Ocean.* 34:1034–1047.

Munn, N.L. 1989. The role of geomorphology in nutrient cycles in headwater mountain streams. Ph. D. diss., Univ. of Georgia, Athens.

Munn, N.L., and J.L. Meyer. 1988. Rapid flow through the sediments of a headwater stream in the southern Appalachians. *Freshwat. Biol.* 20:235–240.

Petersen, R.C., Jr. 1986. *In situ* particle generation in a southern Swedish stream. *Limnol. Ocean.* 31:432–436.

Petersen, R.C., and K.W. Cummins. 1974. Leaf processing in a woodland stream. *Freshwat. Biol.* 4:343–368.

Richey, J.E., J.T. Brock, R.J. Naiman, R.C. Wissmar, and R.F. Stallard. 1980. Organic carbon: oxidation and transport in the Amazon River. *Science* 207:1348–1350.

Rounick, J.S., and M.J. Winterbourn. 1983. The formation, structure and utilization of stone surface organic layers in two New Zealand streams. *Freshwat. Biol.* 13:57–72.

Sherr, E.B. 1988. Direct use of high molecular weight polysaccharide by heterotrophic flagellates. *Nature* 335:348–351.

Stewart, A.J., and R.G. Wetzel. 1982. Influence of dissolved humic material on carbon assimilation and alkaline phosphatase activity in natural algal-bacterial assemblages. *Freshwat. Biol.* 12:369–380.

Thurman, E.M. 1985. Organic Geochemistry of Natural Waters. Dordrecht: Martinus Nijhoff.

Tranvik, L.J., and M.G. Hofle. 1987. Bacterial growth in mixed cultures on dissolved organic carbon from humic and clear waters. *Appl. Env. Microbiol.* 53:482–488.

Tsyban, A.V. 1971. Sea foam as an ecological habitat for bacteria. *Hydrobiol. J.* 7:9–18.

White, D.S., C.H. Elzinga, and S.P. Hendricks. 1987. Temperature patterns within the hyporheic zone of a northern Michigan river. *J. N. Am. Benthol. Soc.* 6:85–91.
Wotton, R.S. 1988. Dissolved organic material and trophic dynamics. *BioScience* 38:172–178.
Zepp, R.G., P.F. Schlotzhauer, and R.M. Sink. 1985. Photosensitized transformations involving electronic energy transfer in natural waters: role of humic substances. *Env. Sci. Tech.* 19:74–81.

Organic Acids in Aquatic Ecosystems
eds. E.M. Perdue and E.T. Gjessing, pp. 301–313
John Wiley & Sons Ltd

Wetlands as the Source of Dissolved Organic Carbon to Surface Waters

H.F. Hemond

Ralph M. Parsons Laboratory, 48–419
Massachusetts Institute of Technology
Cambridge, MA 02139, U.S.A.

Abstract. There is no consensus among researchers regarding the primary source of dissolved organic carbon (DOC) to surface waters. This uncertainty is compounded by an imcomplete understanding of the pathways by which water, and hence DOC, are transported to freshwater streams. A discussion of existing theories is presented and, to stimulate discussion, a hypothesis is presented. Specifically, wetlands bordering stream channels are proposed as the dominant DOC source, while the importance of upland sources is minimized by sorption occurring along lengthy, subsurface flow pathways.

INTRODUCTION

To understand the geochemical significance of dissolved organic carbon (DOC) in freshwater streams, it is necessary to understand the sources and origins of DOC. There is surprisingly little consensus with respect to (*a*) the biochemical pathways which lead to DOC production, (*b*) the locations where it is produced, and (*c*) the fate of DOC (hydrologic transport, biological oxidation, etc.). This lack of consensus arises from many factors, including the complex nature of DOC, differences between DOC-producing systems which have been studied, and knowledge gaps in the hydrologic sciences. In the present discussion, the focus in on question (*b*); it must be answered before the other questions can be addressed.

Much naturally occurring DOC has a very rapid turnover time. Such components of DOC, for example sugars, amino acids, and volatile fatty acids, are not the focus of this discussion; they are rapidly metabolized by the ubiquitous microbiota of natural aquatic systems. Although labile DOC components may influence the chemistry of natural waters and may even be precursors to humic substances, the present emphasis is on DOC that meets the operational definitions of humic substances, and which is relatively recalcitrant and persistent in nature. Humic DOC components make up a majority of the DOC of most rivers and streams, and probably have the larger effects on the acidity of the freshwater chemical environment.

SCOPE AND DEFINITIONS

The origin of dissolved humic substances in freshwater streams is closely associated with the mechanisms of streamflow generation in watersheds. Any discussion of flow paths or wetlands inevitably requires the use of some specific terminology. Because the terminology presently in the literature is not used consistently by all workers, key terms are defined below in the manner in which they will be subsequently used.

Flow Paths

Despite the myriad of hydrologic terms which qualitatively describe the pathways water can follow on a watershed, most of the ideas can be encompassed using the few following terms. Flow paths are defined in the context of watersheds, where precipitation which is deposited on the land eventually travels to surface drainage channels or streams. The following flow paths are conceptual and may or may not be readily measurable or distinguishable from one another.

Overland flow. Overland flow is the process whereby water flows across the surface of the ground, typically toward a stream. The existence and nature of this flow pathway is obvious on an open, impermeable surface such as a parking lot during a rainstorm. Both its existence and its definition are less obvious on a forested watershed, where the surface of the ground becomes difficult to define. Typically, one considers overland flow to include water traveling through, as well as over, the upper loose organic litter layer of a forest soil.

Overland flow is divided into (*a*) Hortonian overland flow and (*b*) saturation (or Dunne) flow. In the former, the existence of standing water at the ground surface is attributed to delivery of water by precipitation at a rate in excess of what can infiltrate into the ground surface, while saturation overland flow occurs because the entire soil profile is saturated from the bottom up and no further pore space is available to accept infiltration.

Interflow. Interflow is a term commonly used to describe lateral waterflow through a temporarily saturated soil layer which is underlain by unsaturated soils (i.e., a perched water table exists). Interflow may occur where a more permeable soil overlies a less permeable soil. At such transitions, downward movement of water in the upper, more permeable layer of soil can deliver water to the interface at a rate in excess of the maximum vertical flow rate in the lower layer, thus leading to the buildup of a saturated zone within the upper layer of soil. Water in the saturated layer then moves laterally in response to the hillslope gradient.

Unsaturated lateral flow. Some workers have proposed that significant amounts of lateral water movement may occur in unsaturated soil layers. The existence of unsaturated flow in soils is an unquestioned fact; the important question is whether unsaturated lateral flow is a quantitatively important process of streamflow generation.

"Groundwater flow." Traditionally, "groundwater flow" means waterflow through soil strata that are more or less continuously saturated over time, at least during a particular season of interest. Sometimes "groundwater flow" is loosely equated to "baseflow." Difficulties exist with both terms; in a broad sense, all subsurface water is groundwater, and "baseflow" by itself is not necessarily restricted to flow through a permanently saturated layer of soil. (For example, flow through fractured bedrock may contribute to baseflow.) For the purposes of this chapter, we will assume that "groundwater flow" refers to lateral flow in a saturated region extending upward from an impermeable base of the soil profile.

Macropores

Macropores are discrete channels, larger than the pore spaces of the immediately adjacent soil mass and often made by roots, soil insects, or even small mammals. Macropores are a prominent feature of most soils. Workers have attributed various roles to macropores; they are held by some to be the dominant conduits of water movement within watersheds. A disputed point is to what extent macropores conduct large amounts of water through otherwise unsaturated regions of soil, thus providing a mechanism for both rapid streamflow response to precipitation events and a route along which water avoids close contact with the soils of a catchment.

Wetland

Wetland is an environment characterized by soils or sediments that are at least intermittently in contact with the atmosphere but which retain a very high water content throughout the year. Within such sediments, anoxic porewaters typically exist, and a characteristic set of wetland biogeochemical processes occurs. These processes typically involve anaerobic metabolism by microbiota and even higher plants. Unlike some definitions, this definition of wetland does not explicitly require that any specific vegetation be present, although the presence of anoxic conditions is normally the result of organic carbon, produced *in situ* by vegetation, acting in concert with impeded gas (O_2) transport due to water saturation.

WETLANDS AS THE PRIMARY SOURCE OF DOC IN FRESHWATERS: A HYPOTHESIS

Proposed for discussion is the following hypothesis: The major sources of humic substances to freshwater streams and rivers are the "stream channel wetlands." This hypothesis applies primarily to glaciated catchments, characterized by relatively shallow soils underlain by bedrock at depths not exceeding a few meters and having high infiltration rates. It does not necessarily apply to catchments having deep, highly weathered, fine-textured soils, where macropore flow and/or shallow interflow are more likely to be important.

Stream channel wetlands include the saturated or near-saturated soils in the immediate vicinity of the stream. They include riparian wetlands as commonly recognized on the basis of vegetation, and possibly the stream sediments themselves. They also include small riparian depressions, abandoned oxbows, and low-gradient stream reaches with braided channels. Diagrammatically, this hypothesis is represented in Fig. 1. Specifically, DOC produced on the upland is held to have minor importance to the stream.

The above hypothesis implies an underlying streamflow generation hypothesis. The key features of this underlying hydrologic hypothesis include the following.

A Limited Role of Overland Flow

Classical Hortonian flow is hypothesized to be uncommon on temperate, glaciated watersheds. Most glaciated soils in humid regions have infiltration capacities well in excess of storms normally experienced (Pearce et al. 1986). Overland flow is rarely reported by hydrologists who have looked for it (Bonell et al. [1981] report an exception). My own observations are consistent with this conclusion, despite the fact that at certain times (most notably during the spring) casual observation would suggest that overland flow does occur. However, such flow is associated with ephemeral stream channels whose location is consistent from time to time and whose origin is believed to be soil saturation from the bottom of the soil profile. It is not extensive over the watershed. If overland flow is infrequent, the opportunity for direct leaching of the uppermost organic layers of soils to the stream is limited to

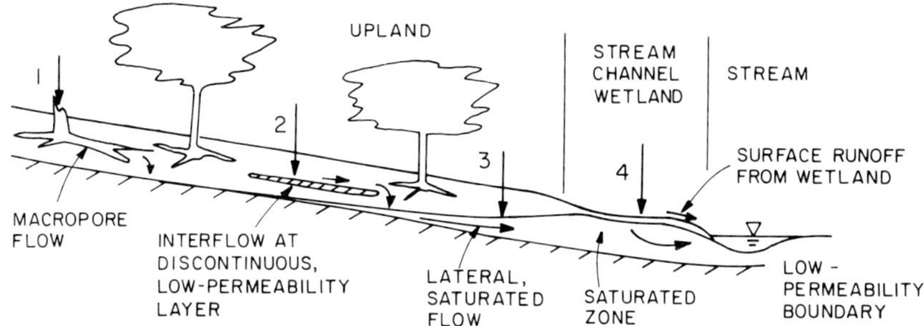

Fig. 1—Possible flow paths for water and DOC to travel to a stream: 1 = macropore flow, 2 = interflow above a permeability contrast, 3 = percolation to the saturated ("groundwater") zone, and 4 = overland flow. It is hypothesized that on glaciated catchments processes 1 and 2 are insufficiently continuous laterally to prevent water from eventually entering the "groundwater" zone, while 4 is restricted to areas bordering streams. Processes 1–3 may all leach DOC from the organic horizon, but this DOC is absorbed by underlying mineral soil. Only DOC leached in the near-channel area actually reaches the stream without severe attenuation.

springtime periods when water tables and streamflow are at their highest. DOC concentrations are typically not at their highest at this time of year.

A Minor Role of Interflow

It is proposed, further, that on most glaciated catchments the importance of interflow as a streamflow generation process is small. This statement is made despite the fact that several workers have reported "interflow" based on the behavior of water collection troughs inserted into the faces of cross-slope trenches (e.g., Whipkey 1965; Weyman 1970; Pilgrim et al. 1978). A concern is that trenches inherently create a major disturbance to the subsurface pathways. Because of the temporal spatial variability of watershed processes, direct and accurate observations of interflow are difficult to obtain. However, forested, glaciated watersheds are typically covered by a highly heterogeneous soil mantle, influenced by variations in slope and soil thickness, and interlaced with structures of biotic origin (e.g., macropores). Under such circumstances it is very difficult to envision a soil layer both of sufficiently low hydraulic conductivity and sufficient horizontal continuity to allow lateral water movement over appreciable (hundreds of meters) distances without encountering high conductivity pathways to an underlying saturated zone. This notion is reinforced by a simple field experiment conducted at Bickford and shown in Fig. 2 (Villars 1988), in which water was applied to a plot at a high rate and was observed to infiltrate to the bottom of the soil profile within tens of minutes.

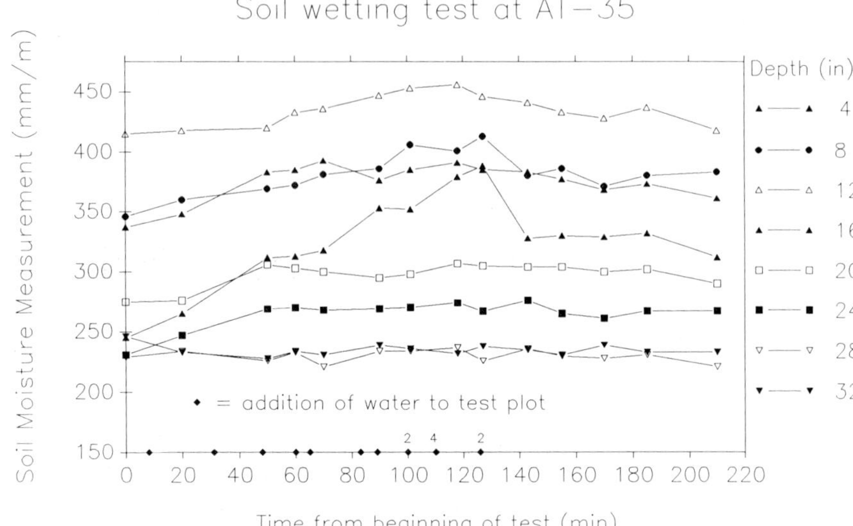

Fig. 2—In an experiment attempting to create a temporary perched water table (necessary for interflow), water was artificially applied to a soil plot at Bickford Watershed as shown, while soil water status was monitored by a neutron probe and nearby piezometer. As shown above, the entire soil profile responded very rapidly to water applications, with no evidence of perching (from Villars 1988).

A Minor Role of Unsaturated Lateral Flow

This part of the flow path hypothesis is based on the shape of the $K-\Psi$ curves for typical soils, in which hydraulic conductivities drop dramatically as soil makes the transition from saturated to unsaturated conditions. Thus, unusual circumstances must be postulated to allow unsaturated flow to exceed in importance saturated lateral flow down a hillslope to the stream. This hypothesis is widely held and is presented as part of the present argument without further evidence. It is noted, however, that Harr (1977) does not discount unsaturated lateral flow as a streamflow generation mechanism.

The "Oldness" of Water

Even on watersheds having very shallow soils, isotopic data almost always show stream water, even at the height of a precipitation event, to be comprised of a large fraction or even a majority of "old" water, i.e., water which was on the watershed before the storm (e.g., Pearce et al. 1986; Bottomley et al. 1984; Sklash and Farvolden 1979). Data from Bickford show that streamflow contains high enough levels of radon to suggest that the water has traveled at substantial depths in the soil (Genereux 1988). Such data suggest that surface runoff is not a dominant streamflow generation process on most catchments.

The Dominance of "Groundwater" Flow

It is proposed that the primary pathway for water from upland regions of a glaciated watershed to reach the stream is by "groundwater" flow as defined earlier. Again, direct evidence is not conclusive, although (*a*) several workers do make a case for "groundwater" flow on the basis of physical observations, (*b*) isotopic data point to a major role of subsurface flow in streamflow generation, and (*c*) the aforementioned rapid water application experiment at Bickford did show rapid saturation of the soil from the bottom. A common objection to this hypothesis is the fact that most small catchments respond very rapidly to inputs of precipitation water. Significantly, streamflow typically reaches its peak toward the end of rainfall and starts to decline as rainfall ceases. How can such rapid response occur if the most rapid mode for transport of water to streams, namely overland flow, is not believed to be important? Freeze (1972) underscores this problem in a series of computer simulations. It has been proposed that the answer lies in the stream channel wetland. Dunne and Black (1970) argued that saturation *overland* flow in stream channel wetlands could entirely account for storm hydrographs on their study site, while Ragan (1968) presents evidence for very rapid *subsurface* response in the near-channel region. In the present hypothesis, both mechanisms are important. Within the stream channel wetland, water content of the soil is high, and addition of even small amounts of precipitation brings about rapid saturation, perhaps converting a water-saturated capillary fringe from a region of negative porewater pressure to a region of positive porewater pressure nearly instantly (Gillham 1984). When this occurs, the subsurface hydraulic gradients towards the stream increase dramatically, leading to rapid subsurface stormflow into the stream. In addition, within the saturated region of the stream channel wetland, saturation overland flow occurs. The early part of a storm response and the entire storm response of a watershed during relatively dry periods of the year are thus attributed to the stream channel wetland. During long storms when upland moisture conditions are sufficiently high, flow from the upland areas by way of the groundwater zone contributes to the hydrograph after the first several hours. The irrigation experiments of Corbett et al. (1975) illustrate clearly the contribution of the upland, after several hours, to storm hydrographs on a glaciated Pennsylvania catchment. The speed of upland delivery of water to the stream is limited by two factors, namely the speed with which water can drain from the unsaturated zone to the groundwater zone and the speed with which saturated flow may travel through the soil profile. Both types of flow are limited by the hydraulic properties of the soil (although it is noted that, because of the heterogeneity of forest soils, the upland's response may be more rapid than simulation models invoking homogeneous soils may suggest [cf. Freeze 1972]).

In the proposed hydrologic scenario for streamflow generation, DOC produced in the stream channel wetland is readily transported to streams by both surface and subsurface routes. Little sorption of DOC occurs in this zone, in part because the soils and stream sediments already have a high organic content. In contrast, DOC produced near the surface of upland soils must be transported long distances to the stream, primarily by way of the groundwater zone. This second route is of less importance because of sorption of DOC by mineral soils.

CHEMICAL EVIDENCE FOR WETLANDS AS THE SOURCE OF DOC

High concentrations of soluble humic materials are found both in soil porewaters from the organic soil horizon and in wetland porewaters. One or both of these locations is usually considered to be the source of dissolved humic substances in streams. Streamflow generated in the wetland provides an unequivocal mechanism for delivery of wetland-produced DOC to streams. By contrast, DOC produced in upland organic soils must be transported tens of meters to the stream, mostly through the saturated "groundwater" zone. This implies an initial, downward movement of DOC-containing water to the mineral-soil dominated "groundwater" horizon.

Generally low concentrations of DOC are observed in "groundwater." For example, Cronan and Aiken (1985) document high concentrations of DOC in porewaters from the upper soil horizons in an Adirondack site and sharply decreasing concentrations of DOC with increasing depth. Transport of DOC at high concentrations from such forest soil sites to drainage streams would require shallow flow through the upper soil horizons (a process which we believe to be important only where saturation overland flow occurs) or rapid flow through deeper macropores. The predominance of vertical downward transport of DOC from the upper soil horizons, and its removal by sorption and ultimate microbial degradation in lower horizons, is consistent with classical theories of soil development (e.g., Paxeus et al. 1985). Such DOC movement aids in the eluviation of metals such as iron and aluminum from upper soil horizons, and deposition of these metals in the B horizon (podzolization).

Evidence from Temporal Variability

DOC concentrations in freshwaters of the northeastern U.S. are generally observed to peak during the growing season (Fig. 3). In the present

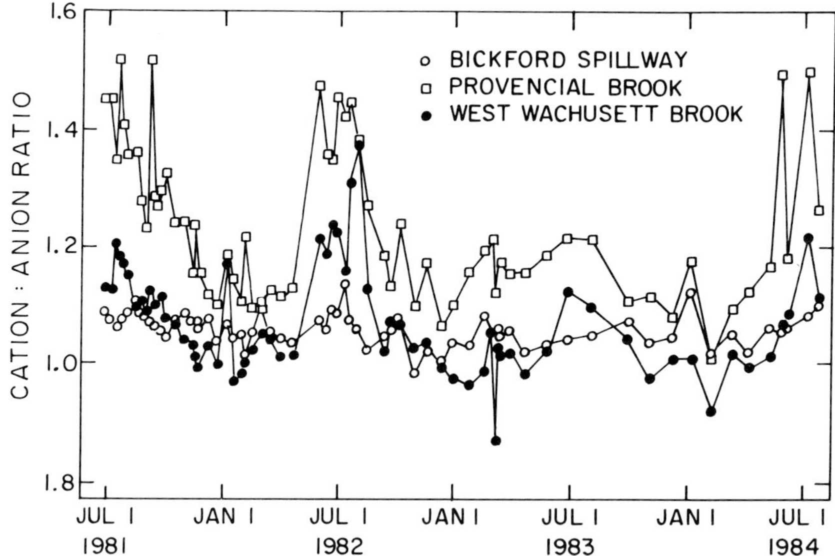

Fig. 3—Pattern of DOC concentrations (as measured by cation:anion ratio) in surface waters at Bickford Reservoir. In upland tributaries, DOC concentrations consistently peak during summertime (from Eshleman and Hemond 1985).

hypothesis, this is interpreted in terms of the greater hydrologic role assumed by stream channel wetlands during the dry season. As upland soil moisture content decreases, groundwater flow from the uplands decreases sharply. The residence time of water in the near-channel wetlands, through which most water must pass before entering the stream, increases. Production of DOC in the wetlands then gives rise to the highest concentrations of DOC in the stream, due to minimum dilution by water from the upland. During dry interstorm periods, when both the overland flow and interflow pathways are unequivocally inoperative, stream DOC is high. Higher biotic activity during the growing period undoubtedly also contributes to higher DOC concentrations during the summer.

During summertime storms, the overland flow and interflow pathways are sometimes held to be responsible for the rises in stream DOC concentrations often seen during these events. We have observed such rises in stream DOC at Bickford to be quite rapid, paralleling rises in stream discharge (Fig. 4). This rapidity is consistent with a stream channel wetland origin. It is not consistent with an upland DOC source, due to the several-hour delays probably associated with water movement from the uplands. In fact, the stream channel wetland can typically account for the total water yield of Bickford during a summer storm.

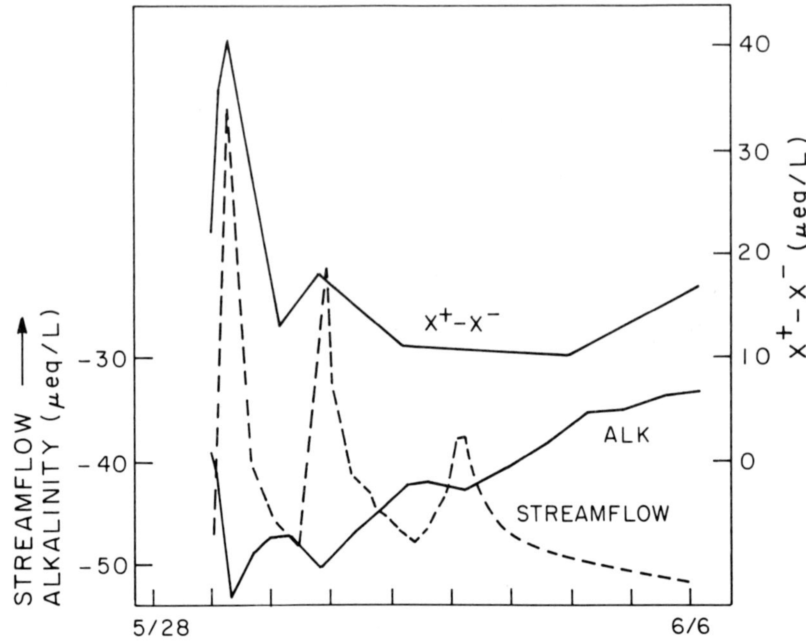

Fig. 4—Alkalinity, DOC, and streamflow at Bickford Reservoir during a series of storm episodes: the drop in stream alkalinity (lower solid line) can be accounted for by corresponding rises in DOC concentrations (the upper solid line, expressed as measured cations minus measured anions). DOC peaks coincide with very little lag to peaks in total streamflow (dashed line). Data from Eshleman and Hemond (1985).

The Existence of Low-DOC Surface Waters

A further argument against the dominance of upland soils as the primary source of DOC to freshwater streams lies in the existence of surface waters draining forested catchments but having very little DOC (e.g., Hubbard Brook [Likens et al. 1977]). If the organic horizon of forest soils (and/or canopy exudation and leaching) were the dominant source of DOC in streams, then the very low levels of DOC in systems such as Hubbard Brook would have to be explained on the basis of differences between the soils at Hubbard Brook and soils horizons elsewhere. To develop such an argument appears problematic, given that the vegetation of the Hubbard Brook watershed is similar to that of other sites which have been studied. One must assume that DOC is produced in the organic horizons at Hubbard Brook much as elsewhere; this DOC is removed as flow travels through mineral soil towards the streams. At Hubbard Brook, very little DOC is subsequently added by the stream channel wetland, whose development may be restricted by topography or by high soil permeability in the vicinity of the streams.

Statistical Relationship with Sulfate Concentrations

Although many details of sulfur cycling on forested catchments are as yet unresolved, it is known that sulfate reduction occurs in wetland areas. Wetland porewaters are typically depleted in sulfate with respect to precipitation. It is hypothesized here that an inverse relationship between DOC and sulfate concentrations exists at many sites and that this relationship reflects the common underlying influence of wetland processes. Such a relationship (with anion deficit used as a measure of DOC) is shown in Fig. 5, which includes data from our own study site at Bickford as well as elsewhere. Perhaps most striking is the consistency in sulfate versus DOC data from several sites. Two "end member" sites are represented by Thoreau's bog (a purely wetland catchment) and Hubbard Brook (a

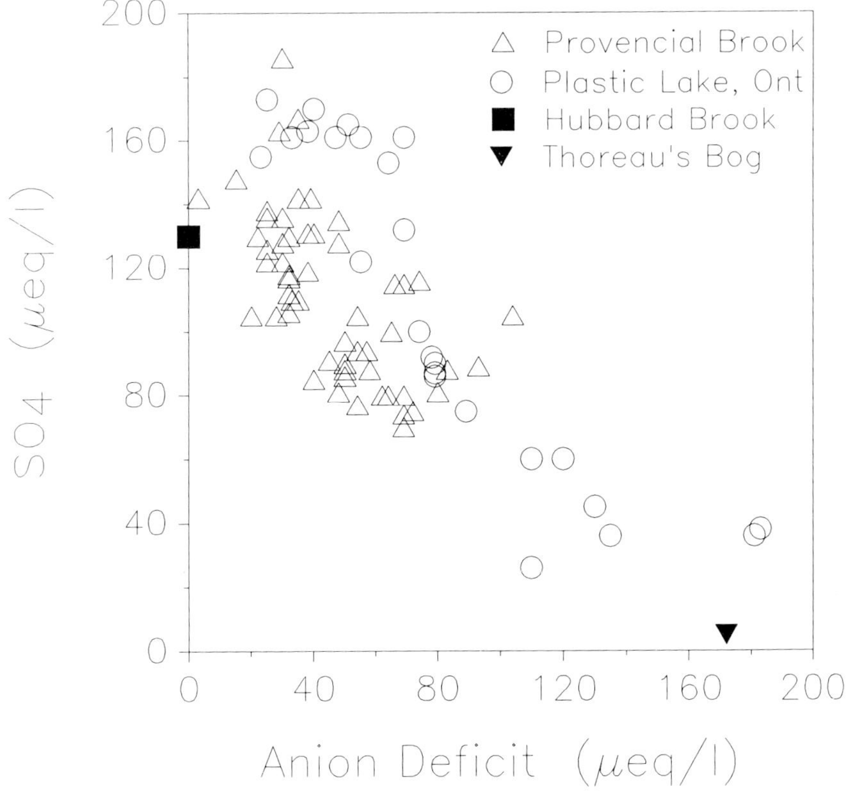

Fig. 5—An observed inverse relationship between sulfate and DOC levels in several natural waters from northeastern U.S. and Canada. Waters closer to the high DOC/low sulfate end member are hypothesized to reflect a greater proportion of wetland drainage. Plastic Lake data are from La Zerte and Dillon (1984); Provencial Brook is on the Bickford catchment.

catchment minimally influenced by wetlands). Along the "mixing line," data points are arrayed not only from the Bickford watershed, but also from a tributary to Plastic Lake (La Zerte and Dillon 1984). We propose that a common mechanistic link is the wetland, which both reduces sulfate and serves as the primary source of DOC.

SUMMARY

To answer the question "Where does the humic DOC in freshwater streams and lakes come from?" requires the application of hydrologic flow path concepts as well as biochemical information. Through a combination of data from the literature, physical reasoning, and patterns seen in DOC distribution, it is proposed here that near-stream wetlands account for the bulk of humic DOC seen in freshwater streams in *glaciated catchments*. This is presented as a plausible, albeit unproven, hypothesis which may help motivate further discussion on the topic.

Acknowledgements. The author thanks Patrick Mulholland for valuable comments on the manuscript. Support for our previous research at Bickford was provided by New England Power Service Co., Empire State Electric Energy Research Corp., American Electric Power Service Co., and the U.S. Geological Survey, Grant 14–08–0001–G1725.

REFERENCES

Bonell, M., D.A. Gilmour, and D.F. Sinclair. 1981. Soil hydraulic properties and their effect on surface and subsurface water transfer in a tropical rainforest catchment. *Hydrol. Sci. Bull.* 26:1–18.

Bottomley, D.J., D. Craig, and L.M. Johnston. 1984. Neutralization of acid runoff by groundwater discharge to streams in Canadian Precambrian Shield watersheds. *J. Hydrol.* 75:1–26.

Corbett, E.S., W.E. Sopper, and J.A. Lynch. 1975. Watershed response to partial area application of simulated rainfall. *Ass. Int. Sci. Hydrol.* 117:63–73.

Cronan, C.S., and G.R. Aiken. 1985. Chemistry and transport of soluble humic substances in forested watersheds at the Adirondack Park, New York. *Geochim. Cosmo. Acta* 49:1697–1705.

Dunne, T., and R. Black. 1970. Partial area contributions to storm runoff in a small New England watershed. *Wat. Res. Res.* 6(5):1296–1311.

Eshleman, K.N., and H.F. Hemond. 1985. The role of organic acids in the acid-base status of surface waters at Bickford Watershed, Massachusetts. *Wat. Res. Res.* 21(10):1503–1510.

Freeze, R.A. 1972. Role of subsurface flow in generating surface runoff. II. Upstream source areas. *Wat. Res. Res.* 8:609–623.

Genereux, D. 1988. Hydrogeochemistry of ^{222}Rn on a forested watershed. M.S. thesis, Massachusetts Institute of Technology, Cambridge, MA.

Gillham, R.W. 1984. The capillary fringe and its effect on water table response. *J. Hydrol.* 67:307–324.

Harr, R.D. 1977. Water flux in soil and subsoil on a steep forested slope. *J. Hydrol.* 33:37–58.

La Zerte, D., and P.J. Dillon. 1984. Relative importance of anthropogenic versus natural sources of acidity in lakes and streams of central Ontario. *Can. J. Fish. Aq. Sci.* 41:1664–1677.

Likens, G.E., F.H. Bormann, R.S. Pierce, J.S. Eaton, and N.M. Johnson. 1977. Biogeochemistry of a Forested Ecosystem. New York: Springer.

Paxeus, N., B. Allard, U. Olofsson, and M. Bengtsson. 1985. Humic substances in ground waters. *Mat. Res. Soc. Symp. Proc.* 50:525–532.

Pearce, A.J., M.K. Stewart, and M.G. Sklash. 1986. Storm runoff generation in humid headwater catchments. I. Where does the water come from? *Wat. Res. Res.* 22(8):1263–1272.

Pilgrim, D.H., D.D. Huff, and R.D. Steele. 1978. A field evaluation of surface and subsurface runoff. II. Runoff processes. *J. Hydrol.* 38:319–341.

Ragan, R.M. 1968. An experimental investigation of partial-area contributions. *Int. Ass. Sci. Hydrol. Publ.* 76:241–249.

Sklash, M.G., and R.N. Farvolden. 1979. The role of groundwater in storm runoff. *J. Hydrol.* 43:45–65.

Villars, M. 1988. Modeling flowpaths and streamflow generation at the Bickford Watershed: applications to acid deposition modeling. M.S. thesis, Massachusetts Institute of Technology, Cambridge, MA.

Weyman, D.R. 1970. Throughflow on hillslopes and its relation to the stream hydrograph. *Int. Ass. Sci. Hydrol. Bull.* 15:25–33.

Whipkey, R.Z. 1965. Subsurface stormflow from forested slopes. *Int. Ass. Sci. Hydrol. Bull.* 10:74–85.

Standing, left to right:
Tom Fisher, Jim Sedell, Michel Meybeck, Cliff Dahm, Dominic Di Toro, Patrick Mulholland

Seated, left to right:
Harry Hemond, Judy Meyer, Mark David, Ingrid Kögel-Knabner

Organic Acids in Aquatic Ecosystems
eds. E.M. Perdue and E.T. Gjessing, pp. 315–329
John Wiley & Sons Ltd

Group Report
What Are the Temporal and Spatial Variations of Organic Acids at the Ecosystem Level?

P.J. Mulholland, Rapporteur
C.N. Dahm
M.B. David
D.M. Di Toro
T.R. Fisher
H.F. Hemond
I. Kögel-Knabner
M.H. Meybeck
J.L. Meyer
J.R. Sedell

INTRODUCTION

The spatial and temporal variation in organic acids and dissolved organic carbon (DOC) in freshwater ecosystems and those mechanisms which effectively regulate this variability were the primary topics for discussion in our group. The approach was to document what is known as a result of field observations and to identify what we believe to be the most important unknowns. The context of these discussions was *variability* in DOC concentration and quality. Fluxes per se, although extremely important to ecosystem function, were not the central issue. We recognize that DOC is a bulk parameter for organic molecules, similar to conductivity for inorganic solutes. It yields information primarily on the most abundant organic compounds, and these are often the compounds of low biological flux within aquatic ecosystems.

In this chapter, the term DOC is used as a surrogate for organic acids. It is recognized that not all DOC molecules are acidic. However, a case for using DOC as an approximate measure of the influence of organic acids can be made on several grounds. These include:

1. Numerous workers have found organic acidity in surface waters to correlate with DOC (Aiken 1985). While the acidity due to DOC may vary, the correlation is strong enough to allow broad trends and patterns of organic acidity to be discussed.

2. Many workers have measured DOC in aquatic systems, and a reasonable data base thus exists. If the data base were to be restricted to those measurements explicitly intended to assess organic acidity (e.g., charge balance analysis, measurement of specific organic acids), far less information would be available.
3. There is no general consensus on what the best measure of organic acidity is. Even "direct" measures are subject to several limitations. Thus, while DOC is a surrogate measure, it may not be much more arbitrary than other methods.

In view of the above, we have chosen to proceed on the assumption that temporal and spatial patterns of DOC will yield a reasonable picture of the variability of organic acids in aquatic ecosystems.

We are concerned with temporal and spatial variability of DOC at the ecosystem level for a variety of reasons. First, DOC serves as an energy source for heterotrophic microbes and contributes to the productivity of higher trophic levels in food webs. Second, DOC plays an important role in regulating the availability of inorganic nutrients to plants and heterotrophic microbes, and strongly influences the mobility of and biological susceptibility to contaminants. Third, the role of DOC in surface water acidification has been an important concern. Fourth, the role of DOC in the global geochemical cycles of several elements (e.g., C, N, P, S, and trace metals) is significant. Finally, DOC has a strong influence on the optical properties of natural water and consequently on photosynthetic processes and photochemically induced redox reactions.

Although the focus of this workshop was on aquatic ecosystems, our considerations were limited for the most part to freshwaters. Lotic (running water) ecosystems and lentic (stationary water) ecosystems with relatively strong influence of terrestrial DOC sources were emphasized. This primarily reflects the lower DOC concentrations and the relatively small variability in these concentrations in the pelagic zones of the ocean and lakes with carbon budgets dominated by algal production. We recognize that variation in DOC quality (physical, chemical) and its influence on chemistry and biology may nonetheless be quite important in these ecosystems, but this topic is the subject of the group report by McKnight et al. (this volume).

VARIABILITY

The vast majority of aquatic ecosystems have DOC concentrations falling within the range of 0.5 to 50 mg/l. In an individual system, temporal variation in DOC concentration is often less than 10 mg/l. Therefore, magnitudes of spatial variability (10^2) and temporal variability (<10) in DOC would be considered small compared to the variability of some solutes such as NH_4^+ or Cl^- (>10^4), but similar to or larger than others such as K^+ or dissolved oxygen (Livingstone 1963; Meybeck 1982).

A central question for consideration is why DOC is not more variable. Several observations might indicate that DOC should be more variable than it is. First, the ultimate source of DOC is net primary production (NPP), for which annual averages vary spatially by at least 10^3 (National Academy of Sciences 1975). Second, the range of hydraulic retention times in aquatic ecosystems ranging from small streams to oceans varies by at least 10^6 (h to >1000 y). Finally, the turnover times of DOC molecules are equally variable, ranging from hours for biological uptake of simple compounds (Seki et al. 1975) to >10^3 y for bulk DOC turnover in deep ocean water (Williams and Druffel 1987).

What limits DOC variability? The most obvious potential limitations are those dealing with sources, transport, and reactivity. Source limitations might include upper limits to organic carbon solubility in soils and upper limits of microbial metabolism. Transport limitations might involve hydrologic throughflow in source zones. Limitations due to DOC reactivity could involve solubility limits in water and very slow biological reactivity for much of the DOC produced. Few studies have examined these processes in the context of limitations to variability, but this approach may be fruitful in gaining a more complete perspective on DOC variability in aquatic ecosystems.

Although the variability of DOC as measured in absolute concentration (mg/l) may be less than some other solutes, the effects of this variability in ecosystems may be much greater. Because of the importance of DOC in many biotic and abiotic processes as mentioned above, a tenfold variability could have a much greater effect on the ecosystem than a similar variability in a more inert ion (e.g., Cl^-, Na^+). Therefore, we hereafter consider the question of DOC variability from the perspective of assessing the primary factors which control the observed variability in space and time.

SOURCES

Potential sources of DOC for aquatic ecosystems are primarily organic-rich soil horizons of terrestrial ecosystems, detrital accumulations within aquatic ecosystems, terrestrial–aquatic interface systems such as floodplains and wetlands, and aquatic primary producers (Fig. 1). Relatively small temporal (diel, seasonal) and spatial variability in DOC concentrations in lakes dominated by autochthonous primary production indicates that this source may be of minor importance to the bulk DOC (Wetzel 1983). However, identification of the relative importance of the other sources in particular systems is generally not possible with present methods.

There is large variability in DOC concentrations in terrestrial soils along

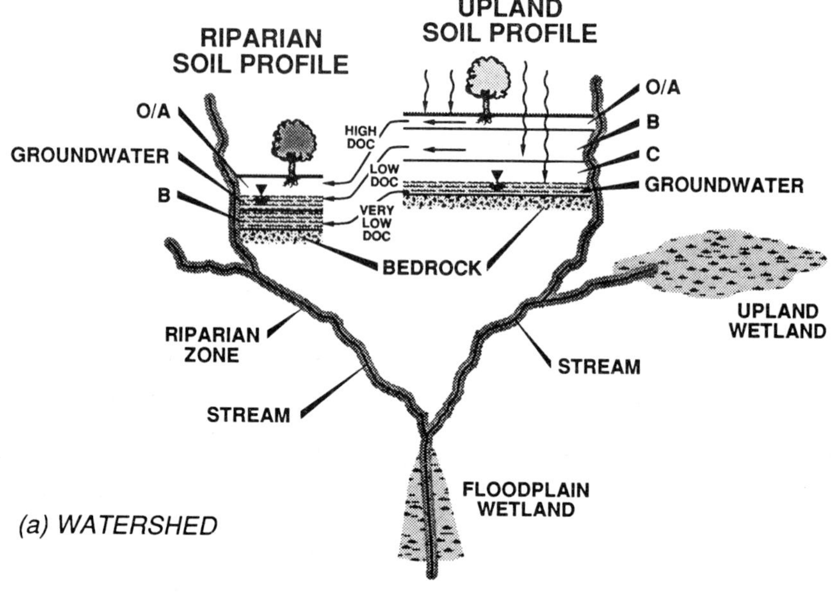

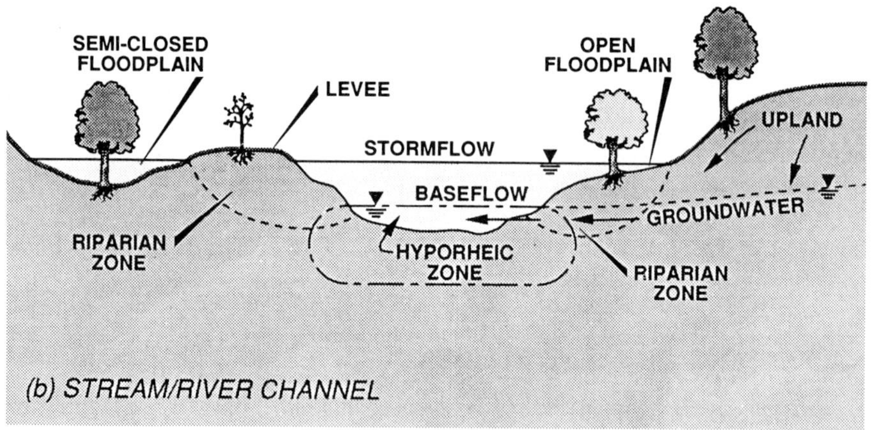

Fig. 1—Potential DOC sources and hydrologic linkages with the stream/river at (*a*) watershed and (*b*) floodplain/channel spatial scales.

the flow path of water (Cronan and Aiken 1985; McDowell and Likens 1988). The highest DOC concentrations are found in surface soil horizons (O and A) as a result of inputs in throughfall and decomposition of surface detrital carbon. Concentrations decline sharply in lower soil horizons (B and C), primarily because of adsorption/co-precipitation in mineral soils with Fe and Al sesquioxides (Dawson et al. 1978; McDowell and Wood 1984). Observations of high rates of microbial respiration in the upper portions of the B horizon indicate that biological decomposition of the adsorbed DOC is important in renewal of sorption sites and maintenance of high rates of adsorption over time. In soil types such as podsols, which lack an iron-rich clay horizon, large quantities of DOC can be produced. This is particularly important in the tropics (Klinge 1965).

Although it is clear that the primary source of DOC in soils is throughfall and litter decomposition, and that this material is largely removed with passage through lower soil horizons, differences in DOC quality in different soil horizons are less well understood (Vance and David 1989). Also, it is unclear whether the DOC of lower soil horizons is always the nonadsorbable fraction from throughfall and O/A horizons, or whether subsequent microbial activity releases previously adsorbed, more hydrophobic DOC.

Studies of organic carbon dynamics in small streams have indicated the potential importance of the riparian and hyporheic zones as DOC sources (Meyer, Hemond, both this volume). Studies in larger rivers have pointed to wetlands and floodplains (Sedell and Dahm, this volume). Sensitive, easily measurable tracers that discriminate among these potential sources of DOC are needed.

Formation of DOC from solid phase precursors occurs via abiotic processes (primarily leaching) and biotic processes (primarily microbial). Up to about 30% of the mass of particulate detritus from terrestrial vegetation is water soluble and leaches rapidly (within 1–2 d) upon contact with water (Petersen and Cummins 1974). However, much of the rapidly leached DOC is quickly taken up by microbes and results in only small variations in DOC concentration in aquatic ecosystems (Lock and Hynes 1976). Thereafter, DOC is produced from plant detritus by slow leaching and microbial decomposition, the latter being the result of incomplete uptake of the soluble products of exoenzymatic activity. DOC is also produced in aquatic ecosystems by release from living algae (Kuserk et al. 1984) and macrophytes (Hough and Wetzel 1975), although these compounds also may be rapidly utilized by microbes resulting in little DOC variation.

A complete understanding of the factors regulating microbial production of DOC from plant detritus in soils and surface waters does not exist. In some environments the process may be first-order (limited by substrate availability); in others inorganic nutrients, or in the case of soils, moisture content, may be important. Temperature will always be a factor. A better understanding of these factors may help to identify upper limits on DOC formation and explain the range of DOC concentrations observed in terrestrial and aquatic environments.

Some evidence suggests that anaerobic environments play an important role in DOC formation in many systems. Concentrations of DOC are routinely higher in environments that have more extensive anaerobic zones, such as saturated soils, wetlands, and interstitial waters of aquatic sediments, compared to concentrations in oxic environments (e.g., Krom and Sholkovitz 1977). Components of the DOC produced by anaerobic metabolism are different in chemical structure from those produced by aerobic processes. For example, volatile fatty acids and carbohydrates occur in higher concentrations under anaerobic conditions (Lovely and Klug 1982; Orem and Hatcher 1987). Aromatic organic compounds may also increase in concentration as microbial attack on ring structures is often slower under anoxic conditions. In general, anaerobic decomposition of particulate organic carbon is less efficient than decomposition under aerobic conditions, with a higher proportion of soluble intermediate metabolites released and a lower proportion of CO_2 produced per unit of particulate organic carbon attacked (Otsuki and Hanya 1972). Some of these soluble metabolites appear to have humic characteristics or to be the precursors of humic materials formed in oxic environments. The distinctive composition of DOC produced under anoxic conditions could influence the subsequent reactivity of this material, both microbially and chemically, when it is eventually routed into oxic surface waters.

TRANSFORMATION AND DEGRADATION

Abiotic processes involving transformation and degradation of DOC in aquatic ecosystems include adsorption/co-precipitation, flocculation, and photochemical transformations. Adsorption/co-precipitation, while extremely important in terrestrial soils, is of less importance in controlling DOC concentrations in most aquatic ecosystems (Dahm 1981). Exceptions may be systems in which substantial iron and aluminum precipitation occurs due to acidic inputs upstream and subsequent neutralization (P.J. Mulholland, unpubl. data), or lakes with high rates of $CaCO_3$ precipitation (Otsuki and Wetzel 1973).

Flocculation of DOC in aquatic ecosystems, as in all flocculation processes, involves a partitioning equilibrium among the larger components of DOC and, unlike precipitation in solubility equilibrium, is enhanced by removal of the aggregated product. Removal of flocculated DOC at solid or air–water interfaces both promotes additional flocculation and alters the composition of the remaining DOC. Some studies (e.g., Petersen 1986) have indicated that flocculation and removal at turbulence-induced air–water interfaces can be important in regulation of DOC in small streams. Its general importance in regulation of DOC concentrations in aquatic ecosystems is unknown but thought to be minimal.

Humic substances play an important role in mediating photochemical production of H_2O_2 and subsequently of OH radicals in aquatic ecosystems.

Chemical transformations of humic substances in the presence of OH radicals include hydroxylation of aromatic components that, in turn, may enhance their biological reactivity. The contribution to DOC variability from photochemically induced DOC transformation and subsequent biological degradation in aquatic ecosystems is unknown but could be significant in low turbidity, high DOC, well-illuminated systems and deserves further study.

Microbial processes play a central role in the transformation and degradation of DOC in all ecosystems, sometimes in conjunction with physical/chemical processes such as adsorption in soils or photochemical reactions in water as already described. Microbial degradation (uptake and oxidation) of DOC occurs primarily at solid interfaces in most aquatic ecosystems (e.g., inorganic sediment surfaces, particulate organic surfaces). This may reflect the importance of the mucillaginous excretions of bacteria attached to surfaces for adsorbing DOC and thus enhancing its biological availability. Free-living microbes in the water column may be important in eutrophic systems or in systems with few particle surfaces (e.g., pelagic zone of lakes or the ocean).

The effect of microbial degradation in regulating DOC concentrations in aquatic ecosystems depends on the magnitude of turnover time (inverse of degradation rate constant) relative to hydraulic retention time (Fig. 2). DOC compounds with short turnover times relative to hydraulic retention times will be maintained at low and relatively constant concentrations; DOC compounds with long turnover times relative to retention times will be largely unaltered, and concentrations will reflect only supply rate. It is the DOC compounds with turnover times similar to the range of hydraulic retention times of most aquatic ecosystems (10^{-1} to 10^{3} d) that will exhibit the maximum variability in concentration resulting from differences or changes in hydraulic residence time and biological degradation rate. Therefore, it is these DOC compounds for which we most need biological reactivity information.

Methods exist for determining degradation rates of very reactive compounds (e.g., additions of simple, radiolabeled compounds known to be present in bulk DOC) and, to some extent, very recalcitrant compounds (e.g., ^{14}C dating of deep ocean DOC). However, there is currently no clear approach for obtaining degradation rates for DOC compounds in the range of about 10^{0} to 10^{-3} d^{-1} (turnover times 10^{0} to 10^{3} d). This methodological gap is probably the range of reactivities most common for bulk DOC in most aquatic ecosystems.

A reactivity classification for sedimentary organic carbon has been proposed (Westrich and Berner 1984). The model envisions multiple components of varying reactivity (the G components) which have turnover times from 10 d to geological time scales. An analogous model for DOC,

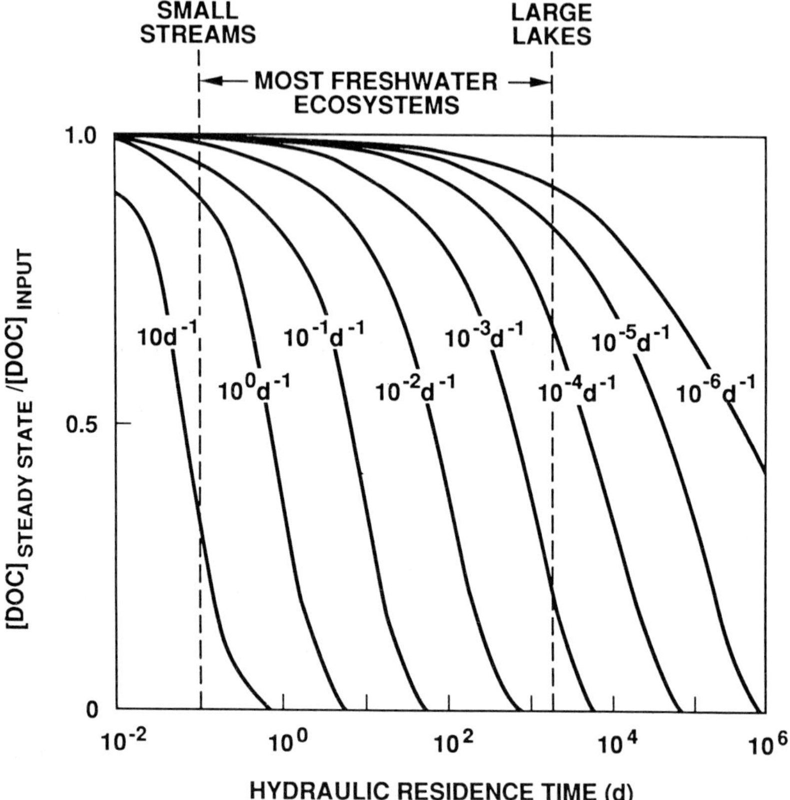

Fig. 2—Relationship between the ratio of the DOC concentration at steady state/DOC concentration in the hydrologic input and hydraulic residence time for DOC compounds of differing degradation rate classes (first-order degradation reactions are assumed). For a given hydraulic residence time, only DOC compounds with degradation rates similar to or greater than the hydraulic residence time will be present in appreciable amounts (relative to their input) at steady state.

ideally with a maximum of about 4 relatively discrete reactivity classes (or perhaps a continuous distribution of reactivity that can be described by a low-order polynomial equation), is needed.

The data necessary for applying the reactivity model could be obtained through long-term microbial degradation studies with bulk DOC, or with fractions of the DOC separated by functional or structural characteristics. These studies should be conducted in conjunction with chemical characterization schemes prior to and at different times during microbial degradation (or minimally, at the end). The challenge is to find easily measurable and widely applicable tracers (biomarkers) indicative of different rates of biological reactivity that will allow partitioning of the bulk DOC pool without performing long-term degradation studies.

HYDROLOGIC CONTROL OF TEMPORAL VARIABILITY

In ecosystems with relatively high hydrologic turnover, temporal and spatial variability are not independent characteristics of the DOC distribution; they are, in fact, closely coupled. For example, a spatial concentration gradient of DOC in a flowing stream is observed as a temporal change in DOC as measured at any given point. Observed variability in aquatic ecosystems is ultimately a result of variation in source/sink rates and physical transport. A temporal/spatial distinction is made here primarily in organizing the discussion, since temporal and spatial variabilities are what we measure in the environment.

Observations of significant temporal variability in DOC concentration (>5 mg/l) have been made in lotic ecosystems at temporal scales ranging from hours to days (storms in small streams), to weeks to months (floods in larger rivers). DOC concentrations are typically positively correlated with discharge and usually show clockwise hysteresis (higher concentrations at equivalent water discharge during the rising stage compared to the falling stage). Often seasonal variation in DOC concentration is also observed, particularly in systems with large seasonal changes in hydraulic residence times (e.g., streams and rivers with very low summer flows and much larger winter–spring flows). Temporal variation in DOC concentration in lakes is generally small, except perhaps for small lakes with large terrestrial or wetland inputs.

On at least the shortest time scales (e.g., hours, days), changes in hydrologic flow paths are almost certainly responsible for observed temporal variability. Temporal variations in the relative importance of flow through different source areas (e.g., upper horizons of terrestrial soils, hyporheic sediments, floodplains, wetlands) to total flow are very important in generating concentration pulses during storms/floods. On longer time scales (e.g., weeks, months), changes in hydraulic residence time of drainage water could also alter DOC formation or degradation in source or sink zones. Again, on longer time scales, hydrologically mediated changes in redox condition in DOC source zones (e.g., saturation of upland, riparian, floodplain soils) are likely to alter the rates of DOC formation and quality of the DOC formed.

The present uncertainty about subsurface water flow paths in both upland and wetlands cannot be overemphasized. Results of some hydrologic studies in small watersheds have identified the riparian zone as being of primary importance in generating streamflow during storms (groundwater ridging model [Sklash and Farvolden 1979]). By contrast, other studies have indicated large amounts of unsaturated lateral flow in upper soil horizons along preferred flow paths (interflow model [Mosley 1979]). Each of these models can account for the observed temporal DOC variations, although the primary source of the DOC would be quite different (riparian vs. upslope soils). Additional efforts at hydrologic modeling coupled with field hydrometric and chemical tracer studies are urgently needed to determine the source of stream DOC during storms. Also, little is known about variations in DOC quality during storms, and more research is needed on this subject.

The hyporheic zone in streams and rivers often contains large accumulations of organic matter that support high rates of microbial metabolism (Meyer 1988). Very little is known concerning its role in releasing DOC as hydraulic gradients and surface water velocities increase during storms. Research is needed to determine the importance of the hyporheic zone as a DOC source compared to upslope and riparian soils.

In larger streams and rivers, floodplain inundation is a potential contributor to the temporal variation in DOC concentration. The role of floodplains in contributing to temporal DOC variation would be dependent on the degree of hydrologic interaction between the river and floodplain. Rivers with extensive levees that hydrologically isolate the floodplain would predictably show much less of a floodplain effect on DOC variations. The relative roles of remote upstream and local floodplain sources in establishing temporal DOC variability could be determined by research focused on rivers with contrasting floodplain interactions.

DRAINAGE GEOMORPHOLOGY AND SPATIAL VARIABILITY

Although surface water DOC concentrations range between 0.5 and 50 mg/l, the range in larger rivers (>100,000 km^2) is considerably less (2-15 mg/l) (Meybeck 1982). Analyses of extensive data sets on spatial variability of DOC concentration in rivers worldwide indicate no relationship between average concentrations and most climatic (e.g., temperature, runoff) or vegetational (e.g., net primary productivity, vegetation type) parameters (Fig. 3). However, positive relationships between concentration and wetland distribution (e.g., fraction of watershed covered by wetland) were observed in regional studies in Finland (Kortelainen and Mannio 1988), New Zealand (Moore 1987), and the southeastern United States (Mulholland and Kuenzler 1979). Observations of much higher spatial variability in DOC concentrations in watersheds or regions with large geomorphic variability (e.g., MacKenzie River) also highlight the potential significance of geomorphological control (M.H. Meybeck, unpubl. data).

Several mechanisms could explain a geomorphologic control of DOC concentration. Factors involved in DOC formation, such as accumulation

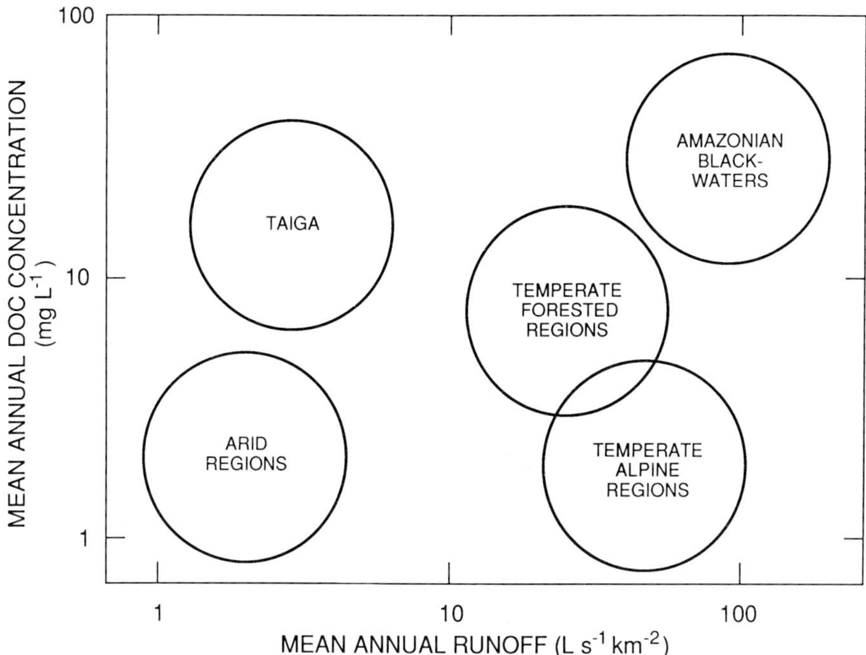

Fig. 3—Mean annual DOC concentration as a function of mean annual runoff for watersheds in different climatic zones.

of particulate organic matter in riparian soils, hydraulic residence time, extensive anaerobic zones in riparian soils and channel sediments, and development of extensive ecotones (e.g., floodplains, braided channels, within-channel debris accumulations) are all geomorphologically controlled. Biological (microbial) degradation of DOC could also be geomorphologically controlled via effects on surface/volume ratios in aquatic ecosystems.

Many geomorphologic features can be readily quantified from maps, photographs, or satellite remote sensing. Identification of relationships between these features and DOC concentrations would be extremely valuable for predicting spatial DOC variation and the effects of human alteration of landscapes on DOC transport. Predictive models with variables that can be quantified from remotely sensed data may be developed for estimating average DOC concentration and transport worldwide.

Other than wetland influences, the role of geomorphology in regulating spatial variability in DOC concentration and quality is largely unknown. Investigations of what geomorphological features are most important as DOC concentration and quality predictors are needed. The relative roles of floodplains, the hyporheic zone, and the water column in DOC formation, transformation, and degradation should be investigated in contrasting geomorphologies (e.g., low-gradient and high-gradient rivers within and between different physiographic provinces).

FUTURE RESEARCH

We summarize future research needs in the form of several questions that need to be answered for a more complete understanding of the spatial and temporal variability in DOC observed in aquatic ecosystems. Each question reflects a hypothesis concerning the relative importance of processes controlling DOC variability. We follow these questions with a list of research studies which would begin to answer them.

Source Questions

- What limits the concentrations of DOC in different source zones, particularly the upper soil horizons of terrestrial environments and wetland ecosystems?
- Do anaerobic environments (including anaerobic microsites in otherwise "oxic" systems) play an important role in the production of DOC with moderate to long turnover times in aquatic ecosystems?
- Is biological degradation of adsorbed DOC the primary mechanism controlling long-term adsorption of DOC in lower soil horizons and the concentrations of DOC found in groundwater?

Transport Questions

- Is temporal variability in DOC primarily a result of hydrologic processes? What are the specific hydrologic flow paths which are most important in transporting DOC from source areas to surface waters?
- Is spatial variability in DOC primarily a result of drainage geomorphology?
- Do upland sources of DOC ever contribute substantially to DOC variability in aquatic ecosystems, and if so, under what conditions?

Reactivity Questions

- Can DOC be fractionated into several (e.g., up to four) relatively discrete classes of biological reactivity rates (decomposition rates)?

Are there easily measurable chemical/structural predictors of reactivity rate?
— Are microbial processes primarily a mechanism of transformation (change in chemical form) and only secondarily a mechanism of degradation of the bulk of DOC compounds present in aquatic ecosystems?
— Are photochemical reactions of DOC an important mechanism for increasing biological degradation of DOC in aquatic ecosystems?

Specific Studies Needed

1. Studies examining the relationship between total organic acids and DOC in a variety of environments.
2. Identification of ecologically meaningful source tracers of DOC (anaerobic vs. aerobic, upland vs. riparian, floodplain vs. hyporheic, terrestrial vs. aquatic).
3. Characterization of microbial processes involved in DOC formation under aerobic and anaerobic conditions and resulting differences (structural quality, quantity) in the DOC produced.
4. Studies conducted in natural ecosystems examining photochemical transformations and subsequent biological degradation.
5. Studies combining streamflow generation modeling with empirical observations of hydrology and DOC dynamics (particularly during storm/flood events).
6. Biological reactivity studies (including technique development) in the full range of natural reactivity rates of bulk DOC, and performed in conjunction with appropriate chemical characterization studies.
7. Studies relating landscape geomorphic features (e.g., channel gradient, hyporheic development, floodplain size, wetland importance) to variations in DOC concentrations and quality (chemical, biological reactivity), including the effect of human alteration of drainage geomorphology.
8. Studies examining longitudinal variation in DOC in rivers with differing hydrologic interaction with floodplains (including a functionally "closed" river such as segments of the Nile River for studies of DOC degradation and transformation).

REFERENCES

Aiken, G.R. 1985. Isolation and concentration technique for aquatic humic substances. In: Humic Substances in Soil, Sediment, and Water, ed. G.R. Aiken, D.M. McKnight, R.L. Wershaw, and P. MacCarthy, pp. 363–385. New York: Wiley.

Cronan, C.S., and G.R. Aiken. 1985. Chemistry and transport of soluble humic substances in forested watersheds of the Adirondack Park, New York. *Geochim. Cosmo. Acta* 49:1697–1705.

Dahm, C.N. 1981. Pathways and mechanisms for removal of dissolved organic carbon from leaf leachate in streams. *Can. J. Fish. Aq. Sci.* 38:68–76.

Dawson, H.J., F.C. Ugolini, B.F. Hrutfiord, and J. Zachra. 1978. Role of soluble organics in the soil processes of a Podzol, Central Cascades, Washington. *Soil Sci.* 126:290–296.

Hough, R.A., and R.G. Wetzel. 1975. The release of dissolved organic carbon from submersed aquatic macrophytes: diel, seasonal, and community relationships. *Verh. Int. Verein. Limnol.* 19:938–948.

Klinge, H. 1965. Podsol soils in the Amazon basin. *J. Soil Sci* 16:95–103.

Kortelainen, P., and J. Mannio. 1988. Natural and anthropogenic acidity sources for Finnish lakes. *Wat. Air Soil Poll.* 42:341–352.

Krom, M.D., and E.R. Sholkovitz. 1977. Nature and reactions of dissolved organic matter in the interstitial waters of marine sediments. *Geochim. Cosmo. Acta* 41:1565–1573.

Kuserk, F.T., L.A. Kaplan, and T.L. Bott. 1984. *In situ* measures of dissolved organic carbon flux in a rural stream. *Can. J. Fish. Aq. Sci.* 41:964–973.

Livingstone, D.A. 1963. Chemical composition of rivers and lakes. U.S. Geological Survey Professional Paper 440-G. Washington, D.C.: GPO.

Lock, M.A., and H.B.N. Hynes. 1976. The fate of dissolved organic carbon derived from autumn-shed maple leaves (*Acer saccharum*) in a temperate hardwater stream. *Limnol. Ocean.* 21:436–443.

Lovely, D.R., and M.J. Klug. 1982. Intermediary metabolism of organic matter in the sediments of a eutrophic lake. *Appl. Env. Microbiol.* 43:552–560.

McDowell, W.H., and G.E. Likens. 1988. Origin, composition, and flux of dissolved organic carbon in the Hubbard Brook Valley. *Ecol. Monogr.* 58:177–195.

McDowell, W.H., and T. Wood. 1984. Podzolization: soil processes control dissolved organic carbon concentrations in stream water. *Soil Sci.* 137:23–32.

Meybeck, M. 1982. Carbon, nitrogen, and phosphorus transport by world rivers. *Am. J. Sci.* 282:401–450.

Meyer, J.L. 1988. Benthic bacterial biomass and production in a blackwater river. *Verh. Int. Verein. Limnol.* 23:1832–1838.

Moore, T.R. 1987. Dissolved organic carbon in forested and cutover drainage basins, Westland, New Zealand. In: Forest Hydrology and Watershed Management, ed. R.H. Swanson, P.Y. Bernier, and P.D. Woodard, pp. 481–487. Wallingford, NZ: IAHS Press.

Mosley, M.P. 1979. Streamflow generation in a forested watershed, New Zealand. *Water Res. Res.* 15:795–806.

Mulholland, P.J., and E.J. Kuenzler. 1979. Organic carbon export from upland and forested wetland watersheds. *Limnol. Ocean.* 24:960–966.

National Academy of Sciences. 1975. Productivity of World Ecosystems. Washington, D.C.: U.S. National Academy of Sciences.

Orem, W.H., and P.G. Hatcher. 1987. Solid-state ^{13}C NMR studies of dissolved organic matter in pore waters from different depositional environments. *Org. Geochem.* 11:73–82.

Otsuki, A., and T. Hanya. 1972. Production of dissolved organic matter from dead green algal cells. II. Anaerobic microbial decomposition. *Limnol. Ocean.* 17:258–264.

Otsuki, A., and R.G. Wetzel. 1973. Interaction of yellow organic acids with calcium carbonate in freshwater. *Limnol. Ocean.* 18:490–493.

Petersen, R.C. 1986. *In situ* particle generation in a southern Swedish stream. *Limnol. Ocean.* 31:432–436.
Petersen, R.C., and K.W. Cummins. 1974. Leaf processing in a woodland stream. *Freshwat. Biol.* 4:343–368.
Seki, H., Y. Yamaguchi, and S. Ichimura. 1975. Turnover rate of dissolved organic materials in a coastal region of Japan at summer stagnation period of 1974. *Arch. Hydrobiol.* 75:297–305.
Sklash, M.G., and R.N. Farvolden. 1979. The role of groundwater in storm runoff. *J. Hydrol.* 43:45–65.
Vance, G.F., and M.R. David. 1989. The effect of acid treatment on dissolved organic carbon retention by a spodic horizon. *Soil Sci. Soc. Am. J.*, in press.
Westrich, J.T., and R.A. Berner. 1984. The role of sedimentary organic matter in bacterial sulfate reduction: the G model tested. *Limnol. Ocean.* 29:236–249.
Wetzel, R.G. 1983. Limnology. Philadelphia: W.B. Saunders.
Williams, P.M., and E.R.M. Druffel. 1987. Radiocarbon in dissolved organic matter in the central North Pacific Ocean. *Nature* 330:246–248.

List of Participants with Fields of Research

ABBT-BRAUN, G.
Engler-Bunte-Institut der Universität
Karlsruhe
Richard-Willstätter-Allee 5
7500 Karlsruhe 1, F.R. Germany

Isolation and characterization of humic substances

BEHMEL, P.
Lehrgebiet Chemie
Universität Göttingen
Von-Siebold-Strasse 2
3400 Göttingen, F.R. Germany

Analytical chemistry of soil organic matter; humic substances chemistry; peat chemistry

BRAUN, A.M.
Institut de Chimie Physique
Ecole Polytechnique Fédérale de
Lausanne
1015 Lausanne, Switzerland

Photochemical and thermal oxidation mechanisms; mechanistic and preparative photochemistry; photochemical technology

CALMANO, W.
Arbeitsbereich Umweltschutztechnik
Technische Universität
Hamburg–Harburg
Eissendorferstrasse 40
2100 Hamburg 90, F.R. Germany

Chemistry of heavy metals in aquatic systems

CHRISTMAN, R.F.
Department of Environmental Science
and Engineering
School of Public Health
University of North Carolina
Chapel Hill, NC 27514, U.S.A.

Humic acid chemistry

DAHM, C.N.
Department of Biology
University of New Mexico
Albuquerque, NM 87131, U.S.A.

Stream ecology; microbial carbon and nutrient cycling

DAVID, M.B.
University of Illinois
Department of Forestry
110 Mumford Hall
Urbana, IL 61801, U.S.A.

Organic acids in forested ecosystems: effects on surface water chemistry

DE HAAN, H.
Limnological Institute
Tjeukemeer Laboratory
De Akkers 47
8536 VD Oosterzee, Netherlands

Biogeochemistry and phytoplankton ecology of humic waters

DI TORO, D.M.
Environmental Engineering
Manhattan College
Bronx, NY 10471, U.S.A.

Water quality modeling

FISHER, T.R.
University of Maryland – CEES
Horn Point Laboratory
Cambridge, MD 21613, U.S.A.

Element cycling; primary productivity; landscape ecology

FRANCKO, D.A.
Department of Botany and Microbiology
Oklahoma State University
Stillwater, OK 74078, U.S.A.

Physiological limnology of algae and aquatic vascular plants; nutrient dynamics

FREDRICKSON, H.L.
GBF – Federal German Institute of Biotechnology
Microbial Ecology Project
Department of Microbiology
Mascheroder Weg 1
3300 Braunschweig, F.R. Germany

Microbial ecology and genetically engineered microbes

FRIMMEL, F.H.
Engler-Bunte-Institut der Universität Karlsruhe
Richard-Willstätter-Allee 5
7500 Karlsruhe 1, F.R. Germany

Environmental aquatic chemistry and technology

GASSEN, H.G.
Institut für Biochemie
Technische Hochschule Darmstadt
Petersenstrasse 22
6100 Darmstadt, F.R. Germany

Biochemistry; molecular cloning

GIGER, W.
Swiss Federal Institute for Water Resources and Water Pollution Control (EAWAG)
8600 Dübendorf, Switzerland

Analysis and behavior of organic chemicals in the environment

GJESSING, E.T.
Norwegian Institute for Water Research NIVA
P.O. Box 68
0808 Oslo 8, Norway

Aquatic chemistry of organic acids and surface water acidification

GLAZE, W.H.
Department of Environmental Science and Engineering
School of Public Health
University of North Carolina
Chapel Hill, NC 27514, U.S.A.

Analytical methods and treatment of organic compounds in water; oxidation of organics

HEDGES, J.I.
School of Oceanography, WB-10
University of Washington
Seattle, WA 98195, U.S.A.

Aquatic organic geochemistry: the sources and fates of organic molecules in lakes, rivers, and the ocean

HEMOND, H.F.
Ralph M. Parsons Laboratory, 48–419
Massachusetts Institute of Technology
Cambridge, MA 02139, U.S.A.

Biogeochemistry

HENRIKSEN, A.
Norwegian Institute for Water Research NIVA
P.O. Box 33, Blindern
0313 Oslo 3, Norway

Acid precipitation; water chemistry

KLUG, M.J.
Kellogg Biological Station
3700 E. Gull Lake Drive
Hickory Corners, MI 49060, U.S.A.

Decomposition of organic matter in naturally occurring anaerobic microbial communities

KÖGEL-KNABNER, I.
Universität Bayreuth
Postfach 10 12 51
8580 Bayreuth, F.R. Germany

Organic matter and humic substances in forest soils

KRAMER, J.R.
McMaster University
Department of Geology
Hamilton, Ontario L8S 4M1, Canada

Organic acids; speciation in watersheds

MANNIO, J.Y.
National Board of Waters and the Environment
PL 250
00101 Helsinki, Finland

Organic and mineral acidity; aluminum and heavy metals in headwater lakes

McKNIGHT, D.M.
U.S. Geological Survey
WRD, MS 408
5293 Ward Road
Arvada, CO 80002, U.S.A.

Biogeochemical processes involving organic material and trace metals

MEYBECK, M.H.
Laboratoire de Géologie Appliquée
Tour 26 - 5ième Etage
Université Pierre et Marie Curie
Place Jussieu
75005 Paris, France

River and estuarine geochemistry

MEYER, J.L.
Zoology Department and Institute of Ecology
University of Georgia
Athens, GA 30602, U.S.A.

Stream ecosystems

MOREL, F.M.M.
Department of Civil Engineering, 48–423
Massachusetts Institute of Technology
Cambridge, MA 02139, U.S.A.

Aquatic chemistry

MULHOLLAND, P.J.
Environmental Sciences Division
Oak Ridge National Laboratory
P.O. Box 2008
Oak Ridge, TN 37831-6036, U.S.A.

Biogeochemistry of nutrients and organic carbon in watersheds; stream ecology

MÜNSTER, U
Max-Planck-Institut für Limnologie
Abt. Mikrobenökologie
Postfach 165
2320 Plön/Holstein, F.R. Germany

Freshwaters: structure and function of DOM; substrate-exoenzyme interactions: amino acids–aminopeptidases, carbohydrates–glycosidases; humic acids: acidity

NEHRKORN, A.H.
Department of Microbiology, FB 2
University of Bremen
P.O. Box 330440
2800 Bremen 33, F.R. Germany

Groundwater microbiology; activated sludge microbiology

NIEMEYER, J.
Abteilung Chemie
Institut für Bodenwissenschaften
Von-Siebold-Strasse 2
3400 Göttingen, F.R. Germany

Interface chemistry in natural systems

ÖHMAN, L.-O.
Department of Inorganic Chemistry
University of Umea
90187 Umea, Sweden

Complexation and precipitation of Al (III) in systems of bio- and geochemical relevance

PERDUE, E.M.
School of Geophysical Sciences
Georgia Institute of Technology
Atlanta, GA 30332, U.S.A.

Acid-base and metal complexation chemistry of organic acids in soil and aquatic environments

PETERSEN, R.C., Jr.
Stream and Benthic Ecology Group
Department of Ecology/Limnology
Box 65, University of Lund
221 00 Lund, Sweden

Limnology; stream ecology; ecotoxicology

SEDELL, J.R.
Pacific Northwest Research Station
FSL Laboratory
3200 Jefferson Way
Corvallis, OR 97331, U.S.A.

Carbon and nutrient cycling in river systems: land/water interactions; geomorphic processes and features determining carbon production and allocations; historical reconstruction of river and wetland systems

SHUMAN, M.S.
Department of Environmental Sciences and Engineering
University of North Carolina
Chapel Hill, NC 27599–7400, U.S.A.

Environmental chemistry; analytical chemistry

SKULBERG, O.M.
Norwegian Institute for Water Research
NIVA
P.O. Box 33, Blindern
0313 Oslo 3, Norway

Experimental hydrobiology; secondary metabolites

STEINBERG, C.E.W.
Fraunhofer-Institut für
Umweltchemie und Ökotoxikologie
P.O. Box 1260
5948 Schmallenberg, F.R. Germany

Eutrophication; oligotrophication (lake therapy); acidification

THURMAN, E.M.
U.S. Geological Survey
4821 Quail Crest Place
Lawrence, KS 66049, U.S.A.

Dissolved organic compounds in natural and polluted waters with special emphasis on both humic substances and contaminate organic compounds, such as agricultural chemicals (herbicides)

TIPPING, E.
Institute of Freshwater Ecology
Windermere Laboratory
Far Sawry
Ambleside, Cumbria LA22 0LP, U.K.

Environmental surface and colloid chemistry; physical chemistry of humic substances

VISSER, S.A.
Department of Soil Science (Agric.)
Comtois Pavillion
Laval University
Ste-Foy
Quebec G1K 7P4, Canada

Physiological effects of fulvic and humic acids of terrestrial and aquatic origin

WEIS, M.
Engler-Bunte-Institut der Universität Karlsruhe
Richard-Willstätter-Allee 5
7500 Karlsruhe 1, F.R. Germany

Metal complexation by organic acids; electrochemical analysis of kinetics and thermodynamics

WERNER, P.W.
DVGW-Forschungsstelle am
Engler-Bunte-Institut
Wasserchemie
Richard-Willstätter-Allee 5
7500 Karlsruhe 1, F.R. Germany

Biodegradation of organic matter; regrowth of drinking water caused by low concentration of organics; remediation of polluted soils and aquifers by biodegradation

WETZEL, R.G.
Department of Biology
University of Michigan
Ann Arbor, MI 48109, U.S.A.

Chemical limnology (particularly dissolved organic compounds), bacterial metabolism, and physiology of aquatic plants

WHITE, D.C.
Institute for Applied Microbiology
University of Tennessee
10515 Research Drive, Suite 300
Knoxville, TN 37932–2567, U.S.A.

Microbial ecology; organic geochemistry

Subject Index

Author Index